Numerical Simulation of Non-Newtonian Flow

RHEOLOGY SERIES

VOLUME 1 **Numerical Simulation of Non-Newtonian Flow** (Crochet, Davies, Walters)

RHEOLOGY SERIES, 1

Numerical Simulation of Non-Newtonian Flow

M.J. CROCHET
Unité de Mécanique Appliquée, Université Catholique de Louvain, Louvain-la-Neuve, Belgium

A.R. DAVIES
Department of Applied Mathematics, University College of Wales, Aberystwyth, U.K.

K. WALTERS
Department of Applied Mathematics, University College of Wales, Aberystwyth, U.K.

ELSEVIER
Amsterdam — Oxford — New York — Tokyo 1984

ELSEVIER SCIENCE PUBLISHERS B.V.
Sara Burgerhartstraat 25,
P.O. Box 211, 1000 AE Amsterdam, The Netherlands

Distributors for the United States and Canada:

ELSEVIER SCIENCE PUBLISHING COMPANY INC.
52, Vanderbilt Avenue
New York, NY 10017

First edition 1984
Second impression 1985

ISBN 0-444-42291-9 (Vol. 1)
ISBN 0-444-42292-7 (Series)

Printed in The Netherlands

To

Brigitte,

Celia

and

Mary

Preface

Two of us (M.J.C. and K.W.) met for the first time in 1974 at a Euromech meeting in Toulouse - the beginning of a happy friendship and collaboration. We invited A.R.D. to join us at a meeting of the Belgian Society of Rheology held at Louvain la Neuve in 1979 under the title "Numerical Simulation of non-Newtonian Flow". At that meeting, the three of us decided to collaborate on a book with the same title. We felt at that time that a need existed for such a book, an impression which has been confirmed in the four years since the initial decision was taken. Activity in the field is increasing and there are encouraging signs that the techniques already developed are being used by industrialists to solve important practical problems. Indeed we have become all too aware of the obvious difficulties of writing a book in a field that is evolving very rapidly.

Some attempt has been made in this book to unify different approaches. For example, two of us (M.J.C. and K.W.) were introduced to non-Newtonian fluid mechanics through different research schools. Accordingly, we have tried in Chapter 2 to carry out a bridge-building exercise, which we hope will be of assistance to newcomers to the field who may be confused by seemingly different approaches to the same subject. Again, in the numerical simulation sections, we consider both finite difference and finite element techniques. This is justified, since both techniques have been employed in the development of the subject. Furthermore, having essentially the same problems solved by both finite difference and finite element techniques may be of help to newcomers to the field who are, as yet, uncommitted to one or the other possibilities.

It will become clear that the approaches in the finite difference and finite element sections are somewhat different. This is due in part to the backgrounds of the major contributors to these sections (A.R.D. for finite differences and M.J.C. for finite elements). In the main, however, it is a reflection of the way in which the two techniques have developed in Newtonian as well as non-Newtonian fluid mechanics.

Collaborating on the book has entailed frequent correspondence and many stimulating meetings in Wales and Belgium (and some points in between), but we are now relieved that the project has been completed and that in future we shall meet in more relaxed style.

Concerning the Belgian connection, special thanks must go to Roland Keunings for his invaluable collaboration and his constant efforts towards better performance of the numerical methods; also to Jules Van Schaftingen for a thorough study of the mixed methods. Jean Meinguet read an early draft of Chapter 8 and made a number of thoughtful suggestions for the improvement of the text.

Roland Keunings and Jean-Marie Marchal gave invaluable assistance in proofreading Chapters 8 - 10. We thank Therese Bodson and Michele Sergant who typed early drafts of the manuscript and Victor Vermeulen and Andre Nackaerts who prepared some of the figures.

Concerning the Welsh connection, we are happy to acknowledge helpful discussions with a number of our friends and colleagues, notably Horst Holstein, Peter Townsend and Mike Webster. Bob Bird of the University of Wisconsin, Madison, made a number of useful comments which improved the final form of Chapter 2. We also thank Pat Evans who assisted in the later stages of the typing of the drafts.

Finally, we are deeply indebted to Robin Evans who prepared the final figures and to Mrs. D. Vincent who expertly typed the camera-ready copy.

M.J. Crochet
A.R. Davies
K. Walters

Contents

Chapter 1

General Introduction

1.1 INTRODUCTION

Materials encountered in industry invariably fall outside the classical extremes of the Newtonian viscous fluid and Hookean elastic solid. When such materials can be classified as fluids, the adjective "non-Newtonian" is usually employed. This book is mainly concerned with non-Newtonian fluids, although we shall find it useful to refer with some regularity to the corresponding Newtonian fluid situation for help and inspiration.

To be precise, we define a non-Newtonian fluid to be one whose behaviour cannot be predicted on the basis of the Navier-Stokes equations. Such fluids may or may not possess a memory of past deformation. If they do, they are called non-Newtonian elastico-viscous liquids or simply elastic liquids.

Examples of non-Newtonian fluids abound in every-day life. Multigrade oils, liquid detergents, paints, printing inks and industrial suspensions all fall within this category as do the polymer solutions and polymer melts used in the plastics processing industries (see, for example, Walters 1980).

In development and use, non-Newtonian fluids often encounter complex geometries: a liquid detergent has to be "squeezed" through a contraction at the exit of a plastic bottle; lubricants have to operate in gears and bearings; molten polymers meet complex geometries (with and without free-surface complications) in injection moulding and similar processes. Many other examples could be cited.

Added complications concern the extreme conditions encountered in many practically important situations. Shear rates of 10^6 sec^{-1} and higher are not uncommon and pressure and temperature can be dominating variables.

It is doubtful whether a *comprehensive* theoretical solution to these practical problems will ever be possible, but the advent of very high-speed computers in the last decade has at least enabled us to make some headway. Certainly, there has been progress in the general area of the numerical prediction of the behaviour of elastic liquids in complex geometries (with and without free surfaces), which is the basic concern of the present book.

It is true that the extension of existing numerical algorithms to include *long-range* memory effects, temperature and pressure variables and more realistic geometries presents challenging problems, but at least the groundwork is now available. This has involved a critical review of numerical simulation as applied to Newtonian fluids and an assessment of how this can be adapted to meet the new challenges of fluids with memory. Not surprisingly, we shall find it necessary to draw heavily on the wealth of literature available in classical fluid mechanics, but our major concern will always be to emphasize the distinctive changes necessary in the non-Newtonian situation.

1.2 RHEOMETRICAL PROPERTIES OF NON-NEWTONIAN FLUIDS

It would be an understatement to say that non-Newtonian elastic liquids manifest material properties which are significantly more complicated than those found in Newtonian viscous liquids. To highlight these differences, it is sufficient for our present purpose to restrict attention to two simple (rheometrical) flows, namely steady simple shear flow and extensional flow.

In a steady simple shear flow with velocity components given by

$$u = \gamma y \, , \qquad v = w = 0 \, , \tag{1.1}$$

where (u,v,w) are the velocity components referred to a rectangular Cartesian coordinate system (x,y,z) and γ is the (constant) velocity gradient or shear rate, the corresponding components of the stress tensor P_{ik} for a Newtonian fluid have the simple form

$$\begin{aligned} P_{xx} - P_{yy} &= 0 \, , \qquad P_{xy} = \gamma\eta \, , \\ P_{yy} - P_{zz} &= 0 \, , \qquad P_{xz} = P_{yz} = 0 \, , \end{aligned} \tag{1.2}$$

where η is the viscosity coefficient which is independent of the shear rate γ.

For non-Newtonian liquids, (1.2) has to be replaced by (see, for example, Walters 1975)

$$\begin{aligned} P_{xx} - P_{yy} &= \nu_1(\gamma) \, , \qquad P_{xy} = \tau(\gamma) = \gamma\eta(\gamma) \, , \\ P_{yy} - P_{zz} &= \nu_2(\gamma) \, , \qquad P_{xz} = P_{yz} = 0 \, . \end{aligned} \tag{1.3}$$

In general, the shear stress τ is not a linear function of γ and we refer to the *apparent viscosity* $\eta(\gamma)$. For most, but not all, non-Newtonian systems, η is found to be a monotonic decreasing function of γ representing so-called 'shear thinning' behaviour. The (η,γ) curve usually has the form shown schematically in Fig. 1.1; falling from a zero shear rate 'first-Newtonian' value η_0 to a 'second-Newtonian' value η_2 at very high shear rates, which can be as much as several orders of magnitude lower than η_0. Experimentally, it is often found that the decreasing part of the curve is very well approximated by a power law of the form

$$\eta(\gamma) = K\gamma^{p-1} \, , \tag{1.4}$$

where K and p are constants.

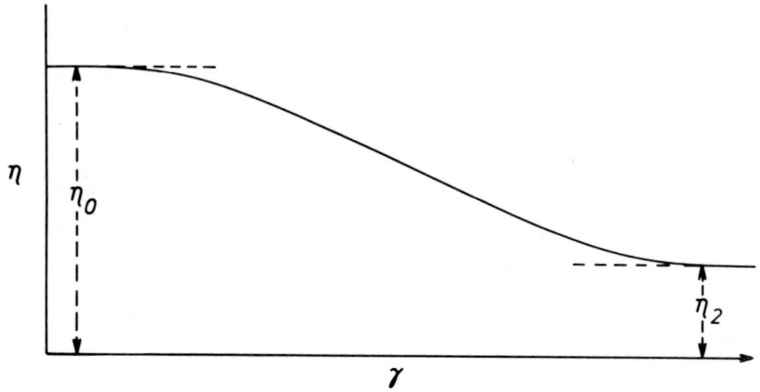

Fig. 1.1 Graph showing typical shear thinning behavior.

Not all non-Newtonian fluids exhibit shear thinning in a steady simple shear flow. The so called Boger (1977/78) fluid (a dilute solution of polyacrylamide in a Maltose syrup/water base) can exhibit a reasonably constant viscosity over a substantial shear rate range (0 to 200 sec^{-1}, say), although in other respects its behaviour is dramatically non-Newtonian. Furthermore, some liquids, notably cornflour suspensions, show *'shear thickening'* behaviour with the viscosity increasing with γ. However, such behaviour is rare.

Normal stress levels in elastic liquids can be high and it is certainly possible for ν_1 and ν_2 to be orders of magnitude higher than the shear stress τ. Available experimental evidence would suggest that for most non-Newtonian systems, ν_2 is smaller than ν_1 $(0.1\ |\nu_1| < |\nu_2| < 0.25\ |\nu_1|)$ and of opposite sign. For the Boger fluid already referred to, however, ν_1 is found to be (approximately) a quadratic function of shear rate γ and ν_2 is zero to within experimental error (Boger 1977/78, Keentok et al 1980).

The normal stress differences give rise to several dramatic demonstrations of non-Newtonian behaviour, including the rod-climbing (or Weissenberg) effect and extrudate swell (die-swell) at the exit of a capillary.

In an extensional flow with velocity components given by

$$u = kx \ , \qquad v = -\frac{k}{2}\, y \ , \qquad w = -\frac{k}{2}\, z \ , \tag{1.5}$$

where k is a constant rate of strain, the corresponding stress distribution can be written in the form (see, for example, Walters 1975, 1980)

$$P_{xx} - P_{yy} = k\eta_E(k) \; ,$$

$$P_{xx} - P_{zz} = k\eta_E(k) \; , \qquad P_{ij} = 0 \quad \text{for} \quad i \neq j \; , \tag{1.6}$$

η_E being the *extensional viscosity*. The ratio η_E/η is called the Trouton ratio. For a Newtonian liquid this ratio is 3. For non-Newtonian liquids, the ratio will be a function of the respective rates of strain k and γ, the relevant γ being chosen as a function of k in an appropriate way. Many elastic liquids are characterized by high Trouton ratios. In the case of polymer melts, for example, this is mainly a result of the reduction in η with γ due to shear thinning (with a fairly constant extensional viscosity). For dilute polymer solutions on the other hand, the very high Trouton ratios arise from the dramatic increase of η_E with k (with nothing like a corresponding fall in the shear viscosity with γ).

1.3 NON-NEWTONIAN FLOW IN COMPLEX GEOMETRIES

The non-Newtonian effects found in simple rheometrical experiments are manifestations of fluid-memory effects. In principle, such effects are now well understood and their theoretical simulation does not involve sophisticated numerical analysis. Rather, conventional rheometrical experiments are to be viewed as providing a foundation set of data which any theoretical modelling process must accommodate, with the accompanying acknowledgment that numerical simulation in non-Newtonian fluid mechanics must involve a study of more complex flows than those encountered in Rheometry. In this book, we shall therefore be largely concerned with the combination of long-range fluid-memory effects on the one hand and complex geometries on the other. The interaction between these factors provides a challenging area of study.

We shall illustrate the general problem by reference to a selected number of flow experiments in which long-range memory effects have resulted in dramatic changes in flow characteristics. By and large, the chosen situations still await satisfactory theoretical solutions.

It is helpful to introduce a non-dimensional parameter W defined by $\lambda U/L$, where U is a characteristic velocity, L a characteristic length and λ is a characteristic (relaxation) time of the fluid (which can be loosely viewed as a measure of the fluid's memory) (see also §3.8).

Circular contraction flows provide our first examples of the general problem (see figure 1.2).

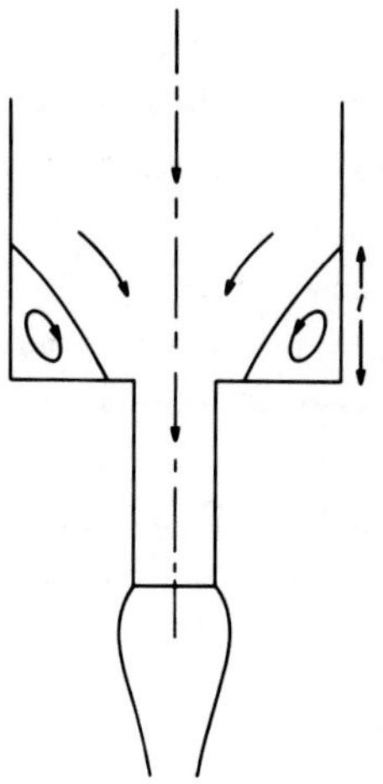

Fig. 1.2 Schematic diagram of a contraction flow.

Fluid is pumped from one fully-developed Poiseuille flow (in the wide capillary) through a contraction zone into a smaller capillary and thence into air where the phenomenon of die-swell occurs.

The second capillary is often long enough to assume a fully-developed Poiseuille flow over at least a part of its length. The corner vortex flow which is usually (but not always) observed is shown schematically in the figure.

Experimentally, it is sometimes found that, beyond a critical value of W, the attachment length l of the vortex increases dramatically with W, giving rise to a so-called 'vortex enhancement' regime, before various forms of instability set in.

A certain pressure drop is required in the contraction zone to force the velocity profile from one Poiseuille flow into another. Other things being equal, it is found that some elastic liquids (very dilute polymer solutions, for example) require a greatly enhanced pressure drop over comparable inelastic liquids, with some indication that a threshold value of W exists below which elasticity has only a small influence and above which the increase in pressure drop with W becomes dramatic (see, for example, Walters and Barnes 1980).

The contraction region is known to contain a non-trivial extensional-flow component, and in qualitative terms the enhanced pressure loss may be attributed to the very high Trouton ratios for elastic liquids already referred to. However, until the very recent work of Keunings and Crochet (1983), all attempts to simulate the pressure drop increase and vortex enhancement were unsuccessful.

Pressure driven flow in the lower capillary is often used to measure the apparent viscosity of non-Newtonian fluids and sometimes to indicate normal stress levels. The second exercise depends critically on the flow in the capillary remaining "fully-developed" up to the exit. There is therefore significant interest in the flow conditions in the capillary-exit zone and also in the related die-swell problem as the extrudate leaves the capillary (see, for example, Boger and Denn 1980).

When the extrudate is drawn down under tension, the situation clearly resembles a fibre-spinning operation. This is another problem of practical importance which is within the range of numerical simulation as the subject commends itself to industrial scientists.

Another set of experiments which provide challenging problems to theoreticians concern the flow of elastic liquids past spheres and cylinders.

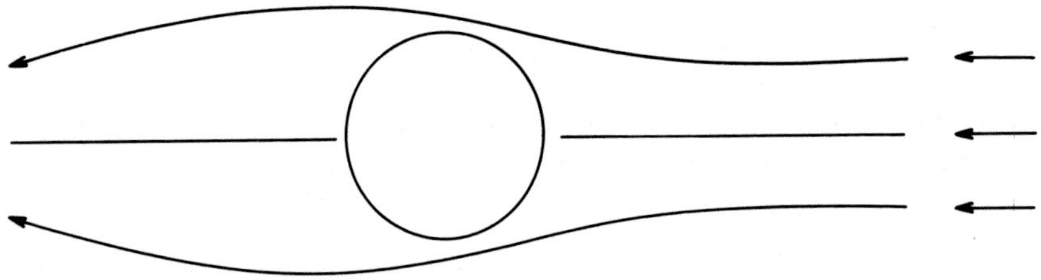

Fig.1.3 Schematic diagram of flow past a cylinder or sphere showing a downstream shift in the streamlines.

Experimental data are available on the drag experienced by the obstacles, and flow visualization techniques provide a means of studying streamline patterns. Our concern here is with the latter experiments (see, for example, Ultman and Denn 1971, Zana et al 1975, Sigli and Coutanceau 1977, Manero and Mena 1981).

Experimental evidence is not conclusive, but it appears that the streamlines can be shifted upstream or downstream depending on the size of the elastic parameter W. For low values of W, the situation illustrated schematically in figure 1.3 pertains, while an upstream shift is sometimes observed at high values of W. Under extreme conditions, there is some evidence of the existence of a stagnant region of fluid in front of the obstacle and this would be consistent with an upstream shift in the streamlines.

Existing analyses predict a downstream shift at low W. As yet no numerical simulations are available to compare with the upstream shift associated with high W.

The above examples are sufficient to indicate that fluid elasticity can produce dramatic changes in flow characteristics. These and many related situations provi an ample practical motivation for the current interest in numerical simulation in non-Newtonian fluid mechanics.

1.4 THE ROLE OF NON-NEWTONIAN FLUID MECHANICS

The basic problem in the numerical simulation of non-Newtonian flow may be stated as follows :

Given the availability of rheometrical data for the test fluid, can one predict the behaviour of the fluid in complex flows such as those discussed in §1.3?

Solution of the problem involves, in the first instance, the construction of suitable rheological equations of state (constitutive equations) for the fluid which are able to simulate (at least qualitatively) the available rheometrical data and which are simple enough to allow computational tractability. In non-Newtonian fluid mechanics, the choice of rheological model depends critically on the type of flow being considered (cf. Chapters 2 and 3) and it is this basic consideration which makes non-Newtonian fluid mechanics basically different from classical fluid mechanics, where the Navier-Stokes equations can be immediately accepted as being valid for all flow situations (cf. Astarita 1976).

Having chosen the most appropriate rheological model, it is then necessary to solve the associated equations in conjunction with the familiar stress equations of motion and the equation of continuity, subject to appropriate boundary conditions. Non-Newtonian fluid mechanics often requires the stress components to be treated as dependent variables along with the velocity components and the pressure - a further complication from the classical situation. Furthermore, it is not in general sufficient to simply take over the boundary conditions of Newtonian fluid mechanics and these have to be adapted and extended to meet the new challenges of fluids with memory. All these points will be explored in depth in subsequent chapters.

The final exercise in the numerical simulation of non-Newtonian flow requires the usual comparison between predictions and experimental data. If the agreement in this regard is not satisfactory, it is of course necessary to look critically at the numerical techniques employed, but it is also open to the investigator to question the original choice of rheological model, so that the fundamental scientific method in non-Newtonian fluid mechanics involves many new features not present in the classical situation. Throughout this book we shall attempt to stress these novel features, since many texts on Newtonian fluid simulation are available and it is not our intention to compete with these on matters of detail (see, for example, Roache 1976, Thomasset 1981, Temam 1979, Girault and Raviart 1979).

Chapter 2

Basic Equations

2.1 INTRODUCTION

The governing equations in non-Newtonian fluid mechanics consist of field equations and constitutive equations. In the isothermal theory, the field equations are the equation of continuity, which is a formal mathematical expression of the principle of conservation of mass, the stress equations of motion, which arise from the application of Newton's second law of motion to a moving continuum (or the principle of balance of linear momentum) and the local expression of the principle of balance of angular momentum. The constitutive equations, or rheological equations of state, relate the stress to the motion of the continuum.

Whereas the field equations are the same for all materials, the constitutive equations will in general vary from one non-Newtonian material to another (and possibly from one type of flow to another). It is this last point which distinguishes non-Newtonian fluid mechanics from classical fluid mechanics, where the use of Newton's viscosity law gives rise to the Navier-Stokes equations which are valid for all Newtonian viscous fluids (see, for example, Astarita 1976).

2.2 FIELD EQUATIONS

We recall for later reference the local form of the field equations (the interested reader may find their derivation from basic principles in, e.g., Schowalter 1978). Unless we specify otherwise, we shall consistently use a rectangular Cartesian coordinate system x_i throughout this book. The components of the velocity vector are denoted by v_i, those of the acceleration vector by a_i; we will make frequent use of the material time derivative D/Dt defined by †

$$\frac{D}{Dt} = \frac{\partial}{\partial t} + v_m \frac{\partial}{\partial x_m} \tag{2.1}$$

in Eulerian coordinates.

Let ρ denote the mass density; the principle of conservation of mass is expressed by the equation

$$\frac{D\rho}{Dt} + \rho \frac{\partial v_m}{\partial x_m} = 0 \; . \tag{2.2}$$

---

†We use standard tensor notation throughout this book. Covariant suffices are written below and contravariant suffices above and the usual summation convention for repeated suffices is assumed. In rectangular Cartesian coordinates, it is not necessary to distinguish between covariant and contravariant suffices.

In most problems encountered in non-Newtonian fluid mechanics, it may be assumed at the outset that the fluid is incompressible; the material derivative of the mass density then vanishes identically, which implies that, from (2.2),

$$\frac{\partial v_m}{\partial x_m} = 0 \ . \tag{2.3}$$

A velocity field which satisfies (2.3) is said to be solenoidal; the conservation of mass is then satisfied identically by expressing the velocity field as the curl of a vector field. This is useful when the flow is two-dimensional or axisymmetric; the velocity components may then be written as spatial derivatives of a scalar function ψ called the stream function. For example in a two-dimensional plane flow, (2.3) reduces to

$$\frac{\partial u}{\partial x} + \frac{\partial v}{\partial y} = 0 \ , \tag{2.4}$$

where we have written $x_1=x$, $x_2=y$, $v_1=u$, $v_2=v$,
and it is possible to define a stream function through

$$u = \frac{\partial \psi}{\partial y} \ , \qquad v = -\frac{\partial \psi}{\partial x} \ . \tag{2.5}$$

Clearly, ψ is arbitrary to the extent of an added function of the time in general, and of an added constant in steady flow.

Let P_{ki} denote the components of the Cauchy stress tensor; P_{ki} denotes the ith component of the force per unit area on a surface normal to the x_k-axis. Also let F_i denote the components of the body force acting per unit mass of the fluid. The stress equations of motion are then given by

$$\frac{\partial P_{ki}}{\partial x_k} + \rho F_i = \rho \frac{Dv_i}{Dt} \ . \tag{2.6}$$

Finally, the principle of balance of angular momentum in the absence of body and surface couples requires that

$$P_{ki} = P_{ik} \ . \tag{2.7}$$

We will assume throughout the rest of this book that the symmetry condition (2.7) is satisfied identically and that the constitutive equations are written accordingly.

2.3 NAVIER STOKES EQUATIONS

It is clear that equations (2.3) and (2.6) are not in themselves sufficient to provide a well-posed problem. We require in addition a relationship between the stress tensor P_{ik} and suitable kinematic variables expressing the motion of the continuum, i.e. we require a set of rheological equations of state. When the fluid is incompressible, the motion of the continuum determines the stress tensor up to an arbitrary isotropic tensor, and we use the decomposition

$$P_{ik} = - p\delta_{ik} + T_{ik} , \tag{2.8}$$

where p is an arbitrary pressure, δ_{ik} are the components of the identity tensor (or Kronecker delta), and T_{ik} are the components of the extra-stress tensor. We note for future reference that the pressure p is left arbitrary and that we will not require that the extra-stress tensor be always traceless.

In the case of an incompressible Newtonian viscous fluid, the rheological equation states that the extra-stress is proportional to the rate-of-deformation tensor d_{ik}, i.e.

$$T_{ik} = 2\eta d_{ik} , \tag{2.9}$$

where η is the shear viscosity and

$$d_{ik} = \frac{1}{2}\left(\frac{\partial v_i}{\partial x_k} + \frac{\partial v_k}{\partial x_i}\right) . \tag{2.10}$$

Since the extra-stress T_{ik} is an explicit function of the rate-of-deformation tensor, it is possible to substitute (2.8)-(2.10) into (2.6); with the use of (2.3) we obtain the 'Navier Stokes' equations :

$$- \frac{\partial p}{\partial x_i} + \eta \frac{\partial^2 v_i}{\partial x_k \partial x_k} + \rho F_i = \rho \frac{Dv_i}{Dt} . \tag{2.11}$$

Any fluid which does not obey the constitutive equation (2.9) is called non-Newtonian. In the next section, we shall discuss the general problem of the formulation of rheological equations of state with particular reference to non-Newtonian *elastic* liquids.

2.4 RHEOLOGICAL EQUATIONS OF STATE. FORMULATION PRINCIPLES

Newcomers to the field of non-Newtonian fluid mechanics often find that the subject of formulating rheological equations of state is a difficult one. Whether this impression is valid or not is to some extent a matter of subjective assessment, but it is certainly true that the newcomer's task has been complicated by the availability of two separate but equivalent ways of approaching the subject, one associated with the name of Oldroyd, Lodge and others (see, for example, Oldroyd 1950, 1958, Lodge 1974) and the other with Coleman, Noll, Rivlin, Ericksen, Green and Truesdell (see, for example, Truesdell and Noll 1965). In the 60's, there developed an unnecessary polarization of attitudes which often led to the mistaken impression that the two approaches were mutually exclusive. The advent of several text books in the 70's has led to the proper acknowledgment of both the convected coordinate ideas of Oldroyd and the concepts embodied in the works of Green, Rivlin, Ericksen, Coleman and Noll (see, for example, Lodge 1974, Huilgol 1975, Bird et al. 1977, Schowalter 1978).

It is not our present intention to write a detailed treatise on the formulation principles of continuum mechanics, but sufficient attention to the subject is clearly necessary to justify the effort that we shall later expend on numerical simulation. At the very least we require a realistic assessment of the standing of the rheological equations which we later attempt to solve, along with the equations of motion and continuity, by numerical techniques. We will simultaneously attempt to draw parallels between and, whenever possible, unify the two approaches to the formulation of rheological equations.

At the basis of the mechanical theory of constitutive equations we find four principles from which we will be able to derive their general form.

i. Principle of determinism of the stress. The stress in a body is determined by the history of motion of that body (Truesdell and Noll 1965). If we wish to calculate the extra-stress T_{ik} at time t, we will assume in general that T_{ik} depends upon the motion of the body for times $t' \leqslant t$.

ii. Principle of local action. In determining the stress at a given material point, the motion outside an arbitrary neighbourhood of the material point may be disregarded (Truesdell and Noll 1965). This principle will have important consequences on the type of kinematic tensors to be used in the constitutive equations.

iii. Principle of coordinate invariance. The relationship between the stress tensor and the history of motion of a body cannot depend on the particular coordinate system used to describe the stress and the history of motion, and rheological equations of state must be endowed with the usual tensorial invariance expected of a physical theory.

iv. Principle of invariance under superposed rigid body motion. The rheological equations must have a significance which is independent of absolute motion in space (Oldroyd 1950). The superposition of a rigid body motion on a given flow history cannot have any effect on the stress field other than that arising from the obvious changes in orientation brought about by the superposed rotational motion.

The observance of these principles leads to the constitutive equations of the so-called simple fluid. In the next section, we describe in some detail how the notion of a simple fluid may be arrived at by means of the two different approaches mentioned above.

2.5 THE SIMPLE FLUID

Let us consider (Fig. 2.1) the motion of a continuum for $t' \leqslant t$, and follow the path of a material point X. With respect to a system of rectangular Cartesian coordinates, the position of X is given by its coordinates x_i' at time $t' \leqslant t$, and x_i at time t. The motion of the continuum is described by the vector equations

$$x_i' = x_i'(X,t') \ , \ t' \leqslant t \ , \tag{2.12}$$

where X represents any material point of the continuum. Since the position of X at time t is given by its coordinates x_i, we will use, instead of (2.12), the vector equation:

$$x_i' = x_i'(x_j,t,t') \ , \ t' \leqslant t \ . \tag{2.13}$$

The right hand side of (2.13) identifies at time t' the position of a material point which occupies the position x_i at time t.

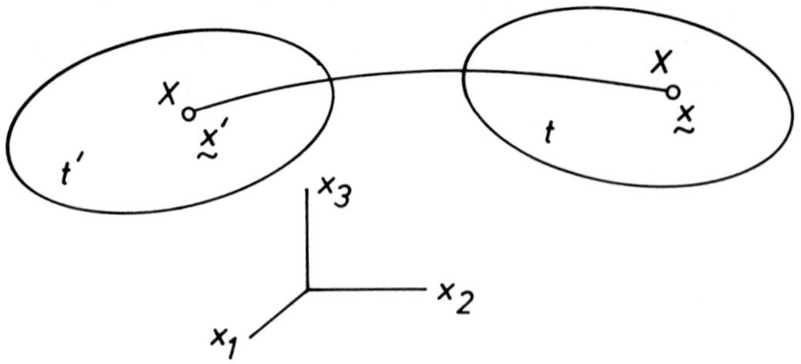

Fig. 2.1 The motion of a continuum.

Assuming that the motion of the continuum is known for $t' \leqslant t$ and given by (2.13), we wish to derive a general form of the rheological equations of state relating the extra-stress tensor T_{ik} to the motion of the continuum.

2.5.1 Convected coordinates

A clear summary of the method of convected coordinates is provided by the following text which is quoted from one of Oldroyd's lesser known references (Oldroyd 1961):

"We assume, simply, that the rheological behaviour of any material element (a part of a moving continuum) is quite independent of the position and the motion of the element as a whole in space, and that the behaviour at any time t may depend on the previous rheological states through which that element has passed, but cannot depend directly on the corresponding states of neighbouring parts of the material. We therefore regard as irrelevant to the problem of formulating equations of state any variable measuring position or translatory or rotatory motion of a material element in space, and any parameter labelling neighbouring material or labelling a time subsequent to the current time t. The easiest method of labelling particles of the material, in a way that does not require reference to where they are in space, is to consider a curvilinear coordinate system embedded in the material and convected with it as it flows or is deformed. If the coordinate surfaces are labelled ξ^j (j=1,2,3), then any particle of material has the same coordinates ξ^j at all times. It follows from what has been said that the equations describing the behaviour at time t of a particular material element at ξ^j can most simply be expressed as relationships between functions of ξ^1,ξ^2,ξ^3 at previous times t' ($-\infty < t' \leqslant t$).

"Of the kinematic variables associated with the material at ξ^j at times t' ($\leqslant t$), all those referring to absolute motion in space may be excluded as irrelevant, so that only those defining the relative distances between parts of a material element (and the way these change with time) are in the present context admissible. A knowledge of the distance ds(t') between an arbitrary pair of neighbouring particles ξ^j and $\xi^j + d\xi^j$ at every instant t' ($\leqslant t$) therefore constitutes complete information about the relevant kinematics, and this information is given by the variable metric tensor of the coordinate system $\gamma_{j\ell}(\xi,t')$ since

$$[ds(t')]^2 = \gamma_{j\ell}(\xi,t')\, d\xi^j\, d\xi^\ell\ . \tag{2.14}$$

In this equation, ξ is written for brevity in place of (ξ^1,ξ^2,ξ^3) ...".

Consider again in Fig. 2.2 the motion of the continuum where we have also indicated coordinate lines moving with the body. It is important to note that the coordinate transformation between the curvilinear system ξ^j and the Cartesian system x_i (which we will also denote by x^i for convenience) will be a function of time. Since the material particle X is now identified by its convected coordinates ξ^j, (2.12) becomes

$$x'^i = x'^i(\xi^k, t') \ , \tag{2.15}$$

and, at time t ,

$$x^i = x^i(\xi^k, t) \ . \tag{2.16}$$

The covariant components of the stress tensor in the ξ^j system at time t are given by

$$\pi_{ik} = \frac{\partial x^j}{\partial \xi^i} \frac{\partial x^\ell}{\partial \xi^k} P_{j\ell} \tag{2.17}$$

and we may in a similar way introduce the contravariant components π^{ik}.

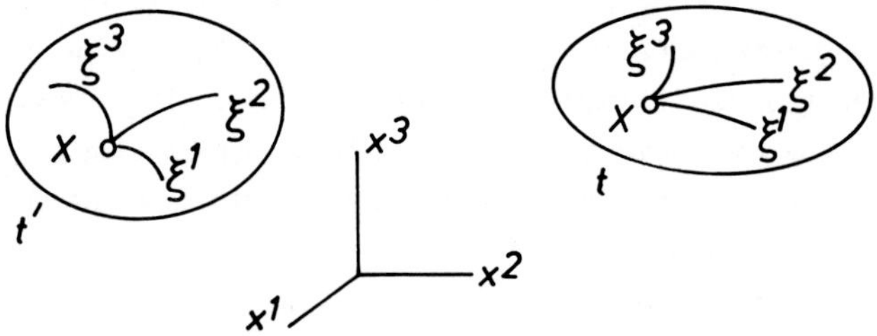

Fig. 2.2 Motion of the body and convected coordinates.

We can now assert that all the kinematic quantities in the constitutive equations for the stress tensor π_{ik} must be derivable from the tensor function

$$\gamma_{j\ell}(\xi, t') \ , \quad t' \leqslant t \ ,$$

where ξ is again written for brevity in place of ξ^1, ξ^2, ξ^3.

A related approach is to consider the *area* of a surface element through an equation of the form (see, for example, Truesdell 1958, White 1964)

$$[dA(t')]^2 = \gamma^{j\ell}(\xi, t') \, d\sigma_j \, d\sigma_\ell \ , \tag{2.18}$$

where $d\sigma_i$ are the convected covariant components of the area vector which do not change with the motion ($d\sigma_i \, d\xi^i$ being the differential volume). The contravariant tensor $\gamma^{j\ell}$ can be obtained from the covariant metric $\gamma_{j\ell}$ through

$$\gamma_{ij}(\xi,t')\ \gamma^{jk}(\xi,t') = \delta_i^k \tag{2.19}$$

and $\gamma^{j\ell}$ can also be selected as a kinematic variable.

We will come back later to a method of obtaining specific constitutive equations by means of algebraic relations between the stress tensor, the metric tensor and its inverse, and their time derivatives. For the sake of generality, we will first consider the rheological equations of the incompressible *simple fluid* :

$$\pi_{ik} = -\,p\gamma_{ik} + \tau_{ik} \ ,$$

$$\tau_{ik} = \mathcal{T}_{ik}[\gamma_{j\ell}(\xi,t') - \gamma_{j\ell}(\xi,t)\ ;\ -\infty<t'\leqslant t] \ , \tag{2.20}$$

where τ_{ik} are the covariant components of the extra-stress tensor at time t, and $\mathcal{T}_{ik}$ is a tensor-valued functional. For the time being, the term "tensor-valued functional" simply means a relationship between a tensor τ_{ik} and arguments which are functions of time. In (2.20) we have chosen $\gamma_{j\ell}(\xi,t') - \gamma_{j\ell}(\xi,t)$ as the kinematic variable instead of $\gamma_{j\ell}(\xi,t')$ alone, since this slight amendment ensures that small deformations correspond to small values of the variable, which actually vanishes for a rigid-body motion.

The problem of formulating rheological equations of state is in principle solved by (2.20). However, to be useful in the solution of flow problems, the rheological equations should preferably be referred to axes fixed in space since they have to be considered in conjunction with the familiar equations of continuity (2.3) and motion (2.6). These are best expressed in terms of fixed coordinates, as are the associated boundary conditions, and, although there have been noteworthy attempts to solve flow problems by recasting all the basic equations in convected coordinates (see, for example, Lodge 1951), there is no doubt that the preferred course of action is to transform the rheological equations to fixed coordinates.

Let $G_{ik}(x,t,t')$ denote the components of the argument of (2.20) in Cartesian coordinates, where x stands for the position x^i of the material point at time t. Using the same transformation of coordinates as in (2.17) we have

$$G_{ik}(x,t,t') = \frac{\partial\xi^j}{\partial x^i}\frac{\partial\xi^\ell}{\partial x^k}\,[\gamma_{j\ell}(\xi,t') - \gamma_{j\ell}(\xi,t)] \ , \tag{2.21}$$

and the Cartesian equivalent of (2.20) is given by

$$P_{ik} = -p\delta_{ik} + T_{ik} \ ,$$

$$T_{ik} = \mathcal{T}_{ik}[G_{j\ell}(x,t,t') \ ; \ -\infty<t'\leq t] \ . \qquad (2.22)$$

Since $\gamma_{j\ell}(\xi,t)$ is the metric tensor at time t, we have

$$\frac{\partial\xi^j}{\partial x^i}\frac{\partial\xi^\ell}{\partial x^k}\gamma_{j\ell}(\xi,t) = \delta_{ik} \ , \qquad (2.23)$$

where δ_{ik} is the metric in the Cartesian system of coordinates. Similarly, since $\gamma_{j\ell}(\xi,t')$ is the metric tensor at time t', we have, with the use of (2.15),

$$\frac{\partial\xi^j}{\partial x'^m}\frac{\partial\xi^\ell}{\partial x'^p}\gamma_{j\ell}(\xi,t') = \delta_{mp} \ . \qquad (2.24)$$

Let us multiply both sides of (2.24) by $\frac{\partial x'^m}{\partial x^i}\frac{\partial x'^p}{\partial x^k}$ and apply the chain rule of differentiation; we obtain

$$\frac{\partial\xi^j}{\partial x^i}\frac{\partial\xi^\ell}{\partial x^k}\gamma_{j\ell}(\xi,t') = \frac{\partial x'^m}{\partial x^i}\frac{\partial x'^p}{\partial x^k}\delta_{mp} = \frac{\partial x'^m}{\partial x^i}\frac{\partial x'^m}{\partial x^k} = C_{ik}(x,t,t') \ . \qquad (2.25)$$

We have thus obtained the right "Cauchy-Green" strain-tensor C_{ik} which may be calculated easily from the representation (2.13) of the motion. The constitutive equations of the simple fluid in Cartesian coordinates are thus given by (2.22) where G_{ik} is the finite deformation tensor defined by

$$G_{ik}(x,t,t') = C_{ik}(x,t,t') - \delta_{ik} \ . \qquad (2.26)$$

We will now, for completeness, show how the constitutive equation of a simple fluid may be obtained without resorting to convected coordinates.

2.5.2 Formal derivation

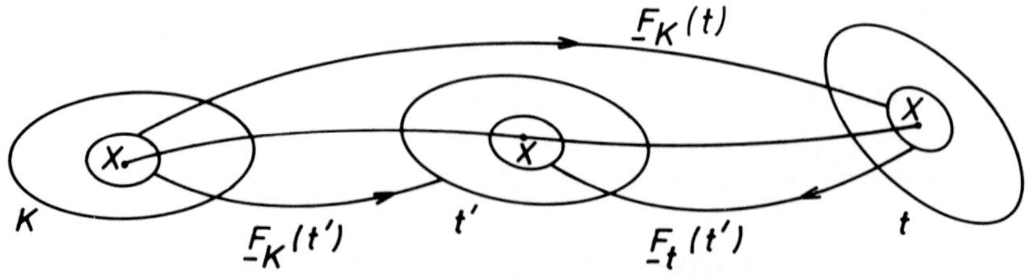

Fig. 2.3 The motion of a neighbourhood of the material point X.

In the present and subsequent sections, we shall occasionally find it easier to use the matrix notation rather than the index notation; it is an easy matter to revert back to rectangular Cartesian coordinates. Moreover, for the sake of brevity, we will omit reference to the material point X in the arguments of the functions.

Let us consider the motion of an arbitrarily small neighbourhood N_X of a material point X in the continuum, and let κ denote some reference configuration of N_X. At time $t' \leqslant t$, the configuration of N_X is fully characterized by the deformation gradient $\underline{F}_\kappa(t')$ which is a linear transformation between the reference configuration κ and the configuration at time t'. For an incompressible material which follows the principles of determinism of the stress and of local action, the Cauchy stress $\underline{P}$ at time t may be written as follows :

$$\underline{P} = - p\underline{I} + \underline{T} \ ,$$

$$\underline{T} = \underline{\mathcal{T}}_\kappa[\underline{F}_\kappa(t') \ ; \ -\infty<t'\leqslant t] \ , \tag{2.27}$$

where $\underline{I}$ is the unit tensor, p is an arbitrary pressure, and $\underline{\mathcal{T}}_\kappa$ is a tensor-valued functional.

The deformation gradient $\underline{F}_\kappa(t')$ may be decomposed as follows (Fig. 2.3):

$$\underline{F}_\kappa(t') = \underline{F}_t(t') \ \underline{F}_\kappa(t) \ , \tag{2.28}$$

where $\underline{F}_t(t')$ stands for the deformation gradient of the configuration at time t' relative to the configuration at time t. Without loss of generality we may re-write the constitutive equation (2.27) as follows :

$$\underline{T} = \underline{\mathcal{T}}_\kappa[\underline{F}_t(t') \ ; \ \underline{F}_\kappa(t) \ ; \ -\infty<t'\leqslant t] \ . \tag{2.29}$$

Let κ' denote another reference configuration related to κ through a linear transformation $\underline{Q}'$ such that

$$\underline{F}_{\kappa'}(t) = \underline{F}_\kappa(t) \ \underline{Q}' . \tag{2.30}$$

If we use κ' as a reference configuration, the constitutive equation (2.29) becomes

$$\underline{T} = \underline{\mathcal{T}}_{\kappa'}[\underline{F}_t(t') \ ; \ \underline{F}_{\kappa'}(t) \ ; \ -\infty<t'\leqslant t] \ . \tag{2.31}$$

We will now express that the material is a fluid by claiming that the response functional $\underline{T}_{\kappa}$ remains the same for any unimodular transformation $\underline{Q}'$ of the reference configuration, i.e. any transformation which preserves the volume of N_X in the reference configuration. With the use of (2.29-31) we obtain the condition

$$\underline{T}_{\kappa}[\underline{F}_t(t') \; ; \; \underline{F}_{\kappa}(t) \; ; \; -\infty<t'\leqslant t]$$

$$= \underline{T}_{\kappa}[\underline{F}_t(t') \; ; \; \underline{F}_{\kappa}(t)\underline{Q}' : \; -\infty<t'\leqslant t] \quad , \tag{2.32}$$

which must be satisfied for any unimodular tensor $\underline{Q}'$. The identity (2.32) will be satisfied only if $\underline{T}_{\kappa}$ depends upon the *determinant* of $\underline{F}_{\kappa}(t)$; since the fluid is incompressible, the determinant equals one, and (2.29) may now be rewritten as

$$\underline{T} = \underline{T}[\underline{F}_t(t') \; ; \; -\infty<t'\leqslant t] \quad . \tag{2.33}$$

We have omitted for $\underline{T}$ the subscript κ since $\underline{T}$ is now independent of the transformations of the reference configuration.

The relative deformation gradient $\underline{F}_t(t')$ may be uniquely decomposed as the product of the relative rotation tensor $\underline{R}_t(t')$, which is orthogonal, and the right relative stretch tensor $\underline{U}_t(t')$, which is symmetric and positive definite,

$$\underline{F}_t(t') = \underline{R}_t(t') \; \underline{U}_t(t') \; . \tag{2.34}$$

Let us now superpose on the motion (2.12) a rigid body motion, comprising translations and rotations which depend upon time. The deformation gradient $\underline{F}_{\kappa}(t')$ is affected as follows by the rigid body motion,

$$\underline{F}^*_{\kappa}(t') = \underline{Q}(t') \; \underline{F}_{\kappa}(t') \quad , \tag{2.35}$$

where $\underline{Q}(t')$ is an orthogonal tensor and $\underline{F}^*_{\kappa}(t')$ is the new value of the deformation gradient. With the help of (2.28) and (2.35) we obtain

$$\underline{F}^*_t(t') = \underline{F}^*_{\kappa}(t') \; [\underline{F}^*_{\kappa}(t)]^{-1} = \underline{Q}(t')\underline{F}_t(t') \; \underline{Q}^T(t) \; , \tag{2.36}$$

where the superscript T denotes the transpose. We may now apply the principle of invariance under superposed rigid body motion by requiring that

$$\underline{T}[\underline{F}^*_t(t') \; ; \; -\infty<t'\leqslant t] = \underline{Q}(t) \; \underline{T}[\underline{F}_t(t') \; ; \; -\infty<t'\leqslant t] \; \underline{Q}^T(t) \tag{2.37}$$

for any orthogonal tensor function $\underset{\sim}{Q}(t')$; (2.37) may be rewritten in the form

$$\underset{\sim}{T}[\underset{\sim}{F}_t(t') ; -\infty<t'\leq t] = \underset{\sim}{Q}^T(t) \underset{\sim}{T}[\underset{\sim}{Q}(t')\underset{\sim}{F}_t(t') \underset{\sim}{Q}^T(t) ; -\infty<t'\leq t] \underset{\sim}{Q}(t) . \quad (2.38)$$

The identity (2.38) must be satisfied in particular if we select for the tensor $\underset{\sim}{Q}(t')$ the form

$$\underset{\sim}{Q}(t') = \underset{\sim}{Q} \underset{\sim}{R}_t^T(t') , \quad (2.39)$$

where $\underset{\sim}{Q}$ is an arbitrary orthogonal tensor. We obtain from (2.38) and (2.34)

$$\underset{\sim}{T}[\underset{\sim}{F}_t(t') ; -\infty<t'\leq t] = \underset{\sim}{Q}^T \underset{\sim}{T}[\underset{\sim}{Q} \underset{\sim}{U}_t(t') \underset{\sim}{Q}^T ; -\infty<t'\leq t]\underset{\sim}{Q} . \quad (2.40)$$

We conclude that $\underset{\sim}{T}$ must be an isotropic tensor-valued functional of the right relative stretch tensor $\underset{\sim}{U}_t(t')$ or, equivalently, of its square $\underset{\sim}{C}_t(t')$ given by

$$\underset{\sim}{C}_t(t') = [\underset{\sim}{U}_t(t')]^2 = \underset{\sim}{F}_t^T(t') \underset{\sim}{F}_t(t') , \quad (2.41)$$

which is precisely the right relative Cauchy-Green strain tensor given by (2.25). We have thus obtained the constitutive equation

$$\underset{\sim}{P} = - p\underset{\sim}{I} + \underset{\sim}{T} , \quad \underset{\sim}{T} = \underset{\sim}{T}[\underset{\sim}{C}_t(t') ; -\infty<t'\leq t] , \quad (2.42)$$

which is essentially equivalent to (2.22), and where $\underset{\sim}{T}$ is an isotropic tensor-valued functional of its argument (see also Lodge and Stark 1972).

2.6 APPROXIMATE CONSTITUTIVE EQUATIONS

Consider now the constitutive equations for the simple fluid in the form given by (2.22). A more suitable time variable in the present context is the time lapse s=t-t'; for convenience we write $G_{j\ell}(s)$ instead of $G_{j\ell}(x,t,t')$ and we have

$$T_{ik} = \mathcal{T}_{ik}[G_{j\ell}(s) , 0\leq s<\infty] . \quad (2.43)$$

As a formal relationship between stress and deformation we may say that (2.43) is completely general and includes all the models which have appeared in the literature to represent isotropic fluids with or without memory. For example, we may identify the functional with a time derivative to recover the Newtonian liquid, and so on. We must also recognize that (2.43) is far too general if we wish to solve specific non-trivial problems, in particular by means of numerical techniques. We will elaborate on two main approaches used for obtaining specific constitutive equations. In the present section, we show how it is possible to obtain asymptotic forms of (2.43) under some special flow conditions and we emphasize their limitations. In the next section, we then follow a more pragmatic approach for generating some simple constitutive equations.

It is important to emphasize that the generality of (2.43) can be inadvertently lost when we proceed to define suitable norms with associated smoothness assumptions. In that sense, the simple fluid with fading memory, developed by Coleman and Noll (1961) is not completely general, since it does not even contain the Newtonian liquid as a special case. We wish however to show the value of the concept of fading memory in the search for narrow-range approximations to the general functional equation (2.43).

Elastic liquids have a fading memory in the sense that the deformation experienced by a fluid element in the distant past must be expected to have a smaller influence on the current stress in that element than those deformations which took place in the recent past. Coleman and Noll (1961) seek to accommodate fading memory through the introduction of a Hilbert-space norm,

$$\| \underline{G}(s) \| = \left[\int_0^\infty |\underline{G}(s)|^2 \, h^2(s) \, ds \right]^{\frac{1}{2}} , \tag{2.44}$$

where $|\underline{G}(s)|$ is the magnitude of the tensor $\underline{G}(s)$ and h is an influence function satisfying

(a) h(s) is defined for $0 \leq s < \infty$ and has positive real values : $h(s) > 0$.

(b) h(s) decays to zero, monotonically for large s, according to
$\lim_{s \to \infty} s^r h(s) = 0 \ , \ r > \frac{1}{2}$.

The norm is clearly designed to give more prominence to small values of s, i.e. the recent past.

2.6.1 Asymptotic form for 'small strain'

The function $\underline{G}(s)$, defined by (2.26), vanishes identically when the neighbourhood of the material point where $\underline{G}(s)$ is defined has not undergone any deformation in the past. The magnitude of the norm (2.44) indicates how much the deformation history of the material element has differed from the rest history for which $\| \underline{G}(s) \|$ vanishes. Coleman and Noll (1961) make certain smoothness assumptions of the functional $\underline{T}$ in (2.43) in the neighbourhood of the rest history and make use of the Fréchet differential, which may be regarded as an extension of the ordinary differential to functionals. The process leads to a set of approximations of the functional $\underline{T}$ which are said to be of order n when the remainder is of order $O(\| \underline{G}(s) \|^{n+1})$ (see also Rivlin 1983).

The first order approximation has an integral representation of the form

$$T_{ik} = \int_0^\infty M_1(s)\ G_{ik}(s)\ ds \quad ; \tag{2.45}$$

for higher orders, the approximation will consist of multilinear functionals which will not in general be endowed with an integral representation. Under certain smoothness assumptions, however, second and third order approximations can be written in the form (cf. Coleman and Noll 1961, Pipkin 1964)

$$T_{ik} = \int_0^\infty M_1(s)\ G_{ik}(s)ds + \int_0^\infty \int_0^\infty M_2(s_1,s_2)\ G_{ij}(s_1)\ G_{jk}(s_2)\ ds_1\ ds_2\ , \tag{2.46}$$

$$\begin{aligned} T_{ik} = & \int_0^\infty M_1(s)\ G_{ik}(s)ds + \int_0^\infty \int_0^\infty M_2(s_1,s_2)\ G_{ij}(s_1)\ G_{jk}(s_2)\ ds_1\ ds_2 \\ & + \int_0^\infty \int_0^\infty \int_0^\infty [M_3(s_1,s_2,s_3)\ G_{j\ell}(s_2)\ G_{\ell j}(s_3)\ G_{ik}(s_1) \\ & + M_4(s_1,s_2,s_3)\ G_{ij}(s_1)\ G_{j\ell}(s_2)\ G_{\ell k}(s_3)]\ ds_1\ ds_2\ ds_3\ , \end{aligned} \tag{2.47}$$

where, from the symmetry of the stress tensor, the kernel functions must satisfy

$$M_2(s_1,s_2) = M_2(s_2,s_1)\ , \quad M_4(s_1,s_2,s_3) = M_4(s_3,s_2,s_1)\ . \tag{2.48}$$

Equations (2.45) are called the equations of finite linear viscoelasticity, (2.46) are called the equations of second-order viscoelasticity and so on.

Equations (2.46)-(2.48) may be regarded as approximations to the general functional equation (2.43) under the condition that $\|\underline{G}(s)\|$ is small. This is typically the case when $|\underline{G}(s)|$ is itself small for all s, and an obvious example would be small amplitude oscillatory shear flow.

The equations of nth order viscoelasticity should be used with a certain degree of caution. A problem arises in implementing the approximations in practical flow problems because the deformation variable G_{ik} is in general a non-linear function of the relevant ordering parameter. As an example, take the case of the steady simple shear flow (cf. (1.1))

$$v_1 = \gamma x_2\ , \qquad v_2 = v_3 = 0\ , \tag{2.49}$$

for which G_{ik} is given by

$$G_{ik} = \begin{bmatrix} 0 & -\gamma s & 0 \\ -\gamma s & \gamma^2 s^2 & 0 \\ 0 & 0 & 0 \end{bmatrix}\ . \tag{2.50}$$

We see that G_{ik} involves both linear *and* quadratic terms in the shear rate γ. If we now consider γ to be the ordering parameter, a possible source of confusion is highlighted. The norm $\| \underline{G}(s) \|$ is of order γ, and *to first order in* γ, (2.45) is certainly a valid approximation; however, the quadratic terms involving γ^2 have no meaning as part of the first order approximation, since the remainder is of the same order. In simple terms, the approximations (2.45)-(2.47) are one order less powerful than a cursory inspection might suggest.

General memory-integral expansions, for which (2.45)-(2.47) represent the early-order terms, have also been discussed by Green, Rivlin and Spencer (1957, 1959,1960). Such expansions can be viewed as approximations to the functional equation (2.43) arising from a procedure analogous to the Taylor-series expansion of an analytic function, or, alternatively, from an application of the Stone-Weierstrass theorem. Alternative memory-integral expansions employing alternative strain (or strain-rate) measures are also available. That developed within a corotational framework has been considered by Goddard and Miller (1966) and given prominence by Bird et al. (1977).

2.6.2 Asymptotic form for 'slow flow'

An alternative approximation procedure can be obtained from a consideration of slow flow, in a sense which must be defined very carefully. If we consider a material point in slow flow, we require that all the kinematic variables associated with it vary slowly. This is an important qualification! For example, we might argue that flow in the neighbourhood of a re-entrant corner is slow, but we could not conclude without caution that other kinematic variables, like $\gamma_{ik}(\xi,t')$, are slowly varying. For the same reason, high frequency small amplitude oscillatory shear flows have to be excluded from this category. Failure to appreciate what is involved in the slow flow approximation has led to more than one instability paradox, where absurd results are obtained through the incorrect use of the approximate equations valid for slow flow. Nothing more will be said on this matter since too much literature space has already been taken up attempting to use slow-flow approximations outside their sphere of validity.

In order to make the definition of slow-flow more precise and to obtain an approximation to the general functional (2.43), Coleman and Noll (1960) introduced the notion of retarded motion. Consider a given deformation history $\underline{G}(s)$, and a further set of histories $\underline{G}_\alpha(s)$ defined by

$$\underline{G}_\alpha(s) = \underline{G}(\alpha s) \quad , \quad 0<\alpha\leq 1 \quad . \tag{2.51}$$

The histories $\underline{G}_\alpha(s)$ are essentially the same as $\underline{G}(s)$ but carried out at a slower rate. It may then be shown that

$$\| \underline{G}_{\alpha}(s) \| = 0(\alpha) \ . \tag{2.52}$$

Let us define the Rivlin-Ericksen (1955) tensor of order n by†

$$\underline{A}^{(n)} = (-1)^n \frac{d^n \underline{G}(s)}{ds^n} \Big|_{s=0} \ . \tag{2.53}$$

By means of the equality

$$\underline{G}_{\alpha}(s) = \sum_{j=0}^{n} \frac{s^j \alpha^j}{j!} (-1)^j \underline{A}^{(j)} + 0(\alpha^{n+1}) \quad , \tag{2.54}$$

Coleman and Noll (1960) have shown that when the norm (2.44) is small through small α, the functional in (2.43) can be approximated by a polynomial function of the derivatives at s=0 of the argument functions of the functional, i.e. the Rivlin-Ericksen tensors $\underline{A}^{(n)}$. A series of approximations can now be obtained, the first, second and third-order approximations being given by (2.55)-(2.57), respectively :

$$T_{ik} = \alpha_1 A^{(1)}_{ik} \quad , \tag{2.55}$$

$$T_{ik} = \alpha_1 A^{(1)}_{ik} + \alpha_2 A^{(2)}_{ik} + \alpha_3 A^{(1)}_{ij} A^{(1)}_{jk} \quad , \tag{2.56}$$

$$T_{ik} = \alpha_1 A^{(1)}_{ik} + \alpha_2 A^{(2)}_{ik} + \alpha_3 A^{(1)}_{ij} A^{(1)}_{jk} + \beta_1 A^{(1)}_{j\ell} A^{(1)}_{\ell j} A^{(1)}_{ik}$$
$$+ \beta_2 A^{(3)}_{ik} + \alpha_5 (A^{(1)}_{ij} A^{(2)}_{jk} + A^{(2)}_{ij} A^{(1)}_{jk}) \ . \tag{2.57}$$

When the functional is such that (2.45)-(2.47) constitute valid approximations, the material constants and the kernel functions are related by the equations

---

† The nth rate-of-strain tensor $\underline{e}^{(n)}$ defined by Oldroyd (1950) is given by $\underline{A}^{(n)} = 2\underline{e}^{(n)}$, $\underline{e}^{(1)}$ being the same as the rate-of-deformation tensor defined in (2.10).

$$\alpha_1 = - \int_0^\infty M_1(s)\ s\ ds \quad , \qquad \alpha_2 = \int_0^\infty M_1(s)\ \frac{s^2}{2}\ ds \ ,$$

$$\beta_2 = - \int_0^\infty M_1(s)\ \frac{s^3}{6} ds \quad , \qquad \alpha_3 = \int_0^\infty \int_0^\infty M_2(s_1,s_2)\ s_1 s_2\ ds_1\ ds_2 \ ,$$

$$\alpha_5 = - \int_0^\infty \int_0^\infty M_2(s_1,s_2)\ \frac{s_1 s_2^2}{2}\ ds_1\ ds_2 \quad ,$$

$$\beta_1 = - \int_0^\infty \int_0^\infty \int_0^\infty [M_3(s_1,s_2,s_3) + \tfrac{1}{2}\ M_4(s_1,s_2,s_3)]\ s_1 s_2 s_3\ ds_1\ ds_2\ ds_3 \ . \qquad (2.58)$$

In an interesting paper, Truesdell (1964) argued from dimensional considerations that the approximations (2.55)-(2.57) must also apply for materials with rapidly fading memory, i.e. for slightly elastic liquids. Such a conclusion is very useful since it highlights the fact that the asymptotic process can be seen as arising from an interplay between material properties on the one hand (i.e. how strong is the memory?) and flow conditions on the other (i.e. how slow is the flow?). In the same paper, Truesdell also offered the more contentious suggestion that there was no reason why equations like (2.56) should not be made the basis of an exact theory, for all motions, in the same sense that the Navier-Stokes equations are regarded as exact for Newtonian liquids.

Inspection of (2.55)-(2.57) shows that the Newtonian liquid is to be found in simple fluid theory as developed by Coleman and Noll, but only as a first-order approximation in slow flow. It is certainly possible to choose kernel functions in (2.45)-(2.47) for any order of approximation to recover the Newtonian liquid, but generalized functions (in this case delta functions) are required in the process and these fall outside the permitted functions in the original Coleman and Noll development. It is possible in principle to modify the functional analysis to accommodate generalized functions and this is under active consideration at the present time (Saut and Joseph 1983).

Before the work of Coleman and Noll on retarded motions, Rivlin and Ericksen (1955) had considered fluids for which the stress could be written as a function of the first n Rivlin-Ericksen tensors. These so-called Rivlin-Ericksen fluids can be viewed as approximations to the simple fluid which are valid under conditions of fading memory and retarded flow, the lower-order approximations being equivalent to (2.55)-(2.57).

2.7 A PRAGMATIC APPROACH TO CONSTITUTIVE EQUATIONS

There are many complex flow problems of theoretical and practical interest which do not meet the conditions necessary to employ the relatively simple approximations (2.45)-(2.47) or (2.55)-(2.57) to the functional equation (2.43).

The interested theoretician is therefore presented with a decision of principle. Since no solution of general validity is possible, does he decide that there is no merit in giving any attention to the flow problem and conclude that it is best left in the file of unsolved and unsolvable problems? Or does he take a more pragmatic stance and do the best he can with the limited information and techniques available? We have little sympathy with the first alternative and choose to adopt the philosophy embodied in the second, even if this means leaving the haven of generality.

We are immediately faced with the problem of deciding how to choose the most appropriate rheological equation of state to suit a given flow problem. Here, there are no simple answers, since there is what one author has called a jungle of constitutive equations to choose from. Certain general points can however be made concerning relatively simple rheological equations of state :

(a) Although most rheological models have their devoted adherents, no one completely satisfactory model is at present available which satisfies the dual requirement of being able to simulate available data from simple rheometrical experiments and at the same time of having the simplicity required to render complex flow problems tractable. Some compromise is required and it has to be acknowledged that the more complex the flow problem, the simpler must be the rheological equations employed.

(b) In what we might call 'engineering applications', we shall have to content ourselves with 'solving' problems by a shrewd combination of analytic solutions for special limiting cases, detailed numerical simulation using simple rheological models, dimensional analysis and physical insight.

It is clearly impracticable to discuss the plethora of simple rheological equations of state which have appeared in the literature. The most we can do is to place them into two basic groups, which we loosely term the single integral models and the Oldroyd/Maxwell models, respectively.

2.7.1 Single integral models

Single integral models may be motivated in different ways. For example, Rivlin and Sawyer (1971) associate such models with the requirement that the effects on the stress at time t of the deformations existing in a fluid element at different times are independent of each other. On the other hand, Bernstein, Kearsley and Zapas (1963) offer thermodynamic arguments to support their particular development. Others see the general single integral models as a logical outcome of the model-building process using the rubber-like liquid model of Lodge (1956) as the starting point.

Such models can be represented in rectangular Cartesian coordinates by the equations

$$T_{ik} = \int_0^\infty [\Phi(I_1,I_2,s)\, G_{ik}(s) + \Psi(I_1,I_2,s)\, G_{ij}(s)\, G_{jk}(s)]\, ds \ , \tag{2.59}$$

where I_1 and I_2 are two independent invariants of $G_{ik}(s)$.

Let us define the tensor $H_{ik}(x,t,t')$ through

$$H_{ik}(x,t,t') = C^{-1}_{ik}(x,t,t') - \delta_{ik} \ , \tag{2.60}$$

where the superscript (-1) denotes the inverse, and use the notation $H_{ik}(s)$ for simplicity. An equivalent form of (2.59) is now

$$T_{ik} = \int_0^\infty [\bar{\Phi}(I_1,I_2,s)\, G_{ik}(s) + \bar{\Psi}(I_1,I_2,s)\, H_{ik}(s)]\, ds \ . \tag{2.61}$$

Seemingly equivalent models using rate of deformation tensors in place of $G_{ik}(s)$ also abound in the literature. Other, more sophisticated refinements may also be found (see, for example, Wagner 1978).

2.7.2 Differential models of the Oldroyd/Maxwell type

The Oldroyd/Maxwell models arise from the reasonable desire to generalize the Maxwell and Jeffrey's models of linear viscoelasticity into a form which ensures their validity under all conditions of motion and stress. At their inception, there was a reasonable expectation that such generalizations would have predictive capacity in a quantitative sense. After all, the simple equation (2.9) representing Newtonian behaviour had been remarkably successful in predicting the behaviour of a number of common fluids and the rheologists of the 50's could be excused some optimism that a similar situation would pertain for elastic liquids. With hindsight, it is now conceded that optimism was largely unjustified and at the present time the Oldroyd/Maxwell models are viewed as providing a *qualitative* description of observed behaviour in simple flows, including stress relaxation after cessation of flow, together with the essential simplicity for use in complex-problem solving (see Chapter 3).

We have seen above that, in convected coordinates, the stress tensor is a functional of the history of the metric tensor. Oldroyd/Maxwell models are generated by means of the assumption that the functional dependence of the stress tensor with respect to the metric tensor is expressed through a tensorially-invariant relation between these tensors and their time derivatives. A typical example would be the rheological equation of state

$$\tau_{ik} + \lambda_1 \frac{D}{Dt} \tau_{ik} = \eta_0 \frac{D}{Dt} \gamma_{ik} \ , \tag{2.62}$$

where D/Dt denotes the material derivative, i.e. a time derivative holding convected coordinates constant.

Before going further we need to know the fixed-component equivalent of the material derivative of a tensor in convected coordinates. This is provided by the following theorem, the proof of which is given in Appendix 1 (cf. Oldroyd 1950).

Theorem The tensor whose convected components are $\frac{D}{Dt}\beta_j^{\;\ell}$ has fixed components

$$\frac{\delta}{\delta t} b_i^{\;k} = \frac{\partial}{\partial t} b_i^{\;k} + v^m \frac{\partial}{\partial x^m} b_i^{\;k} + \frac{\partial v^m}{\partial x^i} b_m^{\;k} - \frac{\partial v^k}{\partial x^m} b_i^{\;m} \quad , \tag{2.63}$$

where v^m represents the components of the velocity vector in the fixed x^i coordinate system, which is not necessarily Cartesian.

We may obtain, similarly, the so-called lower-convected and upper-convected derivatives, which are the fixed-component equivalents of $\frac{D}{Dt}\beta_{j\ell}$ and $\frac{D}{Dt}\beta^{j\ell}$, and are given, respectively, by

$$\frac{\delta b_{ik}}{\delta t} = \frac{\partial b_{ik}}{\partial t} + v^m \frac{\partial b_{ik}}{\partial x^m} + \frac{\partial v^m}{\partial x^i} b_{mk} + \frac{\partial v^m}{\partial x^k} b_{im} \quad , \tag{2.64}$$

$$\frac{\delta b^{ik}}{\delta t} = \frac{\partial b^{ik}}{\partial t} + v^m \frac{\partial b^{ik}}{\partial x^m} - \frac{\partial v^i}{\partial x^m} b^{mk} - \frac{\partial v^k}{\partial x^m} b^{im} \quad . \tag{2.65}$$

If g_{ik} is the metric tensor of the fixed coordinate system x^i, we have (cf. Oldroyd 1950 and (2.53))

$$\frac{\delta^n g_{ik}}{\delta t^n} = A_{ik}^{(n)} = 2e_{ik}^{(n)} \quad , \tag{2.66}$$

and the rate-of-deformation tensor, which we have written for convenience as d_{ik}, is of course given by

$$d_{ik} = \frac{1}{2}\frac{\delta g_{ik}}{\delta t} \quad . \tag{2.67}$$

We shall find it convenient to use the notation

$$\frac{\delta b_{ik}}{\delta t} = \overset{\Delta}{b}_{ik} \quad , \qquad \frac{\delta b^{ik}}{\delta t} = \overset{\nabla}{b}{}^{ik} \quad ; \tag{2.68}$$

so the fixed-component equivalent of (2.62) can be written

$$T_{ik} + \lambda_1 \overset{\Delta}{T}_{ik} = 2\eta_0 \, d_{ik} \, . \tag{2.69}$$

Confining attention now to rectangular Cartesian coordinates to avoid unnecessary confusion over raising and lowering suffices, we may define a general derivative as follows :

$$\overset{\square}{b}_{ik} = (1 - \frac{a}{2}) \overset{\nabla}{b}_{ik} + \frac{a}{2} \overset{\Delta}{b}_{ik} \quad , \tag{2.70}$$

where a is a constant. Clearly, for a=0 we recover the upper-convected derivative and for a=2 the lower-convected derivative. For a=1, we have the corotational derivative for which we reserve the superscript 0 (see, for example, Oldroyd 1958, Bird et al. 1977, Petrie 1979).

Consider now the general Oldroyd model given by

$$T_{ik} + \lambda_1 \overset{\square}{T}_{ik} + \mu_0 T_{jj} d_{ik} - \mu_1 (T_{ij} d_{jk} + T_{kj} d_{ji}) = 2\eta_0 (d_{ik} + \lambda_2 \overset{\square}{d}_{ik} - \mu_2 d_{ij} d_{jk}) \, , \tag{2.71}$$

where η_0 is a constant viscosity coefficient and λ_1, λ_2, μ_0, μ_1, μ_2 are material constants. Several models which are currently in use are contained in (2.71); they are summarized in Table 2.1.

2.7.3 Equivalence between integral and differential models

Before leaving this section, it is of interest to note that some models have a simple integral as well as a differential representation. For example, if in (2.61) we select for the kernel the following form

$$T_{ik} = \frac{\eta_0}{\lambda_1^2} \int_0^{\infty} e^{-\frac{s}{\lambda_1}} \, [(1 - \frac{a}{2}) H_{ik} - \frac{a}{2} G_{ik}(s)] \, ds \; , \tag{2.80}$$

it may easily be shown that, when a = 0, (2.80) is equivalent to (2.72) and that, when a = 2, (2.80) is equivalent to (2.73).

The identity between some simple differential and integral models is an important factor in the evaluation of numerical techniques for solving the flow of Maxwell-type models.

a	λ_1	μ_0	μ_1	λ_2	μ_2	Name	Constitutive equation	Equation No.
0	λ_1	0	0	0	0	Upper-convected Maxwell	$T_{ik} + \lambda_1 \overset{\nabla}{T}_{ik} = 2\eta_0 d_{ik}$	(2.72)
2	λ_1	0	0	0	0	Lower-convected Maxwell	$T_{ik} + \lambda_1 \overset{\Delta}{T}_{ik} = 2\eta_0 d_{ik}$	(2.73)
1	λ_1	0	0	0	0	Corotational Maxwell	$T_{ik} + \lambda_1 \overset{\circ}{T}_{ik} = 2\eta_0 d_{ik}$	(2.74)
a	λ_1	0	0	0	0	Johnson-Segalman (1977) special case of Phan Thien-Tanner (1977)	$T_{ik} + \lambda_1 \overset{\square}{T}_{ik} = 2\eta_0 d_{ik}$	(2.75)
2	λ_1	0	0	λ_2	0	Oldroyd's liquid A (Oldroyd 1950)	$T_{ik} + \lambda_1 \overset{\Delta}{T}_{ik} = 2\eta_0 (d_{ik} + \lambda_2 \overset{\Delta}{d}_{ik})$	(2.76)
0	λ_1	0	0	λ_2	0	Oldroyd's liquid B (Oldroyd 1950)	$T_{ik} + \lambda_1 \overset{\nabla}{T}_{ik} = 2\eta_0 (d_{ik} + \lambda_2 \overset{\nabla}{d}_{ik})$	(2.77)
0	λ_1	μ_0	0	λ_2	0	Four constant Oldroyd (Walters 1979)	$T_{ik} + \lambda_1 \overset{\nabla}{T}_{ik} + \mu_0 T_{jj} d_{ik} = 2\eta_0 (d_{ik} + \lambda_2 \overset{\nabla}{d}_{ik})$	(2.78)
-	0	0	0	0	μ_2	Reiner-Rivlin fluid (special case) (Rivlin 1948)	$T_{ik} = 2\eta_0 d_{ik} - 2\eta_0 \mu_2 d_{ij} d_{jk}$	(2.79)

Table 2.1

2.8 CONSTRAINTS ON RHEOLOGICAL EQUATIONS OF STATE

Once it is conceded that a case exists for employing relatively simple constitutive equations in problem solving, especially in what we may call engineering and industrial applications, we should not be surprised to find that numerous researchers have developed additional constraints which such equations should satisfy, i.e. constraints over and above those arising from the basic formulation principles discussed in §2.4.

First, we would expect any model chosen for numerical simulation studies to represent at least qualitatively the rheometrical behaviour of the real fluids under consideration (see, for example, Walters 1975,1980). This is especially so of the important viscosity/shear rate behaviour in steady shear flow, but also applies to normal stress levels, stress relaxation on cessation of flow, and the ability to produce the abnormally high extensional viscosities found in some dilute polymer solutions. More sophisticated rheometrical tests, such as combined steady and oscillatory shear, may also be available and these often place very severe constraints on rheological models. Indeed, Tanner and Simmons (1967) show from a consideration of orthogonal superposition of steady and oscillatory shear that certain simple Oldroyd/Maxwell models lead to unacceptable instabilities.

A related rheometrical test, namely that associated with non-linear effects in oscillatory shear flow, has also been suggested as providing a severe constraint on constitutive equations. The question posed is the following. Does the departure from linear viscoelastic behaviour depend on the amplitude ε of the motion or the combination $\varepsilon\omega$, where ω is the frequency; i.e. is the departure strain dependent or strain-rate dependent? Available evidence seems to suggest a strain dependent departure from linear behaviour and accordingly Astarita and Marrucci (1974) deprecate the use of strain-rate dependent integral models (but see Goddard 1979).

Another constraint on constitutive equations is provided by the so-called stress-overshoot phenomenon. When a simple shear flow is started from rest in elastic liquids, the shear and normal stresses are often found to overshoot their equilibrium values before reaching a steady state. The constitutive equations employed should be expected to predict such behaviour when it occurs. In addition, Van Es and Christensen (1973) argue that stress overshoot data is sometimes able to rule out certain members of a class of simple integral constitutive equations which do predict stress overshoot.

In the case of very viscous elastic liquids, Lodge has argued that rheometrical tests involving step changes in strain can provide severe constraints on the form of the allowable constitutive equations of the type (2.71) (see, for example, Walters 1980, Lodge 1983).

Microrheological considerations have also been suggested as a possible means of suggesting the most appropriate form of relatively simple constitutive models. For example, Petrie (1979) points out that upper convected models like (2.72) arise in theories of networks of entangled molecules and some dilute solution theories, while the parameter a is required to take non-zero values to accommodate non-affine theories which allow some relative motion between polymer molecules and the observable continuum (see also Johnson and Segalman 1977, Phan Thien and Tanner 1977). Further detailed and useful information on possibilities in the general area of microrheology is provided by the works of Bird et al. (1977), Doi and Edwards (1978) and Curtiss and Bird (1981).

In this section, we have considered a number of possible constraints which may be imposed on simple constitutive equations. It may be argued that, taken together, they present an unacceptably severe constraint, with the resulting equation being too complicated to have predictive ability. This may indeed be so in some cases and we may be required to abandon the less severe to make progress. Deciding which of the constraints are expendable in a given flow situation is to a large extent a matter of experience and subjective judgment, but there is no doubt that the ability to simulate the viscometric functions η and ν_1 is of paramount importance. The prediction of the correct sign and order of magnitude of the second normal stress difference ν_2, extensional viscosity η_E levels, and the existence (or otherwise) of stress overshoot is also advantageous. For this reason, we include in Table 2.2 the relevant rheometrical functions for the models defined in Table 2.1.

2.9 BOUNDARY CONDITIONS

The rheological equations of state have to be solved in conjunction with the stress equations of motion (2.6) and the equation of continuity (2.3). In Newtonian fluid mechanics, the extra stress components can be substituted out of the governing equations yielding the Navier-Stokes equations (2.11). These, together with (2.3) provide four equations in the four unknowns v_i(i=1,2,3) and p. So far as boundary conditions are concerned, it is sufficient to specify the velocity or surface force components over the boundary of the domain of interest, and the pressure at one point when no normal surface force has been specified anywhere on the boundary. For elastic liquids this specification is insufficient on account of fluid memory. If the boundary of the domain contains an entry region, we now need to know the strain history of the fluid entering the domain or, what is equivalent, the knowledge of the stress field on entry to the domain. In practice, the boundary condition requirements are often inadvertently satisfied by assuming "fully-developed flow conditions" on entry to the domain, which essentially implies knowledge of the flow field upstream of the domain of interest.

Name and Equation No.	$\eta(\gamma)$	$\nu_1(\gamma)$	$\nu_2(\gamma)/\nu_1(\gamma)$	$\eta_E(k)$	Stress Overshoot
Upper-convected Maxwell (2.72)	η_0	$2\eta_0\lambda_1\gamma^2$	0	$\frac{2\eta_0}{(1-2\lambda_1 k)}+\frac{\eta_0}{(1+\lambda_1 k)}$	No
Lower-convected Maxwell (2.73)	η_0	$2\eta_0\lambda_1\gamma^2$	-1	$\frac{2\eta_0}{(1+2\lambda_1 k)}+\frac{\eta_0}{(1-\lambda_1 k)}$	No
Corotational Maxwell (2.74)	$\frac{\eta_0}{(1+\lambda_1^2\gamma^2)}$	$\frac{2\eta_0\lambda_1\gamma^2}{(1+\lambda_1^2\gamma^2)}$	$-\frac{1}{2}$	$3\eta_0$	Yes
Johnson-Segalman (1977) special case of Phan Thien -Tanner (1977) (2.75)	$\frac{\eta_0}{[1+2a(1-\frac{a}{2})\lambda_1^2\gamma^2]}$	$\frac{2\eta_0\lambda_1\gamma^2}{[1+2a(1-\frac{a}{2})\lambda_1^2\gamma^2]}$	$-\frac{a}{2}$	$\frac{2\eta_0}{[1-(1-a)2\lambda_1 k]}+\frac{\eta_0}{[1+(1-a)\lambda_1 k]}$	Yes
Oldroyd's liquid A (Oldroyd 1950) (2.76)	η_0	$2\eta_0(\lambda_1-\lambda_2)\gamma^2$	-1	$\frac{2\eta_0(1+2\lambda_2 k)}{(1+2\lambda_1 k)}+\frac{\eta_0(1-\lambda_2 k)}{(1-\lambda_1 k)}$	No
Oldroyd's liquid B (Oldroyd 1950) (2.77)	η_0	$2\eta_0(\lambda_1-\lambda_2)\gamma^2$	0	$\frac{2\eta_0(1-2\lambda_2 k)}{(1-2\lambda_1 k)}+\frac{\eta_0(1+\lambda_2 k)}{(1+\lambda_1 k)}$	No
Four constant Oldroyd (Walters 1979) (2.78)	$\eta_0\left[\frac{1+\lambda_2\mu_0\gamma^2}{1+\lambda_1\mu_0\gamma^2}\right]$	$\frac{2\eta_0(\lambda_1-\lambda_2)\gamma^2}{[1+\lambda_1\mu_0\gamma^2]}$	0	$\frac{3\eta_0[1-\lambda_2 k-2\lambda_1\lambda_2 k^2+3\mu_0\lambda_2 k^2]}{[(1+\lambda_1 k)(1-2\lambda_1 k)+3\mu_0\lambda_1 k^2]}$	Yes
Reiner-Rivlin fluid (Rivlin 1948) (2.79)	η_0	0	$\nu_2=\frac{-\eta_0\mu_2\gamma^2}{2}$	$3\eta_0\left[1-\frac{\mu_2 k}{2}\right]$	No

Table 2.2

APPENDIX 1

Theorem The tensor whose convected components are $\frac{D}{Dt}\beta_j^{\ell}$ has fixed components

$$\frac{\delta}{\delta t} b_i^{\ k} = \frac{\partial}{\partial t} b_i^{\ k} + v^m \frac{\partial}{\partial x^m} b_i^{\ k} + \frac{\partial v^m}{\partial x^i} b_m^{\ k} - \frac{\partial v^k}{\partial x^m} b_i^{\ m} ,$$

where v^m represents the components of the velocity vector in the fixed x^i coordinate system, which is not necessarily Cartesian.

Proof :

Let the components of a mixed second-order tensor be denoted by $\beta_j^{\ \ell}$ in convected coordinates and $b_i^{\ k}$ in fixed coordinates, so that

$$\beta_j^{\ \ell} = \frac{\partial x^i}{\partial \xi^j} \frac{\partial \xi^\ell}{\partial x^k} b_i^{\ k} , \tag{A1.1}$$

or

$$\frac{\partial x^k}{\partial \xi^\ell} \beta_j^{\ \ell} = \frac{\partial x^i}{\partial \xi^j} b_i^{\ k} . \tag{A1.2}$$

We require to determine the tensor $\frac{\delta}{\delta t} b_i^{\ k}$ such that

$$\frac{D}{Dt} \beta_j^{\ \ell} = \frac{\partial x^i}{\partial \xi^j} \frac{\partial \xi^\ell}{\partial x^k} \frac{\delta}{\delta t} b_i^{\ k} . \tag{A1.3}$$

We note that

$$\frac{D}{Dt} \frac{\partial x^i}{\partial \xi^j} = \frac{\partial}{\partial \xi^j} \frac{Dx^i}{Dt} = \frac{\partial v^i}{\partial \xi^j} . \tag{A1.4}$$

We now calculate the material derivative of (A1.2) with respect to t and use (A1.4) to give

$$\frac{\partial v^k}{\partial \xi^\ell} \beta_j^{\ \ell} + \frac{\partial x^k}{\partial \xi^\ell} \frac{D}{Dt} \beta_j^{\ \ell} = \frac{\partial x^i}{\partial \xi^j} \left[\frac{\partial}{\partial t} b_i^{\ k} + v^m \frac{\partial}{\partial x^m} b_i^{\ k} \right] + \frac{\partial v^i}{\partial \xi^j} b_i^{\ k} . \tag{A1.5}$$

We now use (A1.1) and obtain

$$\frac{\partial x^k}{\partial \xi^\ell} \frac{D}{Dt} \beta_j^{\ \ell} = \frac{\partial x^i}{\partial \xi^j} \left[\frac{\partial}{\partial t} b_i^{\ k} + v^m \frac{\partial}{\partial x^m} b_i^{\ k} \right] + \frac{\partial v^m}{\partial x^i} \frac{\partial x^i}{\partial \xi^j} b_m^{\ k} - \frac{\partial v^k}{\partial \xi^\ell} \frac{\partial x^i}{\partial \xi^j} \frac{\partial \xi^\ell}{\partial x^m} b_i^{\ m} , \tag{A1.6}$$

so that

$$\frac{D}{Dt} \beta_j^{\ \ell} = \frac{\partial x^i}{\partial \xi^j} \frac{\partial \xi^\ell}{\partial x^k} \left[\frac{\partial}{\partial t} b_i^{\ k} + v^m \frac{\partial}{\partial x^m} b_i^{\ k} \right] + \frac{\partial v^m}{\partial x^i} \frac{\partial x^i}{\partial \xi^j} \frac{\partial \xi^\ell}{\partial x^k} b_m^{\ k} - \frac{\partial v^k}{\partial x^m} \frac{\partial x^i}{\partial \xi^j} \frac{\partial \xi^\ell}{\partial x^k} b_i^{\ m} . \tag{A1.7}$$

and a comparison of (A1.7) with (A1.3) completes the proof of the theorem.

Chapter 3

Flow Classification

3.1 INTRODUCTION

We have seen that the general characterization of the mechanical behaviour of elastic liquids by means of rheological equations of state is a complex process. It is therefore in order to delineate those flow situations where some simplification is possible and where the resulting flow problems have a measure of tractability. To facilitate this, we attempt a flow classification in which the various flow problems are grouped under five main headings (cf. Crochet and Walters 1983a) :

(i) Flows dominated by the shear viscosity.

(ii) Slow flows (slightly elastic liquids).

(iii) Small deformation flows.

(iv) Nearly-viscometric flows.

(v) Long-range memory effects in complex flows.

Our main concern in the present book is with (v), but it is clearly in order to give some consideration to the other categories, so that the work in subsequent chapters can be placed in a global context.

3.2 FLOWS DOMINATED BY SHEAR VISCOSITY

In an important class of flows, shear viscosity is a dominating influence and it is possible to employ the inelastic 'generalized Newtonian' model

$$T_{ik} = 2\eta(I_2)\, d_{ik} \tag{3.1}$$

with confidence. In (3.1), I_2 is the second invariant of d_{ik} defined in such a way that it collapses to the shear rate γ in a steady simple shear flow; (3.1) is a generally valid equation of state and is in fact a special case of the so-called CEF equation (Criminale, Ericksen and Filbey 1958)

$$T_{ik} = 2\eta(I_2)\, d_{ik} - N_1(I_2)\, \overset{\Delta}{d}_{ik} + 4[N_1(I_2) + N_2(I_2)]\, d_{ij}\, d_{jk} \quad , \tag{3.2}$$

where the 'normal stress coefficients' N_1 and N_2 are given by (cf. (1.3))

$$\nu_1(I_2) = N_1(I_2) I_2^2 \quad ,$$

$$\nu_2(I_2) = N_2(I_2) I_2^2 \quad . \tag{3.3}$$

Equation (3.2) is known to be completely general for the so-called viscometric flows, Poiseuille and Couette flow being the best known examples. In viscometric flows, the determination of the flow field requires at most a knowledge of the shear viscosity function $\eta(I_2)$, i.e. use of (3.1), but (3.2) is needed to provide the associated stress field, so that N_1 and N_2 are important functions in this exercise.

Viscometric flows are not the only ones for which the shear viscosity is a dominating influence. For example, it can be argued that in most (but not all) lubrication analyses, (3.1) can be used with confidence. Furthermore, in engineering analyses of practical problems it is often sufficient to employ (3.1) although some of the finer details of the flow field may be due to viscoelastic effects. As an example, we may quote pressure-driven flow through a straight pipe of non-circular cross section. It is well known that when $\nu_2 \neq 0$ (i.e. when $N_2 \neq 0$), rectilinear flow is not in general possible and some secondary flow in the cross section of the pipe is to be expected (Ericksen 1956, Oldroyd 1965). However, the secondary flow is weak and if one's primary interest is in the flow rate through the pipe, it is usually sufficient to ignore secondary flow effects and employ (3.1) to determine the resulting axial velocity component which determines the flow rate (cf. Dodson, Townsend and Walters 1974).

When (3.1) is employed in the solution of flow problems, the basic equations are necessarily more complicated in detail than the corresponding Navier-Stokes equations for a constant η, but no new conceptual difficulties are encountered and it is true to say that any flow problem which can be solved for a Newtonian viscous fluid is also tractable for the model (3.1) (Crochet and Walters 1983a,b). At the same time, the importance of the generalized Newtonian model in an industrial context justifies more than a passing reference and the subject is considered again in detail in Chapter 9. The treatment is appropriately located in the finite-element section of the book, since the flow geometries arising in most industrial applications are complex and varied; consequently the problems demand the flexibility of the finite-element approach.

Finally, we remark that it is often useful to have some simple means of studying viscoelastic effects in flow problems which are dominated by the shear viscosity. This can be conveniently carried out by using

$$T_{ik} + \lambda \overset{\nabla}{T}_{ik} = 2\eta(I_2)\, d_{ik} \quad , \tag{3.4}$$

where λ is a constant relaxation time. Sometimes, λ is also taken to be a function of I_2, in which case (3.4) becomes the so-called White-Metzner model (White and Metzner 1963). Fluid memory effects can be studied by calculating the influence of λ on the flow characteristics.

3.3 SLOW FLOW (SLIGHTLY ELASTIC LIQUIDS)

We have already seen that the simple fluid model of Coleman and Noll reduces to relatively simple forms under conditions of fading memory and slow flow; equations (2.55), (2.56) and (2.57) are then valid first, second and third-order approximations to the simple fluid in the sense of speed of flow. These equations are explicit in the stress tensor, which can be immediately substituted into the stress equations of motion to yield a flow problem in the velocity components and the pressure. In the appropriate perturbation expansion method, the first-order problem is identical to that for a Newtonian fluid and elastico-viscous effects manifest themselves at second order. However, all the resulting equations have the same general form as those for a Newtonian fluid but are more complex in detail except, of course, at first order. No new techniques are therefore required to solve the flow problem and classical methods can be adapted with little difficulty.

It was pointed out in Chapter 2 that Truesdell (1964) has argued that equations (2.55)-(2.57) can also be applied in the case of slightly elastic liquids flowing in situations which are not necessarily "slow" in the sense of the original retarded-motion expansion. Indeed, it is helpful to associate the ordering process with the non-dimensional parameter $W(= \lambda U/L)$, which was defined in Chapter 1. Use of (2.55)-(2.57) can now be identified with small values of W, which incorporates both the concepts of small characteristic times (corresponding to slightly elastic liquids) and speed of flow, due regard being paid to the important restrictions mentioned in Chapter 2.

Notwithstanding the limited general applicability of the hierarchy equations, we nevertheless advocate a search for analytic solutions for the hierarchy equations as a complementary exercise in attempting the general solution of any complex flow problem. This is not only of assistance in pointing to the structure of the solution to the general problem, but it is also helpful in completing the overall picture. We are not so well endowed with general solutions to complex problems to be able to ignore the helpful information which can be obtained from a consideration of the hierarchy equations, even if these equations are not strictly valid under the conditions pertaining to a given flow problem.

3.4 SMALL-DEFORMATION FLOWS

In Chapter 2, the memory integral expansions (2.45)-(2.47) were argued to be valid approximations to the simple fluid of Coleman and Noll under conditions of small deformation, due regard being paid to the precise range of applicability of each approximation. The most obvious example of a flow in this category is a small amplitude oscillatory-shear flow. "Small" in this context meaning that the amplitude a is small enough to allow a series solution in powers of a.

Very often, the equations of finite linear viscoelasticity only (i.e. (2.45)) are employed and oscillatory flow problems then give rise to governing equations which have essentially the same form as those in classical fluid mechanics with the frequency-dependent complex viscosity function replacing the constant Newtonian viscosity coefficient (Walters 1975).

In practice, it is necessary to specify the kernel functions in the integral expansions to provide quantitative predictions. Usually, one, or at most two, decaying exponential terms are employed. When only one is used in the equations of finite linear viscoelasticity, (2.45) reduces to the simple Maxwell model (cf. §2.7). Use of the higher-order memory integral expansions is hampered by the lack of sufficient experimental information concerning the form of the kernel functions.

3.5 NEARLY-VISCOMETRIC FLOWS

A very popular and important class of flows may be associated with the adjective "nearly-viscometric". These are flows which depart from a viscometric flow like Poiseuille or Couette flow by a small flow field which one needs to quantify in a formal mathematical sense. In the case of Poiseuille flow in a pipe, we can consider the nearly-viscometric flow to be brought about by introducing, for example, (i) a small curvature to the pipe (thus forming an anchor ring), (ii) a small variation to the cross section along the length of the pipe giving rise, for example, to a corrugated pipe, (iii) a pulsatile pressure gradient fluctuating about a non-zero mean (see, for example, Walters 1972).

In the case of Couette flow, one may consider flow between slightly eccentric cylinders in relative rotation or study the general linear stability problem named after Sir G.I. Taylor. Indeed, all linear stability analyses are excellent examples of the application of the idea of a nearly-viscometric flow.

Pipkin and Owen (1967), in an important paper, showed that as many as thirteen kernel functions are needed to describe completely even first order perturbations about a viscometric flow. Not surprisingly, the general theory has not been applied to many problems, pulsatile flow and Taylor stability providing notable exceptions (see, for example, Barnes, Townsend and Walters 1971).

Even for a grouping as relatively simple as nearly-viscometric flows, some compromise has to be made between generality and tractability and numerous rheological equations of state of varying degrees of complexity have been employed. Existing pulsatile-flow analyses show how sensitive solutions can be to the choice of rheological equations and seemingly similar equations can give rise to *qualitatively* different flow predictions (Phan Thien 1978). This provides a salutary warning to those who employ rheological equations which fall short of complete generality. At the same time, we would deprecate a stand which forbids any move outside the haven of "generality" if this is found necessary for reasons of tractability. Here, there is scope for a compromise between bold and careful research if non-Newtonian fluid mechanics is to shed light on an important class of problems.

Whatever rheological equations of state are employed in the solution of nearly-viscometric flows, the resulting flow problems resolve themselves into perturbation problems about a basic viscometric flow. Analytic solutions to the perturbation problems are rarely possible and there has been a significant reliance on numerical solutions in completed work on nearly-viscometric flows.

In conclusion, we remark that in nearly-viscometric flows, the problem is not so much how to solve the governing equations, but rather on what rheological equation should be employed in a specific problem, given that a general description is ruled out for reasons of tractability. At the present time, there are no hard and fast rules to assist in this choice and it is very much a compromise between intuition, common sense and experience.

3.6 HIGHLY ELASTIC LIQUIDS FLOWING IN COMPLEX GEOMETRIES

With the advent of high-speed computers, attention has shifted to problems where liquids with long-range memory (i.e. highly elastic liquids) flow in rather complex geometries. Such situations are of significant practical importance, but we have had to await high speed computers with large store to facilitate exploratory attacks on the various problems.

Sometimes the flow geometries do not involve abrupt changes (such as re-entrant corners) but the problems can still be formidable. For example, squeezing flows and two-roll mill flow are extremely important practical situations which have been reluctant to admit to complete theoretical treatments (see, for example, Lee et al. 1982,1983).

Numerous flows of practical importance involve abrupt geometry changes and sometimes free surfaces, and these bring with them their own particular problems. Converging flow through an abrupt contraction, injection moulding and fibre spinning are just three examples of the many that can be quoted.

In the remainder of the present book, we shall be mainly concerned with the solution of flow problems involving highly elastic liquids flowing in geometries with re-entrant corners; sometimes, free surfaces will also be involved. It is this area which provides examples where numerical simulation in non-Newtonian fluid mechanics is basically different from that in the classical situation.

3.7 GENERAL COMMENTS CONCERNING FLOWS INVOLVING ABRUPT CHANGES IN GEOMETRY

It is generally agreed that for pragmatic reasons (i.e. reasons of tractability) the more complicated the flow problem the simpler has to be the rheological equations of state used in its solution. Here, we are considering very complex flows and by implication the equations have to be rather simple. Workers in the field are aware of the inadequacies of their fluid models but justify current work very simply - they must start somewhere, so they start with simple models, with the hope that their analyses will point the way to a later consideration of more complex and also more realistic models. At the same time, the simple-model analyses can throw considerable light (in a qualitative sense at least) on the flow field and stress field in complex geometries of practical importance.

Most existing work has considered variants of the so-called Maxwell/Oldroyd models, either in their implicit differential or explicit integral form. In the present book, attention will be focused on the simple (upper-convected) Maxwell model with equations of state given by (cf. Chapter 2):

$$T_{ik} + \lambda_1 \overset{\nabla}{T}_{ik} = 2\eta_0\, d_{ik} \quad , \tag{3.5}$$

where η_0 is a constant viscosity coefficient and λ_1 is a constant relaxation time.

The corresponding integral form for the Maxwell model is given by (cf. (2.80))

$$T_{ik} = \frac{\eta_0}{\lambda_1^2} \int_0^\infty e^{-\frac{s}{\lambda_1}} H_{ik}(s)\, ds \quad , \tag{3.6}$$

where the deformation tensor H_{ik} is given by

$$H_{ik} = C_{ik}^{-1}(s) - \delta_{ik} \quad , \tag{3.7}$$

i.e.

$$H_{ik} = \frac{\partial x_i}{\partial x'_m}\frac{\partial x_k}{\partial x'_m} - \delta_{ik} \quad . \tag{3.8}$$

For a steady simple shear flow with velocity components given by

$$u = \gamma y, \qquad v = w = 0 \quad ,$$

the corresponding stress distribution for the Maxwell model is given by

$$P_{xy} = \gamma\eta_0 \quad , \qquad P_{xx} - P_{yy} = 2\eta_0\lambda_1\gamma^2 \quad , \qquad P_{yy} - P_{zz} = 0 \quad , \tag{3.9}$$

which implies a constant apparent viscosity and a quadratic first normal stress difference. This is known to be inadequate for most elastico-viscous systems (where shear thinning can be extremely important) but it may be a fairly realistic model for the Boger (1977/78) fluid in a steady shear flow.

In an extensional flow, with velocity components

$$u = kx \quad , \qquad v = -\frac{ky}{2} \quad , \qquad w = -\frac{kz}{2} \quad , \tag{3.10}$$

the related stress components for the Maxwell model are given by (cf. (1.6))

$$P_{xx} - P_{yy} = k\eta_E(k)$$

$$P_{xx} - P_{zz} = k\eta_E(k) \quad , \qquad P_{ij} = 0 \qquad i \neq j \quad , \tag{3.11}$$

where

$$\eta_E = \frac{3\eta_0}{[1 + \lambda_1 k][1 - 2\lambda_1 k]} \quad . \tag{3.12}$$

Investigation of (3.12) reveals that η_E can take very high values as $\lambda_1 k$ tends to $\frac{1}{2}$ and does in fact become infinite when $\lambda_1 k = \frac{1}{2}$, so that the model is able to simulate the very high extensional viscosity levels found in some dilute polymer solutions.

Workers in the field regard models like (3.5) and (3.6) as useful at the present stage of the development of the subject and argue that later extensions to more complicated (and more realistic) differential and integral constitutive models will not involve many new issues of substance.

From time to time, we shall make reference to the associated Oldroyd B model, given by (cf. (2.77)):

$$T_{ik} + \lambda_1 \overset{\nabla}{T}_{ik} = 2\eta_0[d_{ik} + \lambda_2 \overset{\nabla}{d}_{ik}] \quad , \tag{3.13}$$

where λ_2 is a constant retardation time. The parameter λ_2 is often regarded as expendable and many fluid dynamicists would not anticipate dramatic changes in flow characteristics by the use of (3.13) in place of (3.5). There are however important exceptions, flow in the neighbourhood of a re-entrant corner being an obvious example (Cochrane et al. 1982).

We note finally some important aspects of the solution of complex flow problems for highly elastic liquids not found in classical fluid mechanics. Concerning differential models, the first thing that is immediately apparent from an inspection of model (3.5) is that the stress components are given by implicit differential equations, with the result that these components have to be treated as dependent variables along with the pressure and the velocity components. This makes the situation significantly different from the classical Newtonian case where the equation of continuity and the Navier-Stokes equations lead to four equations in the pressure and the three velocity components.

So far as the solution of flow problems using integral models is concerned, inspection of the relevant constitutive equations (3.6) is revealing since it highlights one of the problems of studying fluids with long-range memory in complex flow situations. Before one is able to solve the flow problem through a determination of the velocity components, it is necessary to obtain the displacement functions x_i' which are of course unknown until the velocity components are known. Some iterative technique is therefore essential if progress is to be made.

3.8 SOME REMARKS ON NON-DIMENSIONAL PARAMETERS

In the solution of flow problems in classical fluid dynamics, it is customary to non-dimensionalize the Navier-Stokes equations using a suitable characteristic velocity U and characteristic length L. The equations then involve one non-dimensional parameter called the Reynolds number R, defined by

$$R = \frac{\rho U L}{\eta_0} , \tag{3.14}$$

where η_0 is the constant viscosity coefficient. It would clearly be in order to attempt a similar non-dimensionalization in the non-Newtonian case. However, in the general problem, this is a very difficult task without any guarantee of complete success (Astarita 1979). At the same time, the process is tractable and meaningful for the simple Maxwell models we shall employ in much of the remainder of this book. The process leads to one additional non-dimensional parameter W defined by

$$W = \lambda_1 \frac{U}{L} , \tag{3.15}$$

where λ_1 is the relaxation time defined in (3.5). W is sometimes called the Weissenberg number, which is regarded by many as a measure of the relative importance of normal and tangential stresses, but this interpretation of W is restricted to flows which are at least approximately viscometric.

In the solution of actual flow problems, the ratio

$$\frac{W}{R} = \frac{\lambda_1 \eta_0}{\rho L^2} \tag{3.16}$$

is often the important non-dimensional variable (see, for example, Thomas and Walters 1964).

Another important non-dimensional parameter in non-Newtonian fluid mechanics is the so-called Deborah number De defined as the ratio of a characteristic time of the fluid (i.e. λ_1) to a characteristic time of the deformation process T, so that

$$De = \frac{\lambda_1}{T} \; . \tag{3.17}$$

Low values of De correspond to fluid-like behaviour and high values to solid-like behaviour. One consequence of this is that even mobile liquid systems with a small characteristic time can appear solid-like in a fast deformation process (see, for example, Walters 1980). Such ideas are of importance within one of the stated objectives of the present book to study highly elastic liquids flowing in complex geometries in which the time scale of the deformation process can often be very small indeed.

Sometimes the Deborah number is given the same definition as W, but it is difficult to see the worth of such a definition in a complex flow situation. Accordingly, we shall avoid confusion by referring to W in (3.15) as the non-dimensional elasticity parameter.

3.9 BASIC EQUATIONS FOR THE FLOW OF A MAXWELL FLUID

Consider a steady two-dimensional flow with velocity components u(x,y) and v(x,y) in the x and y directions, respectively. The physical variables can be written in non-dimensional form through the substitutions

$$x^* = \frac{x}{L} \; , \quad y^* = \frac{y}{L} \; , \quad u^* = \frac{u}{U} \; , \quad v^* = \frac{v}{U} \; ,$$

$$T^*_{ik} = \frac{L}{\eta_0 U} T_{ik} \; , \qquad p^* = \frac{L}{\eta_0 U} p \; , \tag{3.18}$$

to yield governing equations for the upper-convected Maxwell model (3.5) in the form

$$\frac{\partial u}{\partial x} + \frac{\partial v}{\partial y} = 0 \; , \tag{3.19}$$

$$F_x - \frac{\partial p}{\partial x} + \frac{\partial T_{xx}}{\partial x} + \frac{\partial T_{xy}}{\partial y} = R\left[u \frac{\partial u}{\partial x} + v \frac{\partial u}{\partial y}\right] \; , \tag{3.20}$$

$$F_y - \frac{\partial p}{\partial y} + \frac{\partial T_{xy}}{\partial x} + \frac{\partial T_{yy}}{\partial y} = R\left[u \frac{\partial v}{\partial x} + v \frac{\partial v}{\partial y}\right] \; , \tag{3.21}$$

$$T_{xx}\left[1 - 2W\frac{\partial u}{\partial x}\right] + W\left[u\frac{\partial T_{xx}}{\partial x} + v\frac{\partial T_{xx}}{\partial y}\right] - 2WT_{xy}\frac{\partial u}{\partial y} = 2\frac{\partial u}{\partial x} , \qquad (3.22)$$

$$T_{yy}\left[1 - 2W\frac{\partial v}{\partial y}\right] + W\left[u\frac{\partial T_{yy}}{\partial x} + v\frac{\partial T_{yy}}{\partial y}\right] - 2WT_{xy}\frac{\partial v}{\partial x} = 2\frac{\partial v}{\partial y} , \qquad (3.23)$$

$$-WT_{xx}\frac{\partial v}{\partial x} - WT_{yy}\frac{\partial u}{\partial y} + W\left[u\frac{\partial T_{xy}}{\partial x} + v\frac{\partial T_{xy}}{\partial y}\right] + T_{xy} = \frac{\partial u}{\partial y} + \frac{\partial v}{\partial x} , \qquad (3.24)$$

where we have immediately dropped the star notation for convenience of presentation. In (3.20) and (3.21), F_x and F_y are the non-dimensional body forces in the x and y directions, respectively.

Equation (3.19) is obtained from the equation of continuity (2.3); (3.20) and (3.21) are the relevant components of the stress equations of motion (2.6); and (3.22)-(3.24) are determined from the rheological equations of state (3.5).

From (3.19) one can introduce a stream function ψ, defined by (cf. equation (2.5))

$$u = \frac{\partial \psi}{\partial y} , \quad v = -\frac{\partial \psi}{\partial x} , \qquad (3.25)$$

and we define the vorticity ω by

$$\omega = \frac{\partial v}{\partial x} - \frac{\partial u}{\partial y} , \qquad (3.26)$$

so that

$$\omega = -\nabla^2\psi , \qquad (3.27)$$

∇^2 representing the Laplacian operator.

Eliminating the pressure p between (3.20) and (3.21) yields

$$R\left[\frac{\partial \psi}{\partial x}\frac{\partial \omega}{\partial y} - \frac{\partial \psi}{\partial y}\frac{\partial \omega}{\partial x}\right] = \frac{\partial^2 T_{xx}}{\partial x \partial y} + \frac{\partial^2 T_{xy}}{\partial y^2} - \frac{\partial^2 T_{xy}}{\partial x^2} - \frac{\partial^2 T_{yy}}{\partial x \partial y} . \qquad (3.28)$$

The governing equations can be taken to be (3.22)-(3.24), (3.27) and (3.28), yielding five equations in the five unknowns ψ, ω, T_{xx}, T_{xy}, T_{yy}. Alternatively, (3.19)-(3.24) can be used as a system of partial differential equations in terms of the six unknowns u, v, p, T_{xx}, T_{xy}, T_{yy}. For future reference, we point out that in the corresponding axisymmetric case one more constitutive equation together with one more extra stress component variable is required (cf. Chapters 5, 9 and 10).

In handling more complicated models, it is sometimes inappropriate to non-dimensionalize the governing equations, and in Chapter 9, which deals with the generalized Newtonian model, and Chapter 10, which considers more general visco-elastic models like the Phan Thien and Tanner model, the treatment is in dimensional variables. However, the results in Chapter 10 are expressed in terms of a dimensionless elasticity number. For the simple Maxwell model, this is equivalent to the parameter W already defined. For other, more complicated, models, it is sometimes found useful to define a stress ratio S_R given by (§ 10.7)

$$S_R = \frac{\nu_1(\gamma)}{2\eta(\gamma)\gamma} \quad . \tag{3.29}$$

However, this is inappropriate for some models like the Phan Thien-Tanner model for which S_R reaches a maximum for a finite value of γ. In these cases it is appropriate to use a non-dimensional parameter like (3.15) with λ_1 taken as the zero-shear-rate relaxation time.

Chapter 4

An Overview of Numerical Simulation

4.1 INTRODUCTION

The present chapter should be viewed as a bridge between the non-Newtonian fluid mechanics which occupied our attention in Chapters 1-3 and the numerical analysis which will dominate later chapters. The model problem for the upper-convected Maxwell rheological equations has already been outlined in §3.9, at least for the steady-state situation, and it is now convenient to discuss in a general way the various steps in the numerical solution process. These steps are outlined schematically in Fig. 4.1.

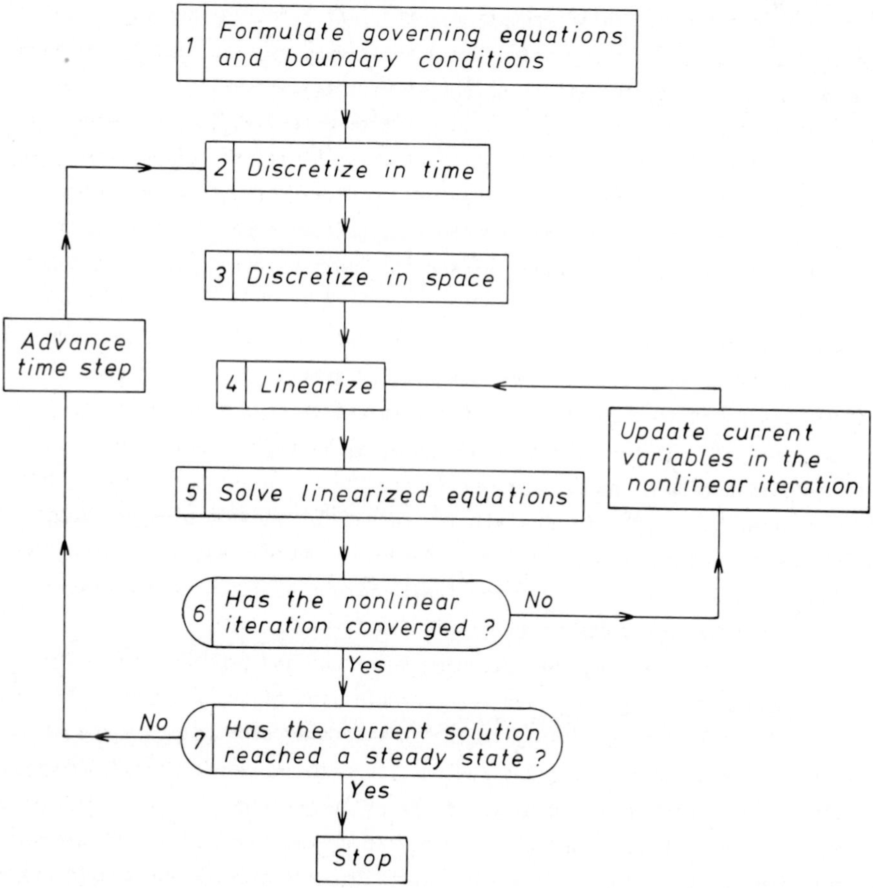

Fig. 4.1 A simple flowchart for numerical simulation.

The basic algorithm describes the simulation of time-dependent, or transient, flows, although most non-Newtonian flows which have been simulated are steady. At the same time, transient flows must not be summarily dismissed, and a part of Chapter 7 is devoted to this topic within the finite difference context.

Steady flows will nevertheless dominate our future discussions and these can be simulated by the appropriate choice of boundary conditions in Step 1 below, with the omission of the time discretization of Step 2, thereby setting $\frac{\partial}{\partial t} \equiv 0$. Alternatively, a steady flow should be obtainable if it exists, as the asymptotic time limit of the transient problem.

4.2 STEP 1 : FORMULATING THE GOVERNING EQUATIONS AND BOUNDARY CONDITIONS

This step was discussed in Chapter 3 within the global context of non-Newtonian fluid mechanics. Recall that the governing equations are those of continuity, momentum and the constitutive equations, the last being defined by the choice of rheological model. In formulating these equations, we have already inferred that an important feature is the selection of physical variables. In two-dimensional flows, it is usual to choose either a $(u,v,p,\underline{T})$ formulation, or a $(\psi,\omega,\underline{T})$ formulation, the (u,v,p) variables being sometimes called *primitive*. We shall see that the choice of variables can dictate the type of partial differential equation (elliptic, hyperbolic, parabolic or mixed) which a governing equation yields (cf. Chapter 6) and this in turn can dictate the choice of boundary conditions.

Boundary conditions clearly depend on the nature of the boundary, i.e. solid (moving or stationary), inflow and outflow, or free surface. The general problem was discussed in §2.9 with particular reference to inflow conditions. It was pointed out that for fluids with memory, a history of past deformation must be specified and this can be achieved by specifying the stress components at the inlet boundary. In principle, there is no need for additional specification of boundary stress components. In the case of free boundaries, extra conditions must be imposed to help to determine their unknown positions computationally and this will be considered in detail in Chapters 9 and 10.

Finally, in formulating our governing equations and boundary conditions, we should mention both *classical* and *weak* formulations of our problem. Classically we talk of a solution as satisfying the equations at each point in a domain, whereas a weak solution satisfies the equations in an average sense. More precisely, a weak formulation consists of taking inner products of the left- and right-hand sides of our equations with suitable test functions, and equating. In numerical work, the weak formulation provides a framework for constructing numerical approximations, and is particularly important in the finite element method.

In the weak formulation of a problem, the boundary conditions may be separated into two kinds: *essential* and *natural*. Essential boundary conditions must be satisfied by the whole space or manifold of admissible functions in which we seek a solution. On the other hand, natural boundary conditions are absorbed into the weak formulation by an integration by parts (cf. Chapter 8).

4.3 STEP 2 : TIME DISCRETIZATION

This step is necessary only when dealing with transient flows and can be omitted for steady flows. The governing partial differential equations are integrated approximately over a finite time-step $\Delta t=k$. It would appear that almost all methods which perform a time integration use finite difference methods for this step. Thus, if we adopt a uniform time-grid $t_n=nk$, $n=0,1,\ldots$, we start from known values of $u,v,p,\underline{T}$ at time $t_0=0$, and set up equations for $u,v,p,\underline{T}$ at the advanced time $t_1=k$. The solution of these equations at space points (x,y) involves Steps 3-6 below. In general, the structure of the equations for $u,v,p,\underline{T}$ at time $t=t_n$ in terms of $u,v,p,\underline{T}$ at time $t=t_{n-1}$ is independent of n. Chapter 7 contains a general discussion of available solutions to time-dependent problems for fluids with memory.

4.4 STEP 3 : SPACE DISCRETIZATION

Space discretization involves the approximation of $u,v,p,\underline{T}$, at prescribed values of x and y on a spatial grid. The required approximation may be generated in several different ways, using finite difference methods, finite element methods, boundary integral methods, multigrid methods or spectral methods. The last three methods have only recently emerged and at present, computational experience with them is limited to Newtonian flows. Accordingly, in the present book, we concentrate entirely on finite difference and finite element methods, both of which are well established and both of which have been applied to non-Newtonian flow problems.

In the finite difference approach, the variables are approximated only at discrete spatial grid-points or nodes (x_j,y_k), and it is precisely these discrete values which are directly calculated. Approximations at an arbitrary space point may then be found by interpolation, if desired. In the finite element method, the variables are approximated by simple (usually algebraic) functions of x and y, defined over finite regions of the (x,y) plane, making up a polygonal or similar grid pattern. In this case, the coefficients in the functional representations of the variables are calculated.

Whatever method is used, the discretized system of governing equations gives rise to a nonlinear algebraic system of equations in the unknowns, be they variable values or coefficients.

The choice between the finite difference and the finite element approach to numerical simulation is in some respects a matter of subjective judgment, but certain important general observations can be made.

The finite difference approach is the easier of the two to implement, especially for newcomers to numerical simulation,and there is an established literature to consult. The background mathematical analysis is also relatively simple. Further, the finite difference technique has advantages with regard to computational time and storage.

Arguments in favour of the finite element approach must include the following. Any initial difficulties associated with this approach are more than compensated by the ease by which a finite element code, once written, can be changed to accommodate new flow situations and geometries. Furthermore, complex geometries can be handled with relative ease, as can free-surface complications, and in these respects, the finite element method has significant advantages over the corresponding finite difference techniques. The finite element method implements natural boundary conditions and this can be viewed as another advantage of the method (cf. Chapter 8).

So far as accuracy and confidence in the computed results are concerned, there is nothing to choose between the finite difference and finite element methods, and it would be inappropriate to use these factors as criteria for, or against,one or the other of the methods.

In the present book we shall consider both the finite difference method (in Chapters 5-7) and the finite element method (in Chapters 8-10) before summarizing the state of the art in Chapter 11.

4.5 STEP 4 : LINEARIZATION

The nonlinear algebraic system must be solved iteratively. At each iteration, the system is *linearized* and the resulting linear system solved to provide an updated approximation to the required solution. In computational fluid dynamics, this linearization is usually performed in one of two ways: either as a step of a Picard-type iteration, or as a step of Newton's method. An example of a Picard-type iteration is the successive decoupling of the governing equations, one variable at a time. This has been the main iteration scheme in finite difference calculations (cf. Chapter 6). Newton's method, which allows a simultaneous update of all variables, is becoming more widely used, particularly in finite element calculations (cf. Chapter 9). There is a growing body of opinion that Newton-type methods are, on the whole, more powerful than simpler fixed-point iterative schemes such as successive decoupling.

It should be noted that, whatever iterative method is used to overcome the nonlinearity, in practical problems there is rarely any applicable theory which will guarantee convergence of the method, nor if convergence is attained, that the solution achieved is unique.

4.6 STEP 5 : SOLUTION OF THE LINEARIZED EQUATION

At each step of the nonlinear iteration loop, one or more systems of linear matrix equations need to be solved. In general the matrices are sparse, i.e. most of the entries are zero. Furthermore, the matrices are structured, in the sense that they are made of blocks which are either null or banded. These remarks are valid whether a Newtonian linearization or successive decoupling is used.

We shall see, particularly in Chapter 6, that several well established methods exist for the efficient solution of sparse matrix equations. It would appear that the most common methods used so far in non-Newtonian flow simulation are based on Gaussian elimination in the context of finite elements, and relaxation methods in the context of finite differences.

4.7 STEP 6 : TERMINATION OF THE NONLINEAR ITERATION LOOP

Steps 4-5 are repeated cyclically until convergence to an approximate solution for the field variables u,v,p,$\underline{T}$ is attained. If the problem under consideration is time-independent, the computed values of u,v,p and $\underline{T}$ represent the solution. If the problem is time-dependent, it is necessary to return to Step 2 to increment the time step, unless of course the computations have reached the final set value of t or have reached a steady state.

All the above stages in the simulation process will be discussed in detail in subsequent chapters.

Chapter 5

Introduction to Finite Differences

5.1 BOUNDARY VALUE PROBLEMS IN ONE AND TWO SPACE DIMENSIONS

In this chapter we give a brief introduction to finite difference methods for boundary value problems in one and two space dimensions. A fuller treatment may be found in the books of Keller (1968,1976) on two-point boundary value problems, and those of Wachspress (1966), Birkhoff (1971), Ames (1977), Smith (1978), Gladwell and Wait (1979), Mitchell and Griffiths (1980) and Meis and Marcowitz (1981), which include material on elliptic boundary value problems.

In Chapters 5 and 6 we treat only time-independent problems. Transient problems will be discussed in Chapter 7.

5.1.1 One space variable

In one dimension the relevant class of problems is that of two-point boundary value problems associated with second order differential equations. Let Ω be the real interval $x_0 \leqslant x \leqslant x_N$, and consider the equation

$$Lu = f(x,u,u') , \qquad x \in \Omega , \tag{5.1}$$

where L is the second order differential operator defined by

$$Lu \equiv au'' + bu' + cu$$

and the prime denotes differentiation with respect to x. In general, the coefficients a, b and c are all functions of x, u and u', but when the coefficients are dependent on x only then L is said to be *linear*. Equation (5.1) is also linear if, in addition, f depends on x only. Generally, both L and (5.1) are *nonlinear*.

A solution u(x) of (5.1) is sought on Ω, subject to two distinct boundary conditions at the end-points x_0 and x_N. Typical boundary conditions are either

$$\text{BC1:} \quad u(x_0) = u_0 , \qquad u(x_N) = u_N , \tag{5.2}$$

or

$$\text{BC2:} \quad u(x_0) = u_0 , \qquad u'(x_N) = 0 . \tag{5.3}$$

As an example, consider

$$u'' + \tfrac{1}{2}\alpha[(u')^2 + u^2] = 1 \;, \qquad 0 \leqslant x \leqslant \tfrac{1}{2}\pi \;, \qquad u(0) = 1 \;, \quad u(\tfrac{1}{2}\pi) = 0 \;. \tag{5.4}$$

When $\alpha = 0$, L is linear and (5.4) has a solution

$$u = 1 - \left(\frac{\pi}{4} + \frac{2}{\pi}\right)x + \tfrac{1}{2}x^2. \tag{5.5}$$

When $\alpha \neq 0$, L is nonlinear, and with $\alpha = 1$ a solution is

$$u = 1 - \sin x \;. \tag{5.6}$$

It can be shown that both these solutions are unique, but this need not have been so. For example, with $\alpha = 1$ and the boundary conditions

$$u(0) = 1 \;, \qquad u'(\tfrac{1}{2}\pi) = 0 \;,$$

the problem has *two* solutions

$$u = 1 \pm \sin x \;. \tag{5.7}$$

When attempting to solve a problem computationally, the question of existence and uniqueness of the theoretical (or exact) solution should always be borne in mind. With particular regard to (5.1) with boundary conditions BC1 or BC2 the following remarks are valid. When L is linear, one of three possibilities holds:

P1 : There is no solution.

P2 : There is one and only one solution.

P3 : There is an infinite continuum of solutions.

When L is nonlinear, there is an additional possibility:

P4 : There are more than one, but countably many solutions.

Thus, in principle, we should attempt to find a numerical solution *only if* in the linear case P2 holds, or in the nonlinear case either P2 or P4 holds. In the case of P4 we can at best expect our numerical algorithm to converge to an approximation of one of the possible solutions.

Unfortunately, in most practical problems, the lack of applicable mathematical theory often makes it impossible to say anything about existence and uniqueness. Numerical solutions, therefore, should not be accepted without careful examination, with reliance on what is known about simpler, related problems, where necessary.

Any *a priori* knowledge of the exact solution should be sought in the numerical solution, and where theoretical considerations fail, physical interpretability in the light of experimental evidence must play an important part in accepting or rejecting a numerical solution. These remarks are, of course, quite general, and not restricted to one-dimensional problems.

5.1.2 Two space variables

In two dimensions, with space variables x and y, let Ω denote a bounded region in the x-y plane, with boundary Γ. Consider the second-order differential operator defined by

$$Lu \equiv a \frac{\partial^2 u}{\partial x^2} + b \frac{\partial^2 u}{\partial x \partial y} + c \frac{\partial^2 u}{\partial y^2} ,$$

where in general a, b and c are all functions of x, y, u, $\partial u/\partial x$ and $\partial u/\partial y$. The partial differential equation

$$Lu = f\left(x,y,u,\frac{\partial u}{\partial x},\frac{\partial u}{\partial y}\right) , \qquad (x,y) \in \Omega , \tag{5.8}$$

is said to be *elliptic* in Ω if

$$b^2 - 4ac < 0, \forall \ (x,y) \in \Omega .$$

On the other hand, equation (5.8) is said to be $\begin{Bmatrix}\text{parabolic}\\ \text{hyperbolic}\end{Bmatrix}$ in some arbitrary (possibly unbounded) region of the x-y plane if $b^2 - 4ac \ (\substack{=\\>})\ 0$ in that region. What distinguishes between the three types of equation is the number of directions at a point (x,y) in the plane along which the integration of the partial differential equation reduces to the integration of an equation involving total differentials only. Elliptic equations possess *no* such directions at a point, whereas parabolic and hyperbolic equations possess one and two, respectively. Where they exist, these directions define *characteristic curves*. What is important is that elliptic equations cannot be solved using step-by-step integration along a characteristic curve, starting from a given initial value, whereas, in principle, such a solution procedure is possible for hyperbolic equations. Moreover, well-posed elliptic problems have their boundary conditions specified on a closed boundary, whereas parabolic and hyperbolic problems do not. Other important properties of elliptic equations, such as maximum principles (cf. §5.4.1), are discussed in Gladwell and Wait (1979).

Three distinct elliptic boundary value problems arise depending on the boundary conditions specified on Γ.

(i) The Dirichlet problem:

$$u = \alpha(x,y) \;, \qquad (x,y) \in \Gamma, \tag{5.9}$$

where α is a prescribed function on Γ.

(ii) The Neumann problem:

$$\frac{\partial u}{\partial n} = \beta(x,y) \;, \qquad (x,y) \in \Gamma, \tag{5.10}$$

where β is prescribed on Γ, and $\partial/\partial n$ denotes partial differentiation along the *outward* normal direction.

(iii) The Robbins problem (mixed boundary condition):

$$\alpha(x,y)u + \beta(x,y)\frac{\partial u}{\partial n} = \gamma(x,y) \;, \qquad (x,y) \in \Gamma, \tag{5.11}$$

where α, $\beta > 0$ on Γ. The positivity constraint may be relaxed to a non-negativity constraint, for example to allow a Dirichlet condition on part of Γ and a Neumann condition on the remainder. Without the constraint the problem may not be well-posed.

The simplest nontrivial example of an elliptic equation is Poisson's equation

$$\nabla^2 u \equiv \frac{\partial^2 u}{\partial x^2} + \frac{\partial^2 u}{\partial y^2} = f(x,y) \;, \qquad (x,y) \in \Omega. \tag{5.12}$$

For a wide class of functions f it is easy to show that the Dirichlet and Robbins problems for Poisson's equation have unique solutions. The corresponding Neumann problem, however, cannot be solved unless the prescribed values β of the normal derivative satisfy the *compatibility condition*

$$\int_\Gamma \beta \, d\Gamma = \int_\Omega f \, d\Omega \;. \tag{5.13}$$

When (5.13) holds the Neumann problem has a solution which is unique apart from an additive arbitrary constant.

Since the elliptic operator ∇^2 is present in the Navier-Stokes equations, both in (u,v,p)- and (ψ,ω)-formulations, it will be instructive to solve (5.12) using finite differences. But we shall start with the one-dimensional equivalent of elliptic boundary value problems, which are the two-point boundary value problems.

5.2 FINITE DIFFERENCE SOLUTION OF TWO-POINT BOUNDARY VALUE PROBLEMS : THE LINEAR CASE

5.2.1 Discretization

Consider the linear second-order equation

$$Lu \equiv u'' + b(x)u' + c(x)u = f(x) , \qquad x \in \Omega , \tag{5.14}$$

subject to

$$\text{BC1: } u(x_0) = u_0 , \qquad u(x_N) = u_N .$$

On the interval $\Omega = [x_0,x_N]$, we place a *uniform* grid or mesh

$$x_j = x_0 + jh , \qquad j = 0,\ldots,N$$

with ***spacing*** $h = (x_N - x_0)/N$. To approximate the solution $u(x)$ on the grid, we define a set of numbers u_j, $j = 0,\ldots,N$, as the solution of a system of finite difference equations which are in some sense an approximation to (5.14). Our notation is that u_j is an *approximation* to $u(x_j)$ when $1 \leq j \leq N-1$, but to satisfy BC1 we have $u_j = u(x_j)$ when $j = 0$ and N.

Perhaps the best-known method of deriving finite difference approximations is based on Taylor series expansions of the solution $u(x)$. Assume that $u \in C^4(\Omega)$, i.e. the fourth derivative $u^{(4)}$ exists and is continuous on Ω. Then

$$u(x_{j+1}) = u(x_j) + hu'(x_j) + \frac{h^2}{2!}u''(x_j) + \frac{h^3}{3!}u'''(x_j) + \frac{h^4}{4!}u^{(4)}(\xi_j^{+}),$$

where $x_j < \xi_j^{+} < x_{j+1}$, and

$$u(x_{j-1}) = u(x_j) - hu'(x_j) + \frac{h^2}{2!}u''(x_j) - \frac{h^3}{3!}u'''(x_j) + \frac{h^4}{4!}u^{(4)}(\xi_j^{-}),$$

where $x_{j-1} < \xi_j^{-} < x_j$, from which it follows that

$$u''(x_j) = \frac{u(x_{j+1}) - 2u(x_j) + u(x_{j-1})}{h^2} - \frac{h^2}{12}u^{(4)}(\xi_j), \tag{5.15}$$

where $x_{j-1} < \xi_j < x_{j+1}$. Notice that we have made use of the continuity of $u^{(4)}$ in writing

$$u^{(4)}(\xi_j^{+}) + u^{(4)}(\xi_j^{-}) = 2u^{(4)}(\xi_j).$$

By terminating the Taylor expansions at the third derivative it follows similarly that

$$u'(x_j) = \frac{u(x_{j+1}) - u(x_{j-1})}{2h} - \frac{h^2}{6} u'''(\eta_j) \ , \tag{5.16}$$

where $x_{j-1} < \eta_j < x_{j+1}$.

Ignoring $O(h^2)$ terms in (5.15) and (5.16) we see that an obvious finite difference approximation of (5.14) is

$$\left. \begin{aligned} L_h u_j &\equiv \left(\frac{u_{j+1} - 2u_j + u_{j-1}}{h^2}\right) + b(x_j)\left(\frac{u_{j+1} - u_{j-1}}{2h}\right) + c(x_j)u_j \\ &= f(x_j) \ , \qquad 1 \leqslant j \leqslant N-1. \end{aligned} \right\} \tag{5.17}$$

Upon multiplication by $-\frac{1}{2}h^2$, (5.17) may be written in the more convenient form

$$-\frac{h^2}{2} L_h u_j \equiv \alpha_j u_{j-1} + \beta_j u_j + \gamma_j u_{j+1} = -\frac{h^2}{2} f(x_j), \qquad 1 \leqslant j \leqslant N-1 \ , \tag{5.18}$$

$$\text{where} \quad \left. \begin{aligned} \alpha_j &= -\tfrac{1}{2}[1 - \tfrac{1}{2}hb(x_j)] \ , \\ \beta_j &= 1 - \tfrac{1}{2}h^2 c(x_j) \ , \\ \gamma_j &= -\tfrac{1}{2}[1 + \tfrac{1}{2}hb(x_j)] \ . \end{aligned} \right\} \tag{5.19}$$

The system (5.18) in matrix notation is

$$\underset{\sim}{\underline{A}}\, \underset{\sim}{u} = \underset{\sim}{r} \ , \tag{5.20}$$

where we have introduced the vectors $\underset{\sim}{u}, \underset{\sim}{r} \in \mathbb{R}^{N-1}$,

$$\underset{\sim}{u} = \begin{pmatrix} u_1 \\ u_2 \\ \cdot \\ \cdot \\ \cdot \\ u_{N-1} \end{pmatrix}, \qquad \underset{\sim}{r} = \begin{pmatrix} r_1 \\ r_2 \\ \cdot \\ \cdot \\ \cdot \\ r_{N-1} \end{pmatrix} = -\frac{h^2}{2} \begin{pmatrix} f(x_1) \\ f(x_2) \\ \cdot \\ \cdot \\ \cdot \\ f(x_{N-1}) \end{pmatrix} - \begin{pmatrix} \alpha_1 u_0 \\ 0 \\ \cdot \\ \cdot \\ 0 \\ \gamma_{N-1} u_N \end{pmatrix}$$

and the $(N-1) \times (N-1)$ matrix

$$\underset{\sim}{A} = \begin{pmatrix} \beta_1 & \gamma_1 & & & & \bigcirc \\ \alpha_2 & \beta_2 & \gamma_2 & & & \\ & \cdot & \cdot & \cdot & & \\ & & \cdot & \cdot & \cdot & \\ & & & \cdot & \cdot & \cdot \\ & & & \alpha_{N-2} & \beta_{N-2} & \gamma_{N-2} \\ & \bigcirc & & & \alpha_{N-1} & \beta_{N-1} \end{pmatrix} . \qquad (5.21)$$

The vector $\underset{\sim}{u}$, whose elements are the finite difference solution of problem (5.14) with BC1, based on the approximation (5.17), must be found by solving the matrix system (5.20). The special structure of the matrix enables us to do this very efficiently using Algorithm 5.1, which we describe in the next section. A matrix of the form (5.21) with non-zero elements only on the diagonal and the two adjacent codiagonals is called a *tridiagonal* matrix.

Example 1. Consider problem (5.4) with $\alpha = 0$. Clearly,

$$b(x) = c(x) = 0 , \qquad f(x) = 1 .$$

Thus in the finite difference approximation (5.18) we have

$$\alpha_j = \gamma_j = -\tfrac{1}{2} , \qquad \beta_j = 1 , \qquad f(x_j) = 1 , \qquad u_0 = 1 , \qquad u_N = 0 .$$

With N = 5, the spacing is $h = \pi/10$, and the grid-points are $x_j = j\pi/10$, $j = 0,\ldots,5$. Equation (5.20) is

$$\begin{pmatrix} 1 & -\frac{1}{2} & 0 & 0 \\ -\frac{1}{2} & 1 & -\frac{1}{2} & 0 \\ 0 & -\frac{1}{2} & 1 & -\frac{1}{2} \\ 0 & 0 & -\frac{1}{2} & 1 \end{pmatrix} \begin{pmatrix} u_1 \\ u_2 \\ u_3 \\ u_4 \end{pmatrix} = \frac{1}{2} \begin{pmatrix} 1 - h^2 \\ -h^2 \\ -h^2 \\ -h^2 \end{pmatrix} = \begin{pmatrix} 0.4507 \\ -0.0493 \\ -0.0493 \\ -0.0493 \end{pmatrix} ,$$

which upon solution by elimination yields

$$\begin{pmatrix} u_1 \\ u_2 \\ u_3 \\ u_4 \end{pmatrix} = \begin{pmatrix} 0.6026 \\ 0.3040 \\ 0.1040 \\ 0.0026 \end{pmatrix} .$$

It is easily verified that each u_j agrees to the number of figures quoted with the values of the theoretical solution (5.5) at $x = x_j$, $j = 1,\ldots,4$.

5.2.2 Solution of tridiagonal systems

The tridiagonal matrix (5.21) is said to be *diagonally dominant* if

$$\left.\begin{aligned} |\beta_1| &\geq |\gamma_1| \,, \\ |\beta_j| &\geq |\alpha_j| + |\gamma_j| \,, \qquad 2 \leq j \leq N-2 \,, \\ |\beta_{N-1}| &\geq |\alpha_{N-1}| \,. \end{aligned}\right\}$$

If, in addition, each $\beta_j \neq 0$ and at least one of the above inequalities is strict, then the matrix $\underline{A}$ is nonsingular and the solution $\underset{\sim}{u}$ of the system (5.20) exists and is unique. Moreover, $\underset{\sim}{u}$ may then be computed in a highly efficient and stable manner using the following algorithm:

Algorithm 5.1
(1) Let $p_1 = \beta_1$.
(2) Generate $p_2,\ldots,p_{N-1}$ from the forward recursion
$$p_j = \beta_j - (\alpha_j\gamma_{j-1})/p_{j-1} \,, \qquad 2 \leq j \leq N-1.$$
(3) Generate $q_1,\ldots,q_{N-1}$ from
$$q_1 = r_1/p_1 \,,$$
$$q_j = (r_j - \alpha_j q_{j-1})/p_j \,, \qquad 2 \leq j \leq N-1.$$
(4) Generate the solution from the backward recursion
$$u_{N-1} = q_{N-1} \,,$$
$$u_j = q_j - (u_{j+1}\gamma_j)/p_j \,, \qquad N-2 \geq j \geq 1.$$

In example 1 we see that $\underline{A}$ is diagonally dominant with strict inequalities in the first and last rows, so that Algorithm 5.1 is applicable.

The algorithm is equivalent to Gauss elimination *without pivoting* [see, for example, Wait (1979)], and is highly efficient with regard to machine arithmetic and storage. If $\underline{A}$ is *not* diagonally dominant, however, the algorithm may become unstable, generating substantial errors in the solution vector $\underset{\sim}{u}$. In this case we would have to employ an alternative, less efficient, *direct method* such as Gauss elimination *with pivoting* [Wait (1979)], the success of which would depend upon $\underline{A}$ being neither singular nor near-singular (ill-conditioned).

5.2.3 Error analysis

In general, it will be of interest to understand the nature of the errors incurred in the finite difference approximation. The *total discretization error* in the numerical solution at the grid-point x_j is defined by

$$e_j = u(x_j) - u_j \,. \tag{5.22}$$

This error occurs as a result of two separate stages of approximation introduced in the transition from (5.14) to (5.17). First, we ignored the $O(h^2)$ terms in

(5.15) and (5.16). This is equivalent to replacing $(Lu)(x_j)$ by $(L_h u)(x_j)$, where

$$L_h u \equiv \frac{u(x+h) - 2u(x) + u(x-h)}{h^2} + b(x)\left[\frac{u(x+h) - u(x-h)}{2h}\right] + c(x)u(x). \qquad (5.23)$$

Secondly, we replaced $(L_h u)(x_j)$ by $L_h u_j$ in writing down equations (5.17). Thus $(Lu)(x_j) \rightsquigarrow (L_h u)(x_j) \rightsquigarrow L_h u_j$.

In the first step we approximated the differential operator L by the difference operator L_h. This gives rise to the *local discretization error* (*local truncation error*) $\tau_j[v]$, which is defined at x_j for any smooth function v by

$$\tau_j[v] = (L_h v)(x_j) - (Lv)(x_j). \qquad (5.24)$$

From the definitions of L and L_h in (5.14) and (5.23), respectively, it follows that

$\tau_j[v] \to 0$ as $h \to 0$ $\forall$ C^2-functions v.

We say that L_h is *consistent* with L. Furthermore, from (5.15) and (5.16) we may deduce

$$\tau_j[u] = -\frac{h^2}{12}[u^{(4)}(\xi_j) + 2b(x_j)u'''(\eta_j)] = O(h^2) , \qquad (5.25)$$

from which we say that L_h has *second-order accuracy* in approximating L for all C^4-functions.

In the second step, we approximated the continuous solution u at the grid-points x_j by the discrete solution $\{u_j\}$ of the system (5.17). We demand that this system is globally stable, i.e., that the solution $\{u_j\}$ depends continuously on the data $\{f(x_j)\}$ as the mesh-size h approaches zero. We say that the difference operator L_h is stable if it is invertible and the inverse operator $(L_h)^{-1}$ is bounded for sufficiently small h. Thus

$$\|(L_h)^{-1}\|_\infty \leqslant C , \qquad (5.26)$$

where C is a constant independent of h. The stability of L_h is easily established for a wide class of coefficient functions b(x) and c(x).

One consequence of stability is that, for given h, the difference equations (5.17) have a unique solution. We may also deduce an upper bound for the discretization error e_j. From (5.22) and (5.24) it follows that

$$L_h e_j = \tau_j[u] , \qquad j = 1,\dots,N-1, \qquad (5.27)$$

with

$$e_0 = e_N = 0 \ . \tag{5.28}$$

Thus the errors e_j satisfy a similar system of difference equations to (5.17), except that the right-hand sides are now the local truncation errors. From (5.26) and (5.27) we obtain the bound

$$|e_j| \leqslant C\left[\max_{1\leqslant k\leqslant N-1} |\tau_k[u]|\right], \qquad 1 \leqslant j \leqslant N-1. \tag{5.29}$$

The most important consequence of stability is that of *convergence*. From (5.25) and (5.29) it follows that

$$e_j = O(h^2) \quad \text{as} \quad h \to 0, \quad x_j \text{ fixed}, \quad u \in C^4. \tag{5.30}$$

Thus at a fixed grid-point x_j the discrete solution u_j converges to the exact solution $u(x_j)$ as the mesh-size approaches zero. Moreover, the *rate of convergence* is $O(h^2)$. We say that the finite difference approximation is convergent, and second-order.

Observe that in deducing convergence, as well as stability, we used the property of consistency implied by (5.25). This result embodies one of the most celebrated theorems of numerical analysis: *Consistency and stability imply convergence.*

In example 1, since the true solution (5.5) is quadratic in x, it follows from (5.25) that the local truncation error vanishes. Consequently the discrete solution should agree exactly with the true solution at the grid-points. In practice, however, rounding errors are incurred as a result of computer storage and arithmetic, and we can never achieve an exact solution. It is meaningless to quote too many decimal places. In most problems, where the truncation errors do not vanish, we refine the mesh to reduce these errors. In such a process the effect of machine rounding error must always be carefully analyzed. In large computational problems, however, it is usually not the rounding errors which dictate the lower limit on mesh-size, but the storage capacity of the computer. This is because the number of variables with which we work increases as the mesh-size is reduced.

5.2.4 Derivative boundary conditions

So far we have considered the boundary conditions BC1 which could be imposed exactly on the discrete solution. We now turn to the problem

$$Lu \equiv u'' + b(x)u' + c(x)u = f(x) \ , \qquad x \in \Omega \ ,$$

subject to

BC2: $u(x_0) = u_0$, $\quad u'(x_N) = 0$.

The derivative boundary condition must be imposed approximately on the discrete solution. We consider two alternatives:

the *one-sided (backward) difference* approximation

$$\frac{u_N - u_{N-1}}{h} = 0 , \tag{5.31}$$

and the *central-difference* approximation

$$\frac{U_{N+1} - u_{N-1}}{2h} = 0 . \tag{5.32}$$

The one-sided difference operator results from the Taylor expansion

$$\frac{u(x_N) - u(x_{N-1})}{h} = u'(x_N) - \frac{h}{2} u''(\xi_N) = -\frac{h}{2} u''(\xi_N) ,$$

where $x_{N-1} < \xi_N < x_N$, giving $O(h)$ truncation error and thus only first-order accuracy. The central-difference operator is of course second-order accurate [(5.16) with $j = N$].

In the first case, (5.31) may be used to eliminate the unknown variable u_N from the difference equations (5.18) which hold for $1 \leqslant j \leqslant N-1$. For $j = N-1$ we then have

$$\alpha_{N-1}u_{N-2} + (\beta_{N-1} + \gamma_{N-1})u_{N-1} = -\frac{h^2}{2} f(x_{N-1}).$$

The (N-1) equations in the (N-1) unknowns $\{u_j\}_{j-1}^{N-1}$ may be written as the tridiagonal system

$$\begin{pmatrix} \beta_1 & \gamma_1 & & & \bigcirc \\ \alpha_2 & \beta_2 & \gamma_2 & & \\ \cdot & \cdot & \cdot & & \\ & \cdot & \cdot & \cdot & \\ & & \cdot & \cdot & \cdot \\ & & \gamma_{N-2} & \beta_{N-2} & \gamma_{N-2} \\ \bigcirc & & & \alpha_{N-1} & (\beta_{N-1} + \gamma_{N-1}) \end{pmatrix} \begin{pmatrix} u_1 \\ u_2 \\ \cdot \\ \cdot \\ \cdot \\ u_{N-2} \\ u_{N-1} \end{pmatrix} = -\frac{h^2}{2} \begin{pmatrix} f(x_1) \\ f(x_2) \\ \cdot \\ \cdot \\ \cdot \\ f(x_{N-2}) \\ f(x_{N-1}) \end{pmatrix} - \begin{pmatrix} \alpha_1 u_0 \\ 0 \\ \cdot \\ \cdot \\ \cdot \\ 0 \\ 0 \end{pmatrix} . \tag{5.33}$$

Having solved (5.33) for the internal points, the numerical solution u_N at the boundary may then be found from (5.31).

In the second case, (5.32) involves the extra unknown u_{N+1} (external to Ω). This may be eliminated by assuming that the difference equation (5.18) holds also for $j = N$, i.e. at the boundary point x_N. Substituting the value $u_{N+1} = u_{N-1}$ into this last equation gives

$$(\alpha_N + \gamma_N)u_{N-1} + \beta_N u_N = -\frac{h^2}{2} f(x_N).$$

We now have N equations in the N unknowns $\{u_j\}_{j=1}^{N}$ resulting in the tridiagonal system

$$\begin{pmatrix} \beta_1 & \gamma_1 & & & \bigcirc \\ \alpha_2 & \beta_2 & \gamma_2 & & \\ \cdot & \cdot & \cdot & & \\ & \cdot & \cdot & \cdot & \\ & & \cdot & \cdot & \cdot \\ & & \alpha_{N-1} & \beta_{N-1} & \gamma_{N-1} \\ \bigcirc & & & (\alpha_N + \gamma_N) & \beta_N \end{pmatrix} \begin{pmatrix} u_1 \\ u_2 \\ \cdot \\ \cdot \\ \cdot \\ u_{N-1} \\ u_N \end{pmatrix} = -\frac{h^2}{2} \begin{pmatrix} f(x_1) \\ f(x_2) \\ \cdot \\ \cdot \\ \cdot \\ f(x_{N-1}) \\ f(x_N) \end{pmatrix} - \begin{pmatrix} \alpha_1 u_0 \\ 0 \\ \cdot \\ \cdot \\ \cdot \\ 0 \\ 0 \end{pmatrix}. \qquad (5.34)$$

Here the internal and boundary values are determined simultaneously. Such an arrangement is also possible in the first case, since we may combine (5.18) and (5.31) to give the $N \times N$ system

$$\begin{pmatrix} \beta_1 & \gamma_1 & & & \bigcirc \\ \alpha_2 & \beta_2 & \gamma_2 & & \\ \cdot & \cdot & \cdot & & \\ & \cdot & \cdot & \cdot & \\ & & \cdot & \cdot & \cdot \\ & & \alpha_{N-2} & \beta_{N-1} & \gamma_{N-1} \\ \bigcirc & & & -1 & 1 \end{pmatrix} \begin{pmatrix} u_1 \\ u_2 \\ \cdot \\ \cdot \\ \cdot \\ u_{N-1} \\ u_N \end{pmatrix} = -\frac{h^2}{2} \begin{pmatrix} f(x_1) \\ f(x_2) \\ \cdot \\ \cdot \\ \cdot \\ f(x_{N-1}) \\ 0 \end{pmatrix} - \begin{pmatrix} \alpha_1 u_0 \\ 0 \\ \cdot \\ \cdot \\ \cdot \\ 0 \\ 0 \end{pmatrix}. \qquad (5.35)$$

<u>Example 2</u>. Let us reconsider problem (5.4), $\alpha = 0$, but this time with boundary conditions

$u(0) = 1$, $\quad u'(\frac{1}{2}\pi) = 0$.

Taking $N = 5$ as before, (5.33) is

$$\begin{pmatrix} 1 & -\frac{1}{2} & 0 & 0 \\ -\frac{1}{2} & 1 & -\frac{1}{2} & 0 \\ 0 & -\frac{1}{2} & 1 & -\frac{1}{2} \\ 0 & 0 & -\frac{1}{2} & \frac{1}{2} \end{pmatrix} \begin{pmatrix} u_1 \\ u_2 \\ u_3 \\ u_4 \end{pmatrix} = \begin{pmatrix} 0.4507 \\ -0.0493 \\ -0.0493 \\ -0.0493 \end{pmatrix}$$

with solution

$$\begin{pmatrix} u_1 \\ u_2 \\ u_3 \\ u_4 \end{pmatrix} = \begin{pmatrix} 0.6052 \\ 0.3091 \\ 0.1117 \\ 0.0130 \end{pmatrix}$$

and boundary value $u_5 = u_4 = 0.0130$. Alternatively, (5.34) is

$$\begin{pmatrix} 1 & -\frac{1}{2} & 0 & 0 & 0 \\ -\frac{1}{2} & 1 & -\frac{1}{2} & 0 & 0 \\ 0 & -\frac{1}{2} & 1 & -\frac{1}{2} & 0 \\ 0 & 0 & -\frac{1}{2} & 1 & -\frac{1}{2} \\ 0 & 0 & 0 & -1 & 1 \end{pmatrix} \begin{pmatrix} u_1 \\ u_2 \\ u_3 \\ u_4 \\ u_5 \end{pmatrix} = \begin{pmatrix} 0.4507 \\ -0.0493 \\ -0.0493 \\ -0.0493 \\ -0.0493 \end{pmatrix}$$

with solution

$$\begin{pmatrix} u_1 \\ u_2 \\ u_3 \\ u_4 \\ u_5 \end{pmatrix} = \begin{pmatrix} 0.5559 \\ 0.2104 \\ -0.0363 \\ -0.1844 \\ -0.2337 \end{pmatrix} .$$

Again it may be verified that the latter solution, based on the central difference approximation of the derivative boundary condition, agrees with the true solution of the problem

$$u = 1 - \frac{\pi}{2}x + \frac{1}{2}x^2$$

to the number of figures quoted. On the other hand, the error in the first approximation is significant, particularly at the boundary point x_5.

A look at the total discretization errors is again informative. In the one-sided difference approximation, (5.27) and (5.28) become

$$L_h e_j = \tau_j[u] , \qquad j = 1,\ldots,N-1 , \tag{5.36}$$

with

$$e_0 = 0 , \qquad e_N - e_{N-1} = -\frac{h^2}{2} u''(\xi_N) . \tag{5.37}$$

For $j \neq N-1$, the truncation errors τ_j vanish, but the second derivative u" does not, which means the e_j satisfying (5.36) and (5.37) are non-zero, $1 \leq j \leq N$. In the central-difference approximation these equations become

$$L_h[e_j] = \tau_j[u] \ , \qquad j = 1,\ldots,N \ ,$$

where the truncation errors all vanish, and

$$e_0 = 0 \ , \qquad e_{N+1} - e_{N-1} = \frac{h^3}{3} u'''(\eta_N) \ ,$$

giving $e_j = 0$, $0 \leq j \leq N$, since u''' vanishes in example 2.

On the whole, the finite difference solution of well-posed linear problems is a straightforward matter, and the analysis of the numerical methods is fairly complete. The same is not true of nonlinear problems which we examine next.

5.3 FINITE DIFFERENCE SOLUTION OF TWO-POINT BOUNDARY VALUE PROBLEMS : THE NONLINEAR CASE

5.3.1 Nonlinear difference equations

Consider the nonlinear second-order equation

$$Lu \equiv u'' - f(x,u,u') = 0 \ , \qquad x \in \Omega \ , \tag{5.38}$$

subject to BC1: $u(x_0) = u_0$, $\quad u(x_N) = u_N$.

Notice that f represents the nonlinear part of the operator L. With the values of u_0 and u_N given, an obvious difference approximation is

$$L_h u_j \equiv \frac{u_{j+1} - 2u_j + u_{j-1}}{h^2} - f\left(x_j,\ u_j,\ \frac{u_{j+1} - u_{j-1}}{2h}\right) = 0 \ , \quad 1 \leq j \leq N-1, \tag{5.39}$$

where the derivatives at interior grid-points have been replaced by central-difference approximations.

Assuming that the partial derivatives $\partial f/\partial u$ and $\partial f/\partial v$ of $f(x,u,v)$ are continuous, the local truncation error $\tau_j[u]$ may be found from Taylor expansions to be

$$\begin{aligned}\tau_j[u] &\equiv L_h u(x_j) - Lu(x_j) \\ &= -\frac{h^2}{12}\left[u^{(4)}(\xi_j) + 2\frac{\partial f}{\partial v}(x_j, u(x_j), u'(\zeta_j))u'''(\eta_j)\right] ,\end{aligned}$$

ξ_j, η_j and $\zeta_j \in (x_{j-1}, x_{j+1})$, so that L_h has second-order accuracy. If also the partial derivatives satisfy

$$0 < \lambda \leq -\partial f/\partial u \leq \mu \ , \qquad |\partial f/\partial v| \leq \nu \ ,$$

for some positive constants λ, μ, ν, then it is easy to establish the stability of the operator L_h for all h satisfying

$$h\nu \leqslant 2 \ . \tag{5.40}$$

More general stability analyses are, however, more difficult.

An important example of f, involving the one-dimensional analogue of the convective term in the Navier-Stokes equations, is given by

$$f(x,u,u') \equiv a(x)u(x)u'(x) + b(x) \ , \tag{5.41}$$

for which the difference equations (5.39) are

$$\frac{u_{j+1} - 2u_j + u_{j-1}}{h^2} = a(x_j)u_j\left(\frac{u_{j+1} - u_{j-1}}{2h}\right) + b(x_j) \ . \tag{5.42}$$

Since $\partial f/\partial v(x,u,u') = a(x)u$, the stability condition (5.40) is simply

$$h\| a(x)u(x)\|_\infty \leqslant 2 \ , \tag{5.43}$$

where $\|\cdot\|_\infty$ denotes the supremum norm on Ω. In computational fluid dynamics this condition is simply a constraint on the *grid Reynolds number*. We discuss in §5.3.4 what can be done if this constraint is violated.

In general, since equations (5.39) are nonlinear, we must employ iterative methods to solve them. The iterative methods which are traditionally used in computational fluid dynamics are all fixed-point methods which work on a contraction mapping principle. Consider the nonlinear system of algebraic equations

$$\underset{\sim}{F}(\underset{\sim}{u}) = \underset{\sim}{0} \ , \qquad \underset{\sim}{F} : \mathbb{R}^M \to \mathbb{R}^M \ , \qquad \underset{\sim}{u} \in \mathbb{R}^M \ . \tag{5.44}$$

It is always possible to rearrange such a system into the form

$\underset{\sim}{u} = \underset{\sim}{G}(\underset{\sim}{u})$, $\underset{\sim}{G} : \mathbb{R}^M \to \mathbb{R}^M$.

A vector $\underset{\sim}{u}^* \in \mathbb{R}^M$ is called a *fixed-point* of $\underset{\sim}{G}$ if

$\underset{\sim}{u}^* = \underset{\sim}{G}(\underset{\sim}{u}^*)$.

Thus a fixed point of $\underset{\sim}{G}$ is a solution of (5.44) and conversely.

The simplest method of solving (5.44) is by the *functional iteration* (or *successive approximation*) :

$$\underset{\sim}{u}_{r+1} = \underset{\sim}{G}(\underset{\sim}{u}_r) , \qquad r = 0,1,2,\ldots, \tag{5.45}$$

where $\underset{\sim}{u}_0$ is taken as some "guess" at the solution $\underset{\sim}{u}^*$. If the sequence $\{\underset{\sim}{u}_r\}$ converges to a limit then this limit is necessarily a fixed-point of $\underset{\sim}{G}$. In practice, the main difficulty is to choose an operator $\underset{\sim}{G}$ such that the iteration converges from the available starting iterate $\underset{\sim}{u}_0$.

A sufficient condition for convergence is that $\underset{\sim}{u}_0 \in \mathcal{D}$ where $\mathcal{D} \subset \mathbb{R}^M$ is a closed region on which $\underset{\sim}{G}$ is a contraction mapping, that is, $\underset{\sim}{G}$ maps $\mathcal{D}$ into itself, and there exists a constant $K < 1$ such that

$$\| \underset{\sim}{G}(\underset{\sim}{u}) - \underset{\sim}{G}(\underset{\sim}{v}) \| \leqslant K \| \underset{\sim}{u} - \underset{\sim}{v} \|, \ \forall \quad \underset{\sim}{u}, \underset{\sim}{v} \in \mathcal{D} .$$

The *contraction mapping theorem* then states that the iteration (5.45) converges to a fixed-point $\underset{\sim}{u}^*$ of $\underset{\sim}{G}$ which is *unique* in $\mathcal{D}$. Since $\mathcal{D}$ is usually unknown in practice, we can rarely be sure in advance whether the starting iterate $\underset{\sim}{u}_0$ is sufficiently near $\underset{\sim}{u}^*$ so that the contraction works. The only test is to try it and see!

We say that the iteration (5.45) has *order* p if the iterates satisfy

$$\| \underset{\sim}{u}_{r+1} - \underset{\sim}{u}^* \| \leqslant C \| \underset{\sim}{u}_r - \underset{\sim}{u}^* \|^p$$

for some constant C. Thus the higher the order the faster the terminal convergence. If $p = 1$ or 2 then the convergence, when it occurs, is said to be linear or quadratic, respectively. The convergence of most simple functional iteration methods is usually linear, whereas Newton-type methods exhibit quadratic terminal convergence at the cost of a slightly more complicated operator $\underset{\sim}{G}$. In practice it often happens that the domain of contraction is larger for low-order methods than for higher. Despite the obvious attraction of Newton-type methods, therefore, simpler iterative methods can also prove useful particularly if the starting iterate $\underset{\sim}{u}_0$ is far from $\underset{\sim}{u}^*$. In the next two sections we discuss both simple functional iteration schemes and Newton's method.

5.3.2 Simple functional iteration

One of the simplest iteration schemes for the nonlinear equations (5.39) is

$$\left.\begin{aligned} &\frac{u_{j+1}^{[r+1]} - 2u_j^{[r+1]} + u_{j-1}^{[r+1]}}{h^2} = f\left(x_j, u_j^{[r]}, \frac{u_{j+1}^{[r]} - u_{j-1}^{[r]}}{2h}\right) \equiv f_j^{[r]} , \\ &\qquad\qquad\qquad\qquad\qquad\qquad\qquad\qquad 1 \leqslant j \leqslant N-1 , \\ &u_0^{[r]} = u_0 , \qquad u_N^{[r]} = u_N ; \qquad r = 0,1,\ldots . \end{aligned}\right\} \tag{5.46}$$

The iteration number r is denoted in square brackets, and the boundary values are fixed for all r to conform with BC1. Upon multiplication by $-\frac{1}{2}h^2$, (5.46) represent the *linearized* system

$$\underset{\sim}{\underline{A}}\underset{\sim}{u}_{r+1} = \underset{\sim}{z}_r , \tag{5.47}$$

where

$$\underset{\sim}{u}_r \equiv \begin{pmatrix} u_1^{[r]} \\ \cdot \\ \cdot \\ \cdot \\ u_{N-1}^{[r]} \end{pmatrix} , \qquad \underset{\sim}{z}_r \equiv -\frac{h^2}{2}\begin{pmatrix} f_1^{[r]} \\ \cdot \\ \cdot \\ \cdot \\ f_{N-1}^{[r]} \end{pmatrix} + \begin{pmatrix} \frac{1}{2}u_0 \\ 0 \\ \cdot \\ \cdot \\ \cdot \\ 0 \\ \frac{1}{2}u_N \end{pmatrix}$$

and the matrix $\underline{A}$ is of tridiagonal form (5.21) with $\alpha_j = \gamma_j = -\frac{1}{2}$, $\beta_j = 1$. The system (5.47) corresponds to the functional iteration

$$\underset{\sim}{u}_{r+1} = \underset{\sim}{G}(\underset{\sim}{u}_r) \equiv \underline{A}^{-1}\underset{\sim}{z}_r , \qquad r = 0,1,\ldots,$$

so that at each iteration step we solve the tridiagonal system (5.47) using Algorithm 5.1. Notice that $\underline{A}$ does not depend on r, which means that the pivot vector $\underset{\sim}{p}$ need only be computed *once* and stored.

An even simpler iteration scheme for (5.39) which avoids the solution of matrix systems is

$$\left.\begin{aligned} &\frac{u_{j+1}^{[r]} - 2u_j^{[r+1]} + u_{j-1}^{[r]}}{h^2} = f\left(x_j, u_j^{[r]}, \frac{u_{j+1}^{[r]} - u_{j-1}^{[r]}}{2h}\right) \equiv f_j^{[r]} , \quad 1 \leq j \leq N-1 , \\ &u_0^{[r]} = u_0 , \qquad u_N^{[r]} = u_N ; \qquad r = 0,1,\ldots, \end{aligned}\right\}$$

which may be rearranged to give

$$\begin{aligned} u_j^{[r+1]} &= \tfrac{1}{2}\left(u_{j+1}^{[r]} + u_{j-1}^{[r]}\right) - \frac{h^2}{2} f_j^{[r]} , \quad 1 \leq j \leq N-1 , \\ &\equiv G_j(\underset{\sim}{u}_r). \end{aligned} \tag{5.48}$$

Depending on the particular form of f, several variants of the schemes (5.46) and (5.48) may also yield linearized systems. The special case (5.41) with its difference equations (5.42) is particularly relevant. We may consider the iterative scheme

$$\frac{u_{j+1}^{[r+1]} - 2u_j^{[r+1]} + u_{j-1}^{[r+1]}}{h^2} = a(x_j)u_j^{[r]}\left(\frac{u_{j+1}^{[r+1]} - u_{j-1}^{[r+1]}}{2h}\right) + b(x_j) , \qquad (5.49)$$

which is the one-dimensional finite difference analogue of an iterative scheme employed by Nickel, Tanner and Caswell (1974), and others, in solving the Navier-Stokes equations by finite elements. Equations (5.49) may be rewritten as the tridiagonal system

$$\alpha_j^{[r]}u_{j-1}^{[r+1]} + \beta_j^{[r]}u_j^{[r+1]} + \gamma_j^{[r]}u_{j+1}^{[r+1]} = -\frac{h^2}{2}\,b(x_j) ,$$

where

$$\left.\begin{aligned} \alpha_j^{[r]} &= -\tfrac{1}{2}[1 + \tfrac{1}{2}ha(x_j)u_j^{[r]}] , \\ \beta_j^{[r]} &= 1 , \\ \gamma_j^{[r]} &= -\tfrac{1}{2}[1 - \tfrac{1}{2}ha(x_j)u_j^{[r]}] . \end{aligned}\right\}$$

In matrix form we have

$$\underline{A}_r \underset{\sim}{u}_{r+1} = \underset{\sim}{z}_r , \qquad (5.50)$$

with

$$\underline{A}_r \equiv \begin{pmatrix} \beta_1^{[r]} & \gamma_1^{[r]} & & \\ \alpha_2^{[r]} & \beta_2^{[r]} & \gamma_2^{[r]} & \\ & \cdot & \cdot & \cdot \\ & & \cdot & \cdot & \cdot \\ & & & \cdot & \cdot & \cdot \\ & & & & \alpha_{N-1}^{[r]} & \beta_{N-1}^{[r]} \end{pmatrix}, \quad \underset{\sim}{z}_r = -\frac{h^2}{2}\begin{pmatrix} b(x_1) \\ \\ \cdot \\ \cdot \\ \cdot \\ \\ b(x_{N-1}) \end{pmatrix} - \begin{pmatrix} \alpha_1^{[r]}u_0 \\ 0 \\ \cdot \\ \cdot \\ \cdot \\ 0 \\ \gamma_{N-1}^{[r]}u_N \end{pmatrix} .$$

Notice that in contrast to (5.47), the system (5.50) has its matrix dependent on r, and consequently must be updated at each iteration. In Algorithm 5.1 both vectors $\underset{\sim}{p}$ and $\underset{\sim}{q}$ need to be recomputed at each step.

An interesting observation may be made concerning the stability of the functional iteration (5.50). The matrix $\underline{A}_r$ is diagonally dominant if and only if

$$|1 \pm \tfrac{1}{2}ha(x_j)u_j^{[r]}| \leq 2 , \qquad 1 \leq j \leq N-1 ,$$

that is,

$$h|a(x_j)u_j^{[r]}| \leq 2 , \qquad 1 \leq j \leq N-1 . \tag{5.51}$$

Notice that (5.51) is the discrete analogue of the stability condition (5.43) for the operator L_h. There is here a close link between the diagonal dominance of the linearized system and the stability of the nonlinear operator. In particular, if inequality (5.51) is violated, the stability of the operator L_h is not guaranteed and the functional iteration may diverge.

The connection between diagonal dominance of the linearized system and nonlinear stability is a special feature of iteration (5.49)-(5.50): we can expect no such connection in general. Diagonal dominance may well play a rôle in the stability of the algorithm for solving the linearized equations, but this will be generally unconnected with the nonlinear stability. Thus, for example, in the iterative scheme (5.46)-(5.47), the diagonal dominance of the matrix $\underline{A}$ has nothing to do with inequality (5.51) which remains the discrete condition for the stability of L_h. In general we can say only that if an unstable algorithm is used to solve the linearized system at each nonlinear iteration step then this may well cause divergence of the nonlinear iterative scheme. Such divergence, however, might also occur for purely independent reasons.

Example 3. To illustrate the main features of simple functional iteration schemes we will now use a few such methods to solve problem (5.4) with $\alpha = 1$. We have

$$f(x,u,u') = 1 - \tfrac{1}{2}[(u')^2 + u^2] ,$$

and we take $n = 5$, $h = \pi/10$, as in the previous examples.

(i) Consider the iteration (5.46)

$$\left.\begin{aligned} &\frac{u_{j+1}^{[r+1]} - 2u_j^{[r+1]} + u_{j-1}^{[r+1]}}{h^2} = 1 - \tfrac{1}{2}\left[\left(\frac{u_{j+1}^{[r]} - u_{j-1}^{[r]}}{2h}\right)^2 + \left(u_j^{[r]}\right)^2\right] , \\ &\qquad\qquad\qquad\qquad\qquad\qquad\qquad\qquad 1 \leq j \leq 4 , \\ &u_0^{[r]} = 1 , \qquad u_5^{[r]} = 0 , \qquad r = 0,1,\ldots . \end{aligned}\right\}$$

As the initial iterate we use a linear approximation. With $c = 2/\pi$, take $u_0(x) = 1 - cx$, yielding $\underset{\sim}{u}_0 = (0.8, 0.6, 0.4, 0.2)^T$. The results of the first few matrix iterations (5.47) are shown in Table 5.1; convergence to 4 decimal places is attained at the 5th iteration.

TABLE 5.1

r	$u_1^{[r]}$	$u_2^{[r]}$	$u_3^{[r]}$	$u_4^{[r]}$
0	0.8000	0.6000	0.4000	0.2000
1	0.6821	0.4113	0.2014	0.0623
2	0.6859	0.4042	0.1840	0.0453
3	0.6884	0.4079	0.1865	0.0461
4	0.6883	0.4081	0.1869	0.0463
5	0.6883	0.4080	0.1869	0.0463

What we have converged to, of course, is the solution of the difference equations $L_h u_j = f_j$. This will differ from the exact solution $u(x_j) = 1 - \sin x_j$ because of non-vanishing discretization error. For comparison, the exact solution at x_j, $j = 1,\ldots,4$, is (0.6910, 0.4122, 0.1910, 0.0489) to 4 decimals.

(ii) Let us replace the previous iterative scheme by

$$\left.\begin{aligned} &\frac{u_{j+1}^{[r+1]} - 2u_j^{[r+1]} + u_{j-1}^{[r+1]}}{h^2} = 1 - \tfrac{1}{2}\left[\left(\frac{u_{j+1}^{[r]} - u_{j-1}^{[r]}}{2h}\right)\left(\frac{u_{j+1}^{[r+1]} - u_{j-1}^{[r+1]}}{2h}\right) + \left(u_j^{[r]}\right)^2\right], \\ &\qquad\qquad\qquad\qquad\qquad\qquad\qquad\qquad\qquad\qquad 1 \leqslant j \leqslant 4, \\ &u_0^{[r]} = 1, \quad u_5^{[r]} = 0; \quad r = 0,1,\ldots . \end{aligned}\right\} \tag{5.52}$$

This reduces to the tridiagonal system

$$\alpha_j^{[r]}u_{j-1}^{[r+1]} + \beta_j^{[r]}u_j^{[r+1]} + \gamma_j^{[r]}u_{j+1}^{[r+1]} = -\frac{h^2}{2}\left[1 - \tfrac{1}{2}\left(u_j^{[r]}\right)^2\right], \quad 1 \leqslant j \leqslant 4,$$

where

$$\alpha_j^{[r]} = -\tfrac{1}{2}[1 - \tfrac{1}{8}(u_{j+1}^{[r]} - u_{j-1}^{[r]})],$$
$$\beta_j^{[r]} = 1,$$
$$\gamma_j^{[r]} = -\tfrac{1}{2}[1 + \tfrac{1}{8}(u_{j+1}^{[r]} - u_{j-1}^{[r]})].$$

The matrix system representing each iteration has its matrix dependent on r. The results of the first few iterations, starting from the same $\underset{\sim}{u}_0$ as previously, are shown in Table 5.2; convergence to 4 decimals is achieved in 4 iterations, making iteration (5.52) slightly more favourable than (5.46).

TABLE 5.2

r	$u_1^{[r]}$	$u_2^{[r]}$	$u_3^{[r]}$	$u_4^{[r]}$
0	0.8000	0.6000	0.4000	0.2000
1	0.6893	0.4165	0.2002	0.0568
2	0.6877	0.7074	0.1867	0.0469
3	0.6882	0.4080	0.1868	0.0463
4	0.6883	0.4080	0.1869	0.0463

A further slight improvement in the rate of convergence in this problem can be achieved by replacing the quadratic term $\left(u_j^{[r]}\right)^2$ in (5.52) by $u_j^{[r+1]}u_j^{[r]}$. The diagonal dominance of our tridiagonal system then becomes mildly exceeded but without affecting stability of Algorithm 5.1. For more general problems, however, it is as well to exercise care in this respect.

(iii) We now consider two iterative methods which do not involve the solution of matrix systems. Iteration (5.48) is

$$\left.\begin{aligned} u_j^{[r+1]} &= \tfrac{1}{2}(u_{j+1}^{[r]} + u_{j-1}^{[r]}) - \frac{h^2}{2}\left[1 - \tfrac{1}{2}\left\{\left(\frac{u_{j+1}^{[r]} - u_{j-1}^{[r]}}{2h}\right)^2 + \left(u_j^{[r]}\right)^2\right\}\right], \\ &\qquad\qquad 1 \leqslant j \leqslant 4, \\ u_0^{[r]} &= 1, \qquad u_5^{[r]} = 0; \qquad r = 0,1,\ldots . \end{aligned}\right\} \tag{5.53}$$

Starting from $\underset{\sim}{u}_0$ as before, we find that 41 iterations are needed to converge to 4 decimals. Even allowing for the much reduced cost of scalar iteration as compared with matrix iteration, the above scheme is not competitive with the methods in (i) and (ii). A considerable improvement over (5.53) can be found using

$$u_j^{[r+1]} = \tfrac{1}{2}\left(u_{j+1}^{[r]} + u_{j-1}^{[r+1]}\right) - \frac{h^2}{2}\left[1 - \tfrac{1}{2}\left\{\left(\frac{u_{j+1}^{[r]} - u_{j-1}^{[r+1]}}{2h}\right)^2 + \left(u_j^{[r]}\right)^2\right\}\right], \qquad 1 \leqslant j \leqslant 4,$$

which takes 21 iterations to converge to 4 decimals.

5.3.3 Newton's method

The basic *Newton* or *Newton-Raphson* method for solving the algebraic system

$$\underset{\sim}{F}(\underset{\sim}{u}) = \underset{\sim}{0}$$

is the functional iteration

$$\underset{\sim}{u}_{r+1} = \underset{\sim}{u}_r - (\nabla \underset{\sim}{F}(\underset{\sim}{u}_r))^{-1} \underset{\sim}{F}(\underset{\sim}{u}_r) , \qquad r = 0,1,\ldots . \tag{5.54}$$

Since the gradient of the fixed-point operator $\underset{\sim}{G}(\underset{\sim}{u}) = \underset{\sim}{u} - (\nabla\underset{\sim}{F}(\underset{\sim}{u}))^{-1}\underset{\sim}{F}(\underset{\sim}{u})$ vanishes at the solution $\underset{\sim}{u} = \underset{\sim}{u}^*$, the method yields quadratic terminal convergence.

Defining the *increment vector* $\delta\underset{\sim}{u}_r \equiv \underset{\sim}{u}_{r+1} - \underset{\sim}{u}_r$, and the *Jacobian matrix* $\underline{J} \equiv \nabla\underset{\sim}{F}$:

$$J_{jk} = \partial F_j/\partial u_k ,$$

the iteration (5.54) is implemented by solving the matrix system

$$\underline{J}_r \delta\underset{\sim}{u}_r = - \underset{\sim}{F}_r \tag{5.55}$$

for the increment, which is then used to update the current iterate:

$$\underset{\sim}{u}_{r+1} = \underset{\sim}{u}_r + \delta\underset{\sim}{u}_r . \tag{5.56}$$

The notation in (5.55) is $\underset{\sim}{F}_r \equiv \underset{\sim}{F}(\underset{\sim}{u}_r)$, $\underline{J}_r \equiv \underline{J}(\underset{\sim}{u}_r)$.

We illustrate the method in example 4, where we solve the nonlinear problem of example 3, i.e. (5.4) with $\alpha = 1$.

Example 4. The algebraic equations are

$$F_j(\underset{\sim}{u}) \equiv \frac{u_{j+1} - 2u_j + u_{j-1}}{h^2} + \tfrac{1}{2}\left[\left(\frac{u_{j+1} - u_{j-1}}{2h}\right)^2 + u_j^2\right] - 1 = 0 , \qquad 1 \leqslant j \leqslant 4 ,$$

$$u_0 = 1 , \qquad u_5 = 0 .$$

The Jacobian matrix is tridiagonal with non-zero elements given by

$$J_{jk}(\underset{\sim}{u}) = \begin{cases} \dfrac{1}{h^2} - \dfrac{1}{4h^2}(u_{j+1} - u_{j-1}) , & j = k+1 , \\ -\dfrac{2}{h^2} + u_j , & j = k , \\ \dfrac{1}{h^2} + \dfrac{1}{4h^2}(u_{j+1} - u_{j-1}) , & j = k-1 . \end{cases}$$

Multiplying (5.55) by $-h^2/2$ we obtain the system

$$\alpha_j^{[r]}\delta u_{j-1}^{[r]} + \beta_j^{[r]}\delta u_j^{[r]} + \gamma_j^{[r]}\delta u_{j+1}^{[r]} = \frac{h^2}{2}F_j^{[r]} ,$$

where

$$\alpha_j^{[r]} = -\tfrac{1}{2}\left[1 - \tfrac{1}{4}\left(u_{j+1}^{[r]} - u_{j-1}^{[r]}\right)\right] ,$$
$$\beta_j^{[r]} = 1 - \frac{h^2}{2} u_j^{[r]} ,$$
$$\gamma_j^{[r]} = -\tfrac{1}{2}\left[1 + \tfrac{1}{4}\left(u_{j+1}^{[r]} - u_{j-1}^{[r]}\right)\right] ,$$

which must be solved subject to the *homogenous* end conditions

$$\delta u_0^{\ r} = \delta u_5^{\ r} = 0 .$$

Each iteration is then completed by the incrementation (5.56). Notice that the homogeneous boundary conditions on $\delta\underset{\sim}{u}$ must be introduced to allow $u_0^{[r+1]} = 1$, $u_5^{[r+1]} = 0$.

Starting from the same $\underset{\sim}{u}_0$ as in example 3, convergence to 4 decimals is obtained in 3 iterations, as shown in Table 5.3. This demonstrates the rapid (quadratic) convergence of the method. We have again used Algorithm 5.1 to solve the linearized equations. The Jacobian matrix is not quite diagonally dominant, but this has no ill effect in the present case.

TABLE 5.3

r	$u_1^{[r]}$	$u_2^{[r]}$	$u_3^{[r]}$	$u_4^{[r]}$
0	0.8000	0.6000	0.4000	0.2000
1	0.6796	0.3953	0.1735	0.0354
2	0.6882	0.4080	0.1868	0.0463
3	0.6883	0.4080	0.1869	0.0463

For large problems in higher dimensions the main criterion in choosing an algorithm for solving the linearized equations of Newton's method is that of speed, for otherwise the benefit of rapid terminal convergence is lost. Where possible, quick elimination methods or preconditioned conjugate-gradient methods based on incomplete factorization are advisable, since these can be adapted to take account of sparsity and structure in the Jacobian. The use of classical iterative methods such as Gauss-Seidel (GS) and successive over-relaxation (SOR), which give rise to the Newton-GS and Newton-SOR methods, will invariably be slower.

The basic Newton method is perhaps one of the most widely used methods for solving nonlinear systems of algebraic equations. For large problems it can prove relatively costly due to the necessity of recomputing the Jacobian at each iteration. In recent years, therefore, much attention has been paid to *modified Newton methods* which use an approximation to the Jacobian which is simpler and less costly to compute, while attempting to preserve quadratic or almost quadratic convergence. The application of such methods to fluid problems has by no means kept pace with their theoretical development and there is considerable scope for advancement in this direction. The book by Wait (1979) and the references contained therein serve as useful introductions to modified Newton methods, while the classical texts of Ortega and Rheinboldt (1970) and Rheinboldt (1974) contain rigorous treatments of the numerical solution of nonlinear algebraic equations generally.

Other methods for nonlinear equations which are worthy of attention, but which have not been widely tried in fluid problems are minimization methods such as conjugate-direction methods [see, for example, Hestenes (1980).] In particular, the conjugate-Newton method of Irons and Elsawaf (1977) has some attractive features and would seem to deserve more attention than it has received.

5.3.4 Stability of finite difference schemes : upwind differences

We now pay further attention to the stability of finite difference formulations. As a model nonlinear equation involving both diffusion and convection terms we consider (5.38) and (5.41) which yield

$$u'' - auu' - b = 0 , \qquad x_0 \leqslant x \leqslant x_N , \tag{5.57}$$

where a and b are functions of x. We shall assume $a > 0$; a^{-1} may represent a variable viscosity and u a velocity.

The conventional central difference (CD) approximation of (5.57) is given by (5.42). Defining

$$\rho_j = \tfrac{1}{2}ha(x_j)u_j ,$$

the CD scheme may be rewritten as

$$-\tfrac{1}{2}h^2L_hu_j \equiv \alpha_ju_{j-1} + \beta_ju_j + \gamma_ju_{j+1} + \tfrac{1}{2}h^2b(x_j) = 0 , \tag{5.58}$$

where

$$\left.\begin{aligned} \alpha_j &= -\tfrac{1}{2}(1 + \rho_j) , \\ \gamma_j &= -\tfrac{1}{2}(1 - \rho_j) , \\ &\text{and} \\ \beta_j &= -(\alpha_j + \gamma_j) . \end{aligned}\right\} \tag{5.59}$$

We have seen that the scheme is stable if $h\|au\|_\infty \leqslant 2$. When a^{-1} is a viscosity and u a velocity, the quantity $h\|au\|_\infty$ is called the *maximum grid Reynolds number* associated with the flow problem, while the quantity $2|\rho_j|$ represents a *local grid Reynolds number*. In terms of ρ_j the stability condition for the CD scheme is $|\rho_j| \leqslant 1$.

By choosing the mesh-size h sufficiently small, we can, in principle, always satisfy the stability condition. In practice, however, mesh refinement is not always feasible because of the demands on computer store. This is normally true only in higher dimensions, but it serves as an impetus for examining alternatives to the CD scheme in one-dimension when the stability condition is violated.

The simplest alternative is to maintain the central difference approximation for the second derivative while using a one-sided difference approximation for the first derivative in (5.57). If we choose the backward difference when $u > 0$ and the forward difference when $u < 0$, that is

$$u'(x_j) \simeq \begin{cases} \dfrac{u_j - u_{j-1}}{h}, & u_j > 0, \\[2ex] \dfrac{u_{j+1} - u_j}{h}, & u_j < 0, \end{cases} \tag{5.60}$$

then the one-sided difference is always on the *upstream* or *upwind* side of x_j. The difference approximation to (5.57) may then be written as (5.58) with (5.59) replaced by

$$\left.\begin{aligned} \alpha_j &= -\tfrac{1}{2}(1 + \rho_j + |\rho_j|) , \\ \gamma_j &= -\tfrac{1}{2}(1 - \rho_j + |\rho_j|) , \\ \beta_j &= -(\alpha_j + \gamma_j) . \end{aligned}\right\} \tag{5.61}$$

Following Roache (1976) we call this scheme the *first upwind differencing* (UD1) method. Roache cites several examples of its use under various names and in different contexts, while Richtmeyer (1957) first credits Lelevier for its application to flow problems. It is not difficult to show that it is an *unconditionally stable* finite difference scheme, i.e., the difference operator L_h has a bounded inverse for all values of $\rho = \frac{1}{2}hau$. In comparison the CD method is conditionally stable ($|\rho| \leqslant 1$).

The disadvantage of UD1 is that the operator L_h is only first-order accurate, with local truncation error

$$\tau_j[u] = \tfrac{1}{2}ha(x_j)|u(x_j)|u''(x_j) + O(h^2). \tag{5.62}$$

The O(h) term in (5.62) is frequently referred to as "false diffusion" or "artificial viscosity" [see, for example, Roache (1976)]. It is larger than the true diffusion term in (5.57) when $|\rho| > 1$.

A second upwind difference scheme (UD2) used by Gentry, Martin and Daly (1966) and Runchal, Spalding and Wolfshtein (1969) displays something near the unconditional stability of UD1 while retaining the second-order accuracy of CD when the spatial variation of the convective term uu' is small. UD2 involves an averaging of velocities on either side of a mesh-point, together with an upwind approximation of derivatives. In (5.57) we take

$$uu' = \tfrac{1}{2}(uu)'$$

with

$$\left.\begin{aligned}
&(uu)'(x_j) \simeq \frac{\bar{u}_R u_R - \bar{u}_L u_L}{h} ,\\
&\text{where}\\
&\bar{u}_R = \tfrac{1}{2}(u_j + u_{j+1}) , \qquad \bar{u}_L = \tfrac{1}{2}(u_j + u_{j-1})\\
&\text{and}\\
&u_R = u_j \text{ when } \bar{u}_R > 0 , \quad u_R = u_{j+1} \text{ when } \bar{u}_R < 0 ,\\
&u_L = u_{j-1} \text{ when } \bar{u}_L > 0 , \quad u_L = u_j \text{ when } \bar{u}_L < 0 .
\end{aligned}\right\} \tag{5.63}$$

In a steady one-dimensional flow there can be no velocity reversal, which means that $\bar{u}_R$ and $\bar{u}_L$ always have the same sign. Defining

$$\rho_j^{\pm} = \tfrac{1}{2}ha(x_j)u_{j\pm1} ,$$

the finite difference approximation of (5.57) may then be written as (5.58) with

$$\left.\begin{aligned}
\alpha_j &= -\tfrac{1}{2}\{1 + \tfrac{1}{4}(\rho_j + \rho_j^-) + \tfrac{1}{4}|\rho_j + \rho_j^-|\} ,\\
\gamma_j &= -\tfrac{1}{2}\{1 - \tfrac{1}{4}(\rho_j + \rho_j^+) + \tfrac{1}{4}|\rho_j + \rho_j^+|\} ,\\
\beta_j &= -(\alpha_j + \gamma_j) + \tfrac{1}{4}(\rho_j^+ - \rho_j^-) .
\end{aligned}\right\} \tag{5.64}$$

The local truncation error is

$$\tau_j[u] = \tfrac{1}{4}ha(x_j)\{\text{sign}\,(u(x_j))\}\{u(x_j)u''(x_j) + (u'(x_j))^2\} + O(h^2) .$$

By comparison with (5.62) we see that when u' is small the false diffusion of UD2 is about one-half that of UD1. Moreover, since $uu'' + (u')^2 = (uu')'$, the UD2 scheme is essentially second-order accurate when the spatial variation of uu' is small.

Another method of Spalding (1972) uses a mixture of central and upwind differences. If $|\rho_j| \leqslant 1$ then Spalding's method (SM) is identical with CD, whereas if $|\rho_j| > 1$ the diffusion term u'' is ignored completely and the convection term uu' approximated as in the UD1 scheme. The resulting finite difference approximation of (5.57) may be written as (5.58) with

$$\left.\begin{aligned} \alpha_j &= -\tfrac{1}{4}\{|1+\rho_j| + |1-\rho_j|\} - \tfrac{1}{2}\rho_j \ , \\ \gamma_j &= -\tfrac{1}{4}\{|1+\rho_j| + |1-\rho_j|\} + \tfrac{1}{2}\rho_j \ , \\ \beta_j &= -(\alpha_j + \gamma_j) \ . \end{aligned}\right\} \tag{5.65}$$

The advantages of SM are that it is unconditionally stable, it has no false diffusion when $|\rho_j| \leqslant 1$, whereas when $|\rho_j| > 1$ the false diffusion is proportional to $(|\rho_j| - 1)$.

Closely related to SM is the exponential fitting method of Allen and Southwell (1955), which has been rediscovered several times in the literature [e.g. Chien (1977)]. The method is locally exact (LE) in the following sense. On each interval $[x_{j-1}, x_{j+1}]$, (5.57) is replaced by its linearized form

$$u'' - a(x_j)u_j u' - b(x_j) = 0 \ , \tag{5.66}$$

which can be solved exactly subject to the boundary conditions $u(x_{j+1}) = u_{j+1}$, $u(x_{j-1}) = u_{j-1}$. The solution is of the form

$$u = A + Bx + Ce^{Kx} \ ,$$

where $K = a(x_j)u_j$, $B = -b(x_j)/K$, and the constants A and C are determined by the boundary conditions. Since

$$u_{j+1} - u_j = Bh + Ce^{Kx_j}(e^{Kh} - 1) \ ,$$

and

$$u_{j-1} - u_j = -Bh + Ce^{Kx_j}(e^{-Kh} - 1) \ ,$$

then by eliminating C, and noting that $Kh = 2\rho_j$, we find

$$-\tfrac{1}{2}\rho_j \coth\rho_j (u_{j+1} - 2u_j + u_{j-1}) + \tfrac{1}{2}\rho_j (u_{j+1} - u_{j-1}) + h^2 b(x_j) = 0 \ . \tag{5.67}$$

Since the difference equation (5.67) is obeyed exactly by the solution of (5.66), then (5.67) is a locally exact finite difference approximation to (5.57) which is of the form (5.58) with

$$\left.\begin{aligned} \alpha_j &= -\tfrac{1}{2}\rho_j(\coth \rho_j + 1) , \\ \gamma_j &= -\tfrac{1}{2}\rho_j(\coth \rho_j - 1) , \\ \beta_j &= -(\alpha_j + \gamma_j) . \end{aligned}\right\} \qquad (5.68)$$

Unfortunately, local exactness does not mean that the local truncation error of (5.67), when applied to (5.57), vanishes identically. When $|\rho_j| \ll 1$, the coefficients in (5.68) tend to those of the CD method (5.59), whereas when $|\rho_j| \gg 1$ the coefficients tend to those of SM ((5.65)). For high grid Reynolds numbers, therefore, the LE and SM methods are virtually identical, with the same false diffusion errors. Again, the LE scheme is unconditionally stable.

Because of the false diffusion errors associated with first-order upwinding and the inherent instability associated with second-order central differences, none of the methods described above are totally satisfactory for high ρ values. In an interesting survey paper, Leonard (1979) refers to several techniques which attempt to retain both accuracy and stability. In particular, Leonard proposes a simple third-order difference scheme which he claims to be optimal in terms of accuracy and stability. These claims are as yet unsubstantiated, and until some satisfactory analysis of higher-order upwinding schemes is forthcoming, it is as well to exercise caution in their use. †

5.4 FINITE DIFFERENCE SOLUTION OF ELLIPTIC BOUNDARY VALUE PROBLEMS : POISSON'S EQUATION

5.4.1 Dirichlet boundary conditions

We introduce finite differences in two space variables by considering the Dirichlet problem associated with Poisson's equation. Let Ω be a closed planar region with boundary Γ. For simplicity, we consider the rectangle $\Omega = \{(x,y): \; x_0 \leqslant x \leqslant x_M \;, \; y_0 \leqslant y \leqslant y_N \}$. We then wish to find $u = u(x,y)$ which satisfies

$$\nabla^2 u \equiv \frac{\partial^2 u}{\partial x^2} + \frac{\partial^2 u}{\partial y^2} = f \;, \qquad (x,y) \in \Omega, \qquad (5.69)$$

with boundary condition

$$u = \alpha \;, \qquad (x,y) \in \Gamma. \qquad (5.70)$$

---

†See, however, Thompson and Wilkes (1982), Wilkes and Thompson (1983).

We cover Ω with a uniform square grid

$$x_j = x_0 + jh \;, \qquad y_k = y_0 + kh \;, \qquad j = 0,\dots,M; \quad k = 0,\dots,N;$$

with grid-spacing $h = (x_M - x_0)/M = (y_N - y_0)/N$. Our main reason for choosing a square grid is to keep the finite difference formulae as simple as possible. It may sometimes be appropriate to use a rectangular grid with different spacings in the x and y directions; such a generalization is straightforward, and reference may be made to the books cited in §5.1.

Let Ω_h denote the set of grid-points inside Ω, and Γ_h denote the set of grid-points on Γ. To approximate the function $u(x,y)$ on the grid we define a set of numbers $u_{j,k}$, $0 \leq j \leq M$, $0 \leq k \leq N$, as the solution of a system of finite difference equations replacing (5.69). Again, our notation is that $u_{j,k}$ approximates $u(x_j,y_k)$ on Ω_h, but on Γ_h we take

$$u_{0,k} = \alpha(x_0,y_k) \;, \qquad u_{M,k} = \alpha(x_M,y_k) \;, \qquad 0 \leq k \leq N \;;$$

$$u_{j,0} = \alpha(x_j,y_0) \;, \qquad u_{j,N} = \alpha(x_j,y_N) \;, \qquad 0 \leq j \leq M \;.$$

Assuming that $u \in C^{4,4}(\Omega)$, i.e., fourth partial derivatives of u exist and are continuous in Ω, then using Taylor expansions with respect to x and y variables respectively, we find (cf. (5.15))

$$\frac{\partial^2 u}{\partial x^2}(x_j,y_k) = \frac{u(x_{j+1},y_k) - 2u(x_j,y_k) + u(x_{j-1},y_k)}{h^2} - \frac{h^2}{12}\frac{\partial^4 u}{\partial x^4}(\xi_j,y_k) \tag{5.71}$$

and

$$\frac{\partial^2 u}{\partial y^2}(x_j,y_k) = \frac{u(x_j,y_{k+1}) - 2u(x_j,y_k) + u(x_j,y_{k-1})}{h^2} - \frac{h^2}{12}\frac{\partial^4 u}{\partial y^4}(x_j,\eta_k) \;, \tag{5.72}$$

where $x_{j-1} < \xi_j < x_{j+1}$, $y_{k-1} < \eta_k < y_{k+1}$. Ignoring $O(h^2)$ terms leads to the *five-point difference approximation* to (5.69) :

$$L_h u_{j,k} \equiv \frac{u_{j+1,k} - 2u_{j,k} + u_{j-1,k}}{h^2} + \frac{u_{j,k+1} - 2u_{j,k} + u_{j,k-1}}{h^2} = f(x_j,y_k),$$

$$1 \leq j \leq M-1 \;, \; 1 \leq k \leq N-1 \;. \tag{5.73}$$

Upon multiplication by $-h^2$, (5.73) may be written in the standard form

$$4u_{j,k} - u_{j+1,k} - u_{j-1,k} - u_{j,k+1} - u_{j,k-1} = -\,h^2 f(x_j,y_k) \;,$$

$$1 \leq j \leq M-1 \;, \; 1 \leq k \leq N-1 \;, \tag{5.74}$$

which represents $(M-1)(N-1)$ linear equations for the same number of unknowns $u_{j,k}$.

Before writing the system (5.74) in matrix form it is important to order the unknown grid-values in a particular way to ensure that the resulting matrix has special structure. We adopt the row-by-row or *natural ordering* of the grid-points by defining the vector $\underset{\sim}{u}$ to be

$$(u_{1,1}, \ldots, u_{M-1,1} ; u_{1,2}, \ldots, u_{M-1,2} ; \ldots, u_{1,N-1}, \ldots, u_{M-1,N-1})^T ,$$

where $(\ )^T$ denotes the transpose. (5.74) may then be written as the matrix system

$$\underline{A}\,\underset{\sim}{u} = \underset{\sim}{r} , \tag{5.75}$$

where $\underline{A}$ is the $(M-1)(N-1) \times (M-1)(N-1)$ matrix

$$\underline{A} = \begin{pmatrix} \underline{C} & -\underline{I} & & & \bigcirc \\ -\underline{I} & \underline{C} & -\underline{I} & & \\ & \cdot & \cdot & \cdot & \\ & & \cdot & \cdot & \cdot \\ & & & \cdot & \cdot & \cdot \\ & & & -\underline{I} & \underline{C} & -\underline{I} \\ \bigcirc & & & & -\underline{I} & \underline{C} \end{pmatrix} . \tag{5.76}$$

$\underline{I}$ is the $(M-1) \times (M-1)$ identity matrix and $\underline{C}$ is the $(M-1) \times (M-1)$ tridiagonal matrix

$$\underline{C} = \begin{pmatrix} 4 & -1 & & & \bigcirc \\ -1 & 4 & -1 & & \\ & \cdot & \cdot & \cdot & \\ & & \cdot & \cdot & \cdot \\ & & & \cdot & \cdot & \cdot \\ & & & -1 & 4 & -1 \\ \bigcirc & & & & -1 & 4 \end{pmatrix} . \tag{5.77}$$

$\underline{A}$ is called a *block tridiagonal* matrix since it is naturally partitioned into $(N-1) \times (N-1)$ blocks each of order $(M-1) \times (M-1)$, and all blocks other than the diagonal and adjacent codiagonals are null. The vector $\underset{\sim}{r}$ in (5.75) has the form

$$(r_{1,1}, \ldots, r_{M-1,1} ; r_{1,2}, \ldots, r_{M-1,2} ; \ldots, r_{1,N-1}, \ldots, r_{M-1,N-1})^T$$

where the elements depend on the functions f and α.

Using the notation $f_{j,k} \equiv f(x_j,y_k)$, $\alpha_{j,k} \equiv \alpha(x_j,y_k)$, then with $M = N = 4$, for example, the system (5.75) is

$$\begin{pmatrix} 4 & -1 & 0 & -1 & 0 & 0 & 0 & 0 & 0 \\ -1 & 4 & -1 & 0 & -1 & 0 & 0 & 0 & 0 \\ 0 & -1 & 4 & 0 & 0 & -1 & 0 & 0 & 0 \\ -1 & 0 & 0 & 4 & -1 & 0 & -1 & 0 & 0 \\ 0 & -1 & 0 & -1 & 4 & -1 & 0 & -1 & 0 \\ 0 & 0 & -1 & 0 & -1 & 4 & 0 & 0 & -1 \\ 0 & 0 & 0 & -1 & 0 & 0 & 4 & -1 & 0 \\ 0 & 0 & 0 & 0 & -1 & 0 & -1 & 4 & -1 \\ 0 & 0 & 0 & 0 & 0 & -1 & 0 & -1 & 4 \end{pmatrix} \begin{pmatrix} u_{1,1} \\ u_{2,1} \\ u_{3,1} \\ u_{1,2} \\ u_{2,2} \\ u_{3,2} \\ u_{1,3} \\ u_{2,3} \\ u_{3,3} \end{pmatrix} = -h^2 \begin{pmatrix} f_{1,1} \\ f_{2,1} \\ f_{3,1} \\ f_{1,2} \\ f_{2,2} \\ f_{3,2} \\ f_{1,3} \\ f_{2,3} \\ f_{3,3} \end{pmatrix} + \begin{pmatrix} \alpha_{0,1} + \alpha_{1,0} \\ \alpha_{2,0} \\ \alpha_{4,1} + \alpha_{3,0} \\ \alpha_{0,2} \\ 0 \\ \alpha_{4,2} \\ \alpha_{0,3} + \alpha_{1,4} \\ \alpha_{2,4} \\ \alpha_{4,3} + \alpha_{3,4} \end{pmatrix} .$$

The discussion of appropriate methods for solving matrix systems of the form (5.75) is left until the next chapter.

The local truncation error of the five-point difference scheme at (x_j,y_k) is

$$\tau_{j,k}[u] = (L_h u)(x_j,y_k) - (\nabla^2 u)(x_j,y_k) , \tag{5.78}$$

i.e.

$$\tau_{j,k}[u] = \frac{h^2}{12}\left[\frac{\partial^4 u}{\partial x^4}(\xi_j,y_k) + \frac{\partial^4 u}{\partial y^4}(x_j,\eta_k)\right] , \tag{5.79}$$

where (5.79) follows from (5.71) and (5.72). Thus $\tau_{j,k}$ is $O(h^2)$, and L_h has second-order accuracy for all $C^{4,4}$-functions.

A bound on the total discretization error, $e_{j,k} = u(x_j,y_k) - u_{j,k}$, may be found by using the fact that the discrete operator L_h gives rise to a *maximum principle* on Ω_h. More precisely, if

$$L_h v_{j,k} \geqslant 0 \tag{5.80}$$

at all points in Ω_h, then $v_{j,k}$ achieves its maximum value on the boundary set Γ_h. We make use of this as follows. From (5.73) and (5.78) we have

$$L_h e_{j,k} = L_h u(x_j,y_k) - L_h u_{j,k} = \tau_{j,k} ,$$

whence

$$|L_h e_{j,k}| \leqslant \tau = \frac{1}{6}h^2 M_4 , \tag{5.81}$$

where

$$M_4 = \sup_{\Omega}\left\{\left|\frac{\partial^4 u}{\partial x^4}\right| , \left|\frac{\partial^4 u}{\partial y^4}\right|\right\} .$$

Choose any function $\phi \geqslant 0$ with the property $L_h\phi_{j,k} = 1$; for example, $\phi = \frac{1}{4}(x^2 + y^2)$. Then

$$L_h(\pm e_{j,k} + \tau\phi_{j,k}) = \pm L_h e_{j,k} + \tau \geqslant 0 ,$$

where the inequality follows from (5.81). Use of the maximum principle then leads to

$$\pm e_{j,k} \leqslant \pm e_{j,k} + \tau\phi_{j,k} \leqslant \max_{\Gamma_h}(\pm e_{j,k} + \tau\phi_{j,k}) = \tfrac{1}{4}(x_M^2 + y_N^2)\tau ,$$

since $e_{j,k} = 0$ on Γ_h. We thus have the error bound

$$|e_{j,k}| \leqslant \frac{1}{24}(x_M^2 + y_N^2)h^2M_4 , \tag{5.82}$$

which establishes the $O(h^2)$ convergence (as $h \to 0$) of the five-point difference scheme, provided $u \in C^{4,4}(\Omega)$.

5.4.2 Mixed boundary conditions

As a special case of the Robbins problem (5.11), in this final section we consider the finite difference solution of Poisson's equation in the same rectangular region Ω as before, subject to a Dirichlet condition on part of the boundary Γ and a Neumann condition on the remainder. More precisely, we consider

$$\nabla^2 u = f , \qquad x_0 \leqslant x \leqslant x_M , \quad y_0 \leqslant y \leqslant y_N ,$$

with

$$u = \alpha(x,y) , \quad \begin{cases} x = x_M , & y_0 \leqslant y \leqslant y_N , \\ y = y_N , & x_0 \leqslant x \leqslant x_M , \end{cases} \tag{5.83}$$

and

$$\frac{\partial u}{\partial n} = \beta(x,y) , \quad \begin{cases} x = x_0 , & y_0 \leqslant y \leqslant y_N , \\ y = y_0 , & x_0 \leqslant x \leqslant x_M . \end{cases} \tag{5.84}$$

On Ω_h we have the five-point difference approximation

$$4u_{j,k} - u_{j+1,k} - u_{j-1,k} - u_{j,k+1} - u_{j,k-1} = -h^2 f_{j,k} , \quad 1 \leqslant j \leqslant M-1 , \quad 1 \leqslant k \leqslant N-1 , \tag{5.85}$$

in which, from (5.83), we take

$$\left.\begin{aligned} u_{M,k} &= \alpha_{M,k} , \quad 0 \leqslant k \leqslant N \\ u_{j,N} &= \alpha_{j,N} , \quad 0 \leqslant j \leqslant M \end{aligned}\right\} . \tag{5.86}$$

To maintain $O(h^2)$ accuracy, we choose central differences to approximate the boundary conditions (5.84). For a typical grid-point (x_0,y_k) on the boundary $x = x_0$, we have

$$\frac{u_{1,k} - u_{-1,k}}{2h} = -\beta_{0,k} , \tag{5.87}$$

while for (x_j,y_0) on the boundary $y = y_0$ we have

$$\frac{u_{j,1} - u_{j,-1}}{2h} = -\beta_{j,0} . \tag{5.88}$$

The minus signs occur on the right of (5.87) and (5.88) because β represents the *outward* normal derivative of u on Γ. As in §5.2.4, the extra unknowns $u_{-1,k}$ and $u_{j,-1}$, which are external to Ω, may be eliminated by assuming that the difference equation (5.85) holds at the boundary points (x_0,y_k) and (x_j,y_0). Eliminating $u_{-1,k}$ between (5.87) and (5.85) with $j = 0$ gives

$$4u_{0,k} - 2u_{1,k} - u_{0,k+1} - u_{0,k-1} = -h^2f_{0,k} + 2h\beta_{0,k} , \quad 1 \leq k \leq N-1 , \tag{5.89}$$

and similarly

$$4u_{j,0} - u_{j+1,0} - u_{j-1,0} - 2u_{j,1} = -h^2f_{j,0} + 2h\beta_{j,0} , \quad 1 \leq j \leq M-1 . \tag{5.90}$$

We note that the corner point (x_0,y_0) is exceptional, since there we have *two* normal derivatives, $\beta^1(x_0,y_0)$ in the x-direction, say, and $\beta^2(x_0,y_0)$ in the y-direction. At $j = k = 0$ we assume that the three equations (5.85), (5.87) and (5.88) hold simultaneously. We can then eliminate both $u_{-1,0}$ and $u_{0,-1}$ to give

$$4u_{0,0} - 2u_{1,0} - 2u_{0,1} = -h^2f_{0,0} + 2h(\beta^1_{0,0} + \beta^2_{0,0}) . \tag{5.91}$$

In contrast to the pure Dirichlet problem of the previous section, in which all the boundary values of u are given, the above finite difference treatment of mixed boundary conditions introduces the values of u along the boundaries $x = x_0$ and $y = y_0$ as unknowns in the difference equations. Adopting the natural ordering of grid-points, the vector $\underset{\sim}{u}$ which we wish to determine is then

$$(u_{0,0}, u_{1,0}, \ldots , u_{M-1,0} ; u_{0,1}, \ldots , u_{M-1,1} ; \ldots ; u_{0,N-1}, \ldots , u_{M-1,N-1})^T .$$

In assembling the difference equations (5.85), (5.89), (5.90) and (5.91) into matrix form, the coefficient matrix is rendered symmetric if we multiply (5.89) and (5.90) by the factor $\frac{1}{2}$, and (5.91) by $\frac{1}{4}$. We then obtain the block tri-diagonal system

$$\underline{A}\,\underset{\sim}{u} = \underset{\sim}{r} ,$$

where $\underline{A}$ is the MN × MN matrix

$$\underline{A} = \begin{pmatrix} \frac{1}{2}\underline{E} & -\underline{K} & & & \bigcirc \\ -\underline{K} & \underline{E} & -\underline{K} & & \\ \cdot & \cdot & \cdot & & \\ & \cdot & \cdot & \cdot & \\ & & \cdot & \cdot & \cdot \\ \bigcirc & & -\underline{K} & \underline{E} & -\underline{K} \\ & & & -\underline{K} & \underline{E} \end{pmatrix} , \qquad (5.92)$$

$\underline{E}$ being the M × M tridiagonal matrix

$$\underline{E} = \begin{pmatrix} 2 & -1 & & & \bigcirc \\ -1 & 4 & -1 & & \\ \cdot & \cdot & \cdot & & \\ & \cdot & \cdot & \cdot & \\ & & \cdot & \cdot & \cdot \\ \bigcirc & & -1 & 4 & -1 \\ & & & -1 & 4 \end{pmatrix} , \qquad (5.93)$$

and $\underline{K}$ the M × M diagonal matrix

$$\underline{K} = \begin{pmatrix} \frac{1}{2} & & & & & & \bigcirc \\ & 1 & & & & & \\ & & 1 & & & & \\ & & & \cdot & & & \\ & & & & \cdot & & \\ \bigcirc & & & & & \cdot & \\ & & & & & & 1 \end{pmatrix} . \qquad (5.94)$$

The elements of $\underset{\sim}{r}$ depend on the functions f, α and β.

For example, with M = N = 3, we have the system

$$\begin{pmatrix} 1 & -\frac{1}{2} & 0 & -\frac{1}{2} & 0 & 0 & 0 & 0 & 0 \\ -\frac{1}{2} & 2 & -\frac{1}{2} & 0 & -1 & 0 & 0 & 0 & 0 \\ 0 & -\frac{1}{2} & 2 & 0 & 0 & -1 & 0 & 0 & 0 \\ -\frac{1}{2} & 0 & 0 & 2 & -1 & 0 & -\frac{1}{2} & 0 & 0 \\ 0 & -1 & 0 & -1 & 4 & -1 & 0 & -1 & 0 \\ 0 & 0 & -1 & 0 & -1 & 4 & 0 & 0 & -1 \\ 0 & 0 & 0 & -\frac{1}{2} & 0 & 0 & 2 & -1 & 0 \\ 0 & 0 & 0 & 0 & -1 & 0 & -1 & 4 & -1 \\ 0 & 0 & 0 & 0 & 0 & -1 & 0 & -1 & 4 \end{pmatrix} \begin{pmatrix} u_{0,0} \\ u_{1,0} \\ u_{2,0} \\ u_{0,1} \\ u_{1,1} \\ u_{2,1} \\ u_{0,2} \\ u_{1,2} \\ u_{2,2} \end{pmatrix} = -h^2 \begin{pmatrix} \frac{1}{4}f_{0,0} \\ \frac{1}{2}f_{1,0} \\ \frac{1}{2}f_{2,0} \\ \frac{1}{2}f_{0,1} \\ f_{1,1} \\ f_{2,1} \\ \frac{1}{2}f_{0,2} \\ f_{1,2} \\ f_{2,2} \end{pmatrix} + h \begin{pmatrix} \frac{1}{2}(\beta^1_{0,0}+\beta^2_{0,0}) \\ \beta_{1,0} \\ \beta_{2,0} \\ \beta_{0,1} \\ 0 \\ 0 \\ \beta_{0,2} \\ 0 \\ 0 \end{pmatrix} + \begin{pmatrix} 0 \\ 0 \\ \frac{1}{2}\alpha_{3,0} \\ 0 \\ 0 \\ \alpha_{3,1} \\ \frac{1}{2}\alpha_{0,3} \\ \alpha_{1,3} \\ \alpha_{3,2}+\alpha_{2,3} \end{pmatrix} .$$

Using the maximum principle as before, we obtain a bound on the total discretization error

$$|e_{j,k}| \leqslant \frac{1}{24}(x_M^2 + y_N^2)h^2M_4 + \frac{1}{6}h^2M_3 , \tag{5.95}$$

where

$$M_3 = \sup_{\Gamma} \left\{ \left|\frac{\partial^3 u}{\partial x^3}\right|, \left|\frac{\partial^3 u}{\partial y^3}\right| \right\} .$$

The second term in (5.95) arises because $e_{j,k}$ does not vanish everywhere on Γ_h; the central difference formulae (5.87) and (5.88) give rise to $O(h^2)$ errors in approximating the normal derivatives. As before, therefore, the overall convergence is $O(h^2)$ as $h \to 0$.

Chapter 6

Finite-Difference Simulation: Differential Models

6.1 INTRODUCTION

In the present chapter we extend the basic methods, introduced in Chapter 5, to solve the governing equations for the flow of a non-Newtonian fluid. We shall only consider differential constitutive equations, leaving the treatment of integral models to the next chapter. Furthermore, it will be useful to restrict attention in the main to one model fluid; for this purpose we take the Maxwell fluid. In particular, we shall study methods for the steady incompressible two-dimensional flow of a Maxwell fluid, the basic equations for which were introduced in §3.9.

In the first instance we shall make the following simplifying assumptions:

(i) The flow region Ω is planar and is either a rectangle or a union of rectangles whose sides are parallel to the coordinate axes. In general, when Ω is planar but non-rectangular, we assume that we may transform Ω into a region of the required form. Conformal transformation techniques for this purpose are referenced in §6.6.3.

(ii) The boundary Γ is solid or the union of a solid boundary, Γ_s, with a well-defined entry boundary, Γ_{in}, exit boundary, Γ_{out}, and possibly an axis of symmetry, Γ_{sym}, i.e.

$$\Gamma = \Gamma_s \cup \Gamma_{in} \cup \Gamma_{out} \cup \Gamma_{sym} .$$

The solid boundary may have a moving part, e.g. a moving plate, but otherwise all boundaries are stationary. Free surfaces are excluded, but see §6.6.3.

(iii) Γ_{in} and Γ_{out} are sufficiently far away from any obstacle, barrier, or abrupt change in geometry, so that the entry and exit flows may be regarded as "fully-developed". In particular, the exit length must be sufficiently long to allow the stresses to relax before Γ_{out} is reached.

Simple examples of flow regions are given in Fig. 6.1(a-c). In (a) the entry and exit flows could be of Poiseuille type if the flow is pressure driven, or alternatively of Couette type if the solid boundary $\Gamma_{s,1}$ moves with constant velocity parallel to the axis, thus generating the flow. In (b) an axis of symmetry forms part of the boundary, the full geometry representing a planar contraction. In (c), where the flow in a cavity is generated by a moving plate, there is no entry or exit.

$$\Gamma_s = \overset{4}{\underset{m=1}{\cup}} \Gamma_{s,m}$$
$$\Gamma = \Gamma_s \cup \Gamma_{in} \cup \Gamma_{out}$$

(a)

$$\Gamma_s = \overset{3}{\underset{m=1}{\cup}} \Gamma_{s,m}$$
$$\Gamma = \Gamma_s \cup \Gamma_{in} \cup \Gamma_{out} \cup \Gamma_{sym}$$

(b)

$$\Gamma = \Gamma_s = \overset{4}{\underset{m=1}{\cup}} \Gamma_{s,m}$$

(c)

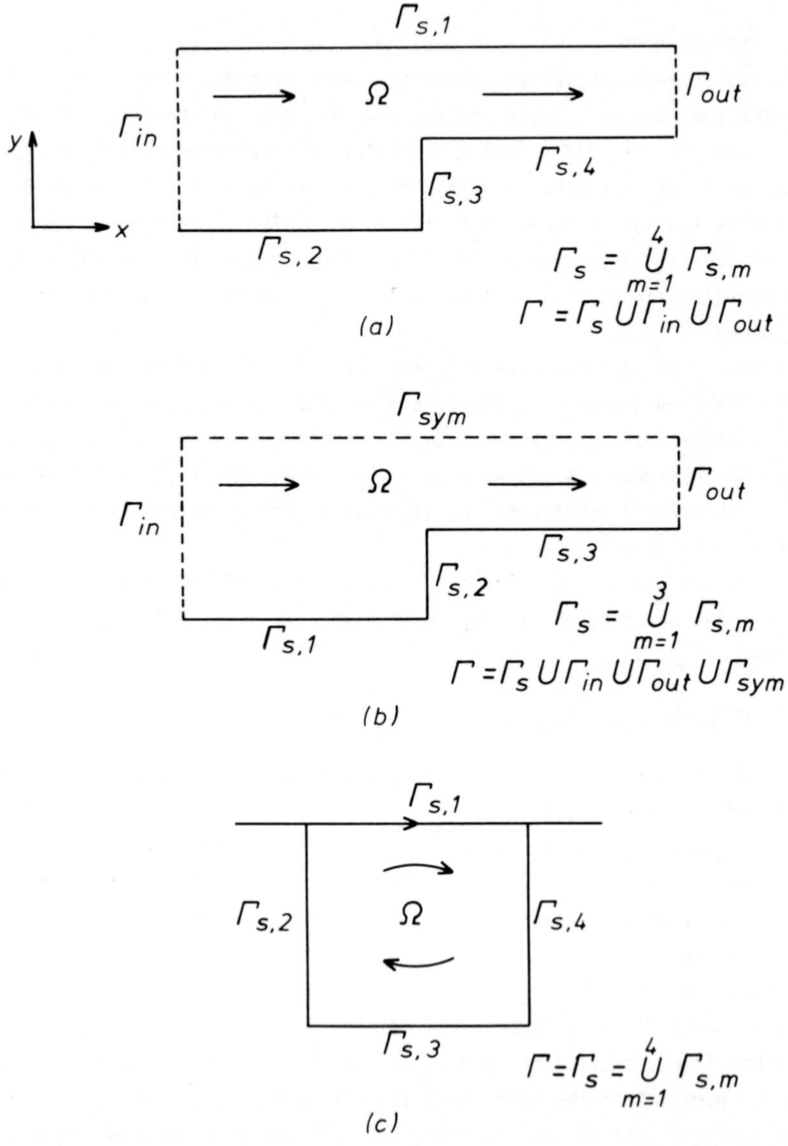

Fig. 6.1 Some typical flow geometries.

Historically, the finite difference solution of the Navier-Stokes equations has been dominated by the choice of stream function and vorticity (ψ,ω) as variables. This is because the discretized form of the coupled elliptic equations for ψ and ω can be solved on a conventional rectangular grid. It took much longer for finite difference methods in terms of the primitive variables (u,v,p) to be developed; their success mainly depends on iteration over interlocking velocity and pressure grids. At the time of writing, finite difference attempts at simulating non-Newtonian flow have been made only in terms of (ψ,ω), with stresses as additional variables. We therefore restrict attention to this formulation in the bulk of this chapter.

Recall from §3.9 that taking non-dimensional variables $(\psi,\omega,\underline{T})$ and non-dimensional parameters R and W, the equations governing the flow of the Maxwell fluid are

$$\nabla^2\psi = -\omega \ , \tag{6.1}$$

$$R\left(\frac{\partial\psi}{\partial x}\frac{\partial\omega}{\partial y} - \frac{\partial\psi}{\partial y}\frac{\partial\omega}{\partial x}\right) = \frac{\partial^2}{\partial x\partial y}(T^{xx} - T^{yy}) + \left(\frac{\partial^2}{\partial y^2} - \frac{\partial^2}{\partial x^2}\right)T^{xy} \ , \tag{6.2}$$

and

$$\left.\begin{aligned}
&T^{xx}\left(1 - 2W\frac{\partial u}{\partial x}\right) + W\left(u\frac{\partial T^{xx}}{\partial x} + v\frac{\partial T^{xx}}{\partial y}\right) - 2WT^{xy}\frac{\partial u}{\partial y} = 2\frac{\partial u}{\partial x} \ ,\\
&T^{xy} + W\left(u\frac{\partial T^{xy}}{\partial x} + v\frac{\partial T^{xy}}{\partial y}\right) - WT^{xx}\frac{\partial v}{\partial x} - WT^{yy}\frac{\partial u}{\partial y} = \frac{\partial u}{\partial y} + \frac{\partial v}{\partial x} \ ,\\
&T^{yy}\left(1 - 2W\frac{\partial v}{\partial y}\right) + W\left(u\frac{\partial T^{yy}}{\partial x} + v\frac{\partial T^{yy}}{\partial y}\right) - 2WT^{xy}\frac{\partial v}{\partial x} = 2\frac{\partial v}{\partial y} \ .
\end{aligned}\right\} \tag{6.3}$$

It will be notationally convenient in this and the next chapter to use upper indices to denote tensor components while reserving lower indices to designate grid points, as in the previous chapter.

Note that the left-hand side of (6.2) consists only of convected vorticity terms. A vorticity diffusion term may be introduced by means of the substitution

$$T^{ik} = S^{ik} + 2d^{ik} \ , \tag{6.4}$$

where S^{ik} denotes the non-Newtonian contribution to the extra-stress tensor. Equation (6.2) may then be written in the elliptic form

$$\nabla^2\omega + R\left(\frac{\partial\psi}{\partial x}\frac{\partial\omega}{\partial y} - \frac{\partial\psi}{\partial y}\frac{\partial\omega}{\partial x}\right) = \frac{\partial^2}{\partial x\partial y}(S^{xx} - S^{yy}) + \left(\frac{\partial^2}{\partial y^2} - \frac{\partial^2}{\partial x^2}\right)S^{xy} \ . \tag{6.5}$$

(Notice that the right-hand side of (6.5) vanishes identically in the Newtonian case.)

Since in the numerical simulation of viscoelastic flows we are most often concerned with low Reynolds numbers, the left-hand side of (6.5) is usually diffusion dominated. It is generally accepted (see, for example, Hughes 1979) that diffusion dominated equations are easier to solve numerically than those dominated by convection, in the sense that stability and accuracy of the discrete representations are more easily ensured. It is therefore appropriate to adopt (6.5) in preference to (6.2). In the case of high R-values, however, we note that Gatski and Lumley (1978a,b) have succeeded in solving (6.2) as it stands.

The substitution (6.4) was first introduced by Perera and Walters (1977a,b), and independently by Leal (1979). Alternative substitutions have been proposed in the context of other constitutive equations (see Townsend, 1980a, for second order fluids, and Holstein 1981, Teifenbruck and Leal 1982, for Oldroyd-type fluids).

6.1.1 Boundary conditions

Our basic problem is to solve the coupled system of equations (6.1), (6.3) and (6.5) for the five variables ψ, ω, S^{xx}, S^{xy} and S^{yy}, subject to the appropriate boundary conditions. Equations (6.1) and (6.5), which are elliptic in ψ and ω respectively, require either Dirichlet or Neumann conditions on Γ. In contrast, the system (6.3) (with (6.4)) is first-order hyperbolic in S^{xx}, S^{xy} and S^{yy}, and it is sufficient to specify extra-stress values at entry (Γ_{in}) only. Parts of Γ joining onto Γ_{in} will be characteristics (possibly singular) of the system, and therefore their boundary stress values are determinable, in principle, from the information at the relevant points of intersection with Γ_{in}. We note that the specification of extra-stress at entry must be consistent with the complete past history of deformation prior to entry, a feature not shared by the memoryless Newtonian fluid. If there is no entry flow (cf. Fig. 6.1c) then the boundary stresses must be determined by iteration using the fixed velocity boundary conditions and iterative determination of boundary velocity gradients.

We may split up the various boundary conditions as follows :

(i) *Entry.* ψ and ω are found by integrating and differentiating the appropriate fully-developed velocity profile. $\underset{\sim}{S}$ may be found by solving (6.3) (with (6.4)) directly with the appropriate values of velocities and velocity gradients.

(ii) *Exit.* ψ and ω may be found as in (i), or alternatively by setting the normal derivatives $\partial\psi/\partial n$, $\partial\omega/\partial n$ to zero. Dirichlet exit conditions would appear to be the more widely used in the finite difference simulation of viscoelastic flows, but they can induce small oscillations in streamlines near Γ_{out} when the elasticity parameter W increases. These are usually safely ignored. They arise because a smooth "computational" fully-developed exit flow emerges from the discrete calculations, which differs from the exact exit flow. The Dirichlet condition on Γ_{out} will attempt to correct this smooth exit flow, resulting in oscillation. The alternative Neumann condition will generally preserve the smoothness of the exit flow, but a slower rate of convergence of the iterative procedure may result (see, for example, Townsend 1980b).

(iii) *Solid boundaries.* The no-slip velocity condition ensures that

$$\psi = \text{constant} , \qquad \frac{\partial\psi}{\partial n} = \text{constant} , \tag{6.6}$$

where the second constant vanishes if the boundary is at rest. The first of (6.6) is a Dirichlet condition for ψ, whereas the second may be used in generating a computational boundary condition for ω (see §6.2.3). Special provision must be made for treating re-entrant corners, where ω is singular (§6.2.4).

(iv) *Axes of symmetry.* Here we have Dirichlet conditions on ψ and ω given by

$$\psi = \text{constant} , \qquad \omega = 0 . \tag{6.7}$$

The first of (6.7) holds since any axis of symmetry is necessarily a streamline; the vorticity vanishes as a result of symmetry in the velocity field.

6.2 DISCRETIZATION

We now discuss the finite difference approximation of each of the governing equations. Consider a square grid of spacing h imposed on the flow region Ω, thereby defining a set of grid points $\{(x_j,y_k) : x_j = x_0 + jh , y_k = y_0 + kh\}$. As in §5.4 we let Ω_h and Γ_h denote the sets of internal and boundary grid points, respectively.

The Poisson equation (6.1) may be approximated by the five-point difference scheme

$$4\psi_{j,k} - \psi_{j+1,k} - \psi_{j-1,k} - \psi_{j,k+1} - \psi_{j,k-1} = h^2\omega_{j,k} , \quad (x_j,y_k) \in \Omega_h , \tag{6.8}$$

where, if we assume Dirichlet conditions on Γ, we have

$$\psi_{j,k} = \psi(x_j,y_k) , \qquad (x_j,y_k) \in \Gamma_h . \tag{6.9}$$

In (6.8), $\psi_{j,k}$ and $\omega_{j,k}$ are approximations to $\psi(x_j,y_k)$ and $\omega(x_j,y_k)$, respectively. If the grid points are *naturally ordered* then the matrix system (6.8) has block tridiagonal structure (cf. §5.4.1).

Let a_{mn} denote the (m,n)th element of a general $N \times N$ matrix $\underline{A}$. Then $\underline{A}$ is said to be *diagonally dominant* if

$$|a_{mm}| \geq \sum_{\substack{n=1 \\ n \neq m}}^{N} |a_{mn}| , \qquad \forall \quad m=1, \ldots, N. \tag{6.10}$$

Any block tridiagonal matrix system of the form

$$K_0 u_{j,k} - K_1 u_{j+1,k} - K_2 u_{j-1,k} - K_3 u_{j,k+1} - K_4 u_{j,k-1} = f_{j,k} \quad , \tag{6.11}$$

where each of the coefficients K depends on (j,k), will therefore possess a diagonally dominant matrix if

$$|K_0| \geq |K_1| + |K_2| + |K_3| + |K_4| , \qquad \forall \ (j,k) . \tag{6.12}$$

The system (6.8) clearly has a diagonally dominant matrix; moreover there is strict inequality for rows corresponding to grid points (x_j, y_k) of depth one in from the boundary, as a result of (6.9) (see §5.4.1).

We saw in the previous chapter that diagonal dominance can play two rôles :

(i) It can guarantee the stability of certain algorithms for solving matrix systems (in particular Algorithm 5.1 for tridiagonal systems).

(ii) It can be a sufficient condition for the stability of certain discrete operators L_h (possibly nonlinear). This is especially relevant for the upwind difference schemes described in §5.3.4.

Diagonal dominance will play similar rôles for the block tridiagonal systems which we shall generate in this and later sections.

6.2.1 The vorticity equation

In discretizing the left-hand side of (6.5) we again use the five-point difference approximation for the Laplacian :

$$\nabla^2 \omega(x_j, y_k) \simeq - \frac{1}{h^2} (4\omega_{j,k} - \omega_{j+1,k} - \omega_{j-1,k} - \omega_{j,k+1} - \omega_{j,k-1}) ,$$

and we take central differences for the ψ-derivatives :

$$\left. \begin{aligned} \frac{\partial \psi}{\partial x}(x_j, y_k) &\simeq \frac{\psi_{j+1,k} - \psi_{j-1,k}}{2h} , \\ \frac{\partial \psi}{\partial y}(x_j, y_k) &\simeq \frac{\psi_{j,k+1} - \psi_{j,k-1}}{2h} . \end{aligned} \right\} \tag{6.13}$$

Assume for the present that the derivatives on the right of (6.5) can also be approximated to $O(h^2)$, so that

$$F(x_j,y_k) \approx F_{j,k} \quad , \tag{6.14}$$

where

$$F \equiv \frac{\partial^2}{\partial x \partial y}(S^{xx} - S^{yy}) + \left(\frac{\partial^2}{\partial y^2} - \frac{\partial^2}{\partial x^2}\right) S^{xy} \; . \tag{6.15}$$

Then a fully second-order accurate scheme for (6.5) is obtained if the ω-derivatives on the left-hand side are approximated by central differences as in (6.13). We call this a CD scheme. Introducing the variables

$$\alpha_{j,k} = \tfrac{1}{4}R(\psi_{j+1,k} - \psi_{j-1,k}) \; , \qquad \beta_{j,k} = \tfrac{1}{4}R(\psi_{j,k+1} - \psi_{j,k-1}) \; , \tag{6.16}$$

the scheme may be written

$$K_0\omega_{j,k} - K_1\omega_{j+1,k} - K_2\omega_{j-1,k} - K_3\omega_{j,k+1} - K_4\omega_{j,k-1} = -\,h^2F_{j,k}, (x_j,y_k) \in \Omega_h \; , \tag{6.17}$$

where

$$\left.\begin{aligned} &K_1 = 1 - \beta_{j,k} \; , \qquad K_2 = 1 + \beta_{j,k} \; , \\ &K_3 = 1 + \alpha_{j,k} \; , \qquad K_4 = 1 - \alpha_{j,k} \; , \\ &\text{and} \\ &K_0 = K_1 + K_2 + K_3 + K_4 = 4 \; . \end{aligned}\right\} \tag{6.18}$$

The CD scheme is easily shown to be diagonally dominant if and only if

$$\rho_{j,k} \equiv \max\,(|\alpha_{j,k}|,|\beta_{j,k}|) \leqslant 1 \; , \; \forall \;\; (x_j,y_k) \in \Omega_h \; . \tag{6.19}$$

The quantity $\max\limits_{\Omega_h} \{2_{j,k}\}$ is usually referred to as the *maximum grid Reynolds number*. In practice, satisfaction of condition (6.19) either restricts Reynolds numbers to extremely low values or demands an excessively fine mesh. If the condition is greatly violated then either unrealistic oscillations are set up in the numerical solution (see, for example, Spalding 1972) or iterative methods for solving the matrix system fail to converge.

When (6.19) is violated, the simplest approach is to use the first upwind difference scheme (UD1) (cf. §5.3.4). An ω-derivative on the left-hand side of (6.5) is approximated to $O(h)$ by a backward difference if the coefficient velocity is positive, and by a forward difference if negative :

$$\frac{\partial\omega}{\partial x}(x_j,y_k) \simeq \begin{cases} \dfrac{\omega_{j,k} - \omega_{j-1,k}}{h}, & \beta_{j,k} > 0 \\[2ex] \dfrac{\omega_{j+1,k} - \omega_{j,k}}{h}, & \beta_{j,k} < 0 \end{cases}$$

$$\frac{\partial\omega}{\partial y}(x_j,y_k) \simeq \begin{cases} \dfrac{\omega_{j,k} - \omega_{j,k-1}}{h}, & -\alpha_{j,k} > 0 \\[2ex] \dfrac{\omega_{j,k+1} - \omega_{j,k}}{h}, & -\alpha_{j,k} < 0 \ . \end{cases}$$

The other terms in (6.5) are kept to $O(h^2)$, as in CD. The UD1 scheme may then be written compactly as (6.17), with

$$\left.\begin{array}{ll} K_1 = 1 + |\beta_{j,k}| - \beta_{j,k} , & K_2 = 1 + |\beta_{j,k}| + \beta_{j,k} , \\ K_3 = 1 + |\alpha_{j,k}| + \alpha_{j,k} , & K_4 = 1 + |\alpha_{j,k}| - \alpha_{j,k} , \\ K_0 = K_1 + K_2 + K_3 + K_4 = 4 + 2(|\alpha_{j,k}| + |\beta_{j,k}|) . & \end{array}\right\} \qquad (6.20)$$

The scheme is unconditionally diagonally dominant, but is only a first-order approximation to (6.5). The $O(h)$ term in the local truncation error is

$$\tfrac{1}{2} Rh \left\{ \left|\frac{\partial\psi}{\partial y}\right| \frac{\partial^2\omega}{\partial x^2} + \left|\frac{\partial\psi}{\partial x}\right| \frac{\partial^2\omega}{\partial y^2} \right\} . \qquad (6.21)$$

This is the *false diffusion* error, which for high Reynolds number may become larger than the true diffusion term.

As mentioned previously, most often in viscoelastic flow simulations we are concerned with low R-values; the false diffusion error (6.21) is then usually not serious. Many workers have used UD1 exclusively (or its equivalent) for the vorticity equation for various constitutive models (see, for example, Crochet and Pilate 1975; Perera and Walters 1977a,b; Davies et al 1979; Townsend 1980a,b). A natural improvement is to use CD when condition (6.19) is satisfied *locally*, with a switch to UD1 when the condition is violated. This idea has been used by Cochrane et al (1981, 1982) and Walters and Webster (1982). The switch causes an abrupt onset of false diffusion, albeit small. A continuous switch-over is available in Spalding's method (SM of §5.3.4); here the false diffusion error in UD1 when $\rho_{j,k} > 1$ is reduced by ignoring the true diffusion term. The compact form of SM applied to (6.5) may be written as (6.17), with

$$\left.\begin{aligned} K_1 &= s(\beta_{j,k}) - \beta_{j,k}\ , \qquad K_2 = s(\beta_{j,k}) + \beta_{j,k}\ , \\ K_3 &= s(\alpha_{j,k}) + \alpha_{j,k}\ , \qquad K_4 = s(\alpha_{j,k}) - \alpha_{j,k}\ , \\ K_0 &= K_1 + K_2 + K_3 + K_4 = 2[s(\alpha_{j,k}) + s(\beta_{j,k})]\ , \end{aligned}\right\} \tag{6.22}$$

where the function s(x) is defined by

$$s(x) = \tfrac{1}{2}(|1+x| + |1-x|)\ . \tag{6.23}$$

In Newtonian test calculations ($F_{j,k} \equiv 0$), Richards and Crane (1979) have found SM to be superior to the upwinding schemes UD1, UD2 and LE of §5.3.4.

Perhaps the approach which comes nearest to attaining both second-order accuracy and unconditional stability is the *deferred correction* method advocated by Dennis and Chang (1969). The idea is quite simple. Let $L\{\omega_{j,k}\}$ and $M\{\omega_{j,k}\}$ denote, respectively, the left-hand sides of (6.17) when the formulae of (6.18) and (6.20) are substituted. Then

$$L\{\omega_{j,k}\} = M\{\omega_{j,k}\} + C\{\omega_{j,k}\}\ ,$$

where

$$C\{\omega_{j,k}\} = -\ 2(|\alpha_{j,k}|+|\beta_{j,k}|)\omega_{j,k} + |\beta_{j,k}|(\omega_{j+1,k} - \omega_{j-1,k}) + |\alpha_{j,k}|(\omega_{j,k+1} - \omega_{j,k-1}). \tag{6.24}$$

Let $\{\omega_{j,k}^{[0]}\}$ denote the approximation obtained from UD1, i.e. by solving

$$M\{\omega_{j,k}^{[0]}\} = -\ h^2 F_{j,k}\ . \tag{6.25}$$

From it we can obtain the correction $C\{\omega_{j,k}^{[0]}\}$. Now obtain successive iterates defined by

$$M\{\omega_{j,k}^{[r+1]}\} + C\{\omega_{j,k}^{[r]}\} = -\ h^2 F_{j,k}\ , \qquad r=0,1,\ldots\ . \tag{6.26}$$

If this sequence converges, the limit solution satisfies

$$L\{\omega_{j,k}\} = -\ h^2 F_{j,k}\ ,$$

i.e. the CD scheme. Thus CD accuracy is obtained through a sequence of iterates each of which is a diagonally-dominant calculation. If the sequence does not converge, or the limit is greatly different from the UD1 solution $\{\omega_{j,k}^{[0]}\}$, then the latter must be looked upon as a highly suspect approximation to the solution of the vorticity equation.

It would be impracticable to request too many iterative corrections in (6.26), and usually only one is used. In this case, if each of (6.25) and (6.26) is to be solved by the same iterative method, instead of using two sequences of inner iterations it is possible to replace them by one sequence. This replacement can be made in a variety of ways and a few details are given in §6.3.2. Dennis and Chang (1970), Veldman (1973), and Richards and Crane (1978) describe calculations of this kind for Newtonian problems. For calculations on second-order fluids, Pilate and Crochet (1977) have used this type of deferred correction, with extensions. Richards and Crane (1979) also discuss deferred correction methods for upwinding schemes other than UD1.

We next describe how the right-hand side of (6.5) may be approximated to $O(h^2)$. The partial derivatives of form $\partial^2S/\partial x^2$ and $\partial^2S/\partial y^2$, respectively, are replaced by the usual central difference formulae

$$\frac{1}{h^2}(S_{j+1,k} - 2S_{j,k} + S_{j-1,k}) \quad \text{and} \quad \frac{1}{h^2}(S_{j,k+1} - 2S_{j,k} + S_{j,k-1}) \ . \tag{6.27}$$

The mixed partial derivative $\partial^2S/\partial x\partial y$ is usually replaced by one of the three $O(h^2)$ approximations

$$\left.\begin{aligned}
&\frac{1}{4h^2}(S_{j+1,k+1} - S_{j+1,k-1} - S_{j-1,k+1} + S_{j-1,k-1}) \ , && (6.28a)\\
&\frac{1}{2h^2}(S_{j+1,k+1} + 2S_{j,k} + S_{j-1,k-1} - S_{j,k+1} - S_{j+1,k} - S_{j,k-1} - S_{j-1,k}) \ , && (6.28b)\\
&\frac{1}{2h^2}(S_{j,k+1} + S_{j+1,k} + S_{j,k-1} + S_{j-1,k} - S_{j-1,k+1} - 2S_{j,k} - S_{j+1,k-1}) \ , && (6.28c)
\end{aligned}\right\}$$

based on the 4-point and 7-point computational molecules shown in Fig. 6.2(a-c), respectively. Formula (6.28a) is normally used, but then special precautions need to be taken when differencing across a re-entrant corner where the stress is singular (§6.2.5). Some workers (e.g. Paddon, 1979; Holstein, 1981) prefer to avoid a re-entrant corner grid point by using (6.28b or c).

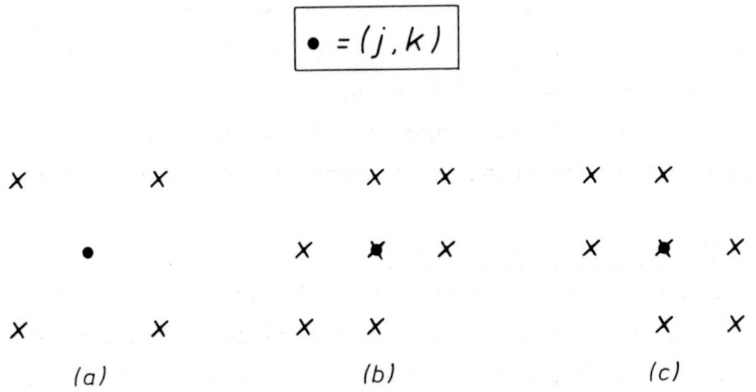

Fig.6.2 Computatational molecules for mixed second derivative.

Combining (6.27) and (6.28a) in (6.14) gives

$$
\begin{aligned}
h^2F_{j,k} = {} & \tfrac{1}{4}(S^{xx}_{j+1,k+1} - S^{xx}_{j+1,k-1} - S^{xx}_{j-1,k+1} + S^{xx}_{j-1,k-1} \\
& - S^{yy}_{j+1,k+1} + S^{yy}_{j+1,k-1} + S^{yy}_{j-1,k+1} - S^{yy}_{j-1,k-1}) \\
& - S^{xy}_{j+1,k} - S^{xy}_{j-1,k} + S^{xy}_{j,k+1} + S^{xy}_{j,k-1} \,.
\end{aligned}
\tag{6.29}
$$

Similar expressions result from combining (6.27) and (6.28b or c).

To complete this section on the vorticity equation, we briefly mention the approach of Gatski and Lumley (1978a,b) who do not use the transformation (6.4) but solve (6.2) as it stands. Their work is on an Oldroyd (1950) B fluid, but this does not affect the form of (6.2). They use $O(h)$ one-sided differences for all terms on the left-hand side of (6.2) but $O(h^2)$ central differences for the stress derivatives on the right-hand side. Forward differences replace the ψ-derivatives in (6.13), but otherwise the same upwind differences for ω as in UD1 is used. Clearly the scheme is slightly less accurate than UD1. Of greater interest, however, is that Gatski and Lumley approximate the normal stresses T^{xx} and T^{yy} at staggered half-grid points only, $\{(x_{j+\frac{1}{2}}, y_{k+\frac{1}{2}}):j,k=0,1,\ldots\}$; the shear stress is approximated at the usual grid points. Thus, for normal stresses, (6.28a) is replaced by

$$
\frac{1}{h^2}\,(T_{j+\frac{1}{2},k+\frac{1}{2}} - T_{j+\frac{1}{2},k-\frac{1}{2}} - T_{j-\frac{1}{2},k+\frac{1}{2}} + T_{j-\frac{1}{2},k-\frac{1}{2}}) \,. \tag{6.30}
$$

A combination of formulae (6.27) for shear stress derivatives and (6.30) for normal stress derivatives ensures that no re-entrant corners are encountered in approximating the right-hand side of (6.2). This approach is clearly applicable to the right-hand side of (6.5) also.

Gatski and Lumley's published results are for high Reynolds number flows. In particular, their treatment of the vorticity equation is inapplicable when $R = 0$.

6.2.2 The constitutive equations

In the discretization of the stream function and vorticity equations we are able to make use of methods which are well-documented and tested in Newtonian computational fluid dynamics. Clearly, for the hyperbolic constitutive equations (6.3), this is not possible since there is no Newtonian counterpart. Consequently, numerical methods for solving hyperbolic constitutive equations are still at an early stage of development. At present there is essentially only one finite difference scheme available, and this has only first-order accuracy. Attempts at developing second-order accurate methods have not yet proved successful (Tiefenbruck and Leal 1982). There is therefore a definite need for more research in this area in the immediate future.

Before discussing the method of discretization, there is a basic choice to be made as to how the transformed stress-tensor $\underline{S}$ in (6.4) should be computed. We can either solve (6.3) for the components of $\underline{T}$, and then perform the transformation (6.4) numerically (Method 1); alternatively we may substitute (6.4) into (6.3) to give a system of hyperbolic equations for $\underline{S}$ which we then solve directly (Method 2). Although both methods are *mathematically* equivalent, they need not be *numerically* equivalent because of the different discretization errors at each stage.

Method 1 We may write (6.3) in the form

$$\left.\begin{aligned} A_1T^{xx} + W\,LT^{xx} &= 2BT^{xy} + F_1 \\ A_2T^{xy} + W\,LT^{xy} &= CT^{xx} + BT^{yy} + F_2 \\ A_3T^{yy} + W\,LT^{yy} &= 2CT^{xy} + F_3 \end{aligned}\right\}, \tag{6.31}$$

$$\left.\begin{aligned} &\text{where } L \equiv \frac{\partial\psi}{\partial y}\frac{\partial}{\partial x} - \frac{\partial\psi}{\partial x}\frac{\partial}{\partial y}, \\ &A_1 = 1 - 2W\frac{\partial^2\psi}{\partial x\partial y}, \quad A_2 = 1, \quad A_3 = 1 + 2W\frac{\partial^2\psi}{\partial x\partial y}, \\ &B = W\frac{\partial^2\psi}{\partial y^2}, \quad C = -W\frac{\partial^2\psi}{\partial x^2}, \\ &F_1 = 2\frac{\partial^2\psi}{\partial x\partial y}, \quad F_2 = \frac{\partial^2\psi}{\partial y^2} - \frac{\partial^2\psi}{\partial x^2}, \quad F_3 = -2\frac{\partial^2\psi}{\partial x\partial y} \quad . \end{aligned}\right\} \tag{6.32}$$

In discretizing the operator L, care must again be taken to ensure stability of the resulting matrix system. Diagonal dominance is achieved if first-order upwind differencing is used when $A_m \geq 0$, with first-order downwind differencing when $A_m < 0$. Downwinding is the exact opposite of upwinding, i.e. a forward difference if the coefficient velocity is positive, a backward difference if negative.

Introducing the variables

$$\left.\begin{aligned} \alpha^A_{j,k} &= \tfrac{1}{4}W(\psi_{j+1,k} - \psi_{j-1,k})\ \text{sign}\ (A)\ , \\ \beta^A_{j,k} &= \tfrac{1}{4}W(\psi_{j,k+1} - \psi_{j,k-1})\ \text{sign}\ (A)\ , \end{aligned}\right\} \tag{6.33}$$

the discretized form of (6.31) may then be written

$$K_0^{A_1}\, T^{xx}_{j,k} - K_1^{A_1}\, T^{xx}_{j+1,k} - K_2^{A_1}\, T^{xx}_{j-1,k} - K_3^{A_1}\, T^{xx}_{j,k+1} - K_4^{A_1}\, T^{xx}_{j,k-1} = h^2[2B_{j,k}\, T^{xy}_{j,k} + F_{1,j,k}]\ \text{sign}\ (A_1)\ , \tag{6.34a}$$

$$K_0^{A_2}\, T^{xy}_{j,k} - K_1^{A_2}\, T^{xy}_{j+1,k} - K_2^{A_2}\, T^{xy}_{j-1,k} - K_3^{A_2}\, T^{xy}_{j,k+1} - K_4^{A_2}\, T^{xy}_{j,k-1} = h^2[C_{j,k}\, T^{xx}_{j,k} + B_{j,k}\, T^{yy}_{j,k} + F_{2,j,k}], \tag{6.34b}$$

$$K_0^{A_3}\, T^{yy}_{j,k} - K_1^{A_3}\, T^{yy}_{j+1,k} - K_2^{A_3}\, T^{yy}_{j-1,k} - K_3^{A_3}\, T^{yy}_{j,k+1} - K_4^{A_3}\, T^{yy}_{j,k-1} = h^2[2C_{j,k}\, T^{xy}_{j,k} + F_{3,j,k}]\ \text{sign}\ (A_3)\ , \tag{6.34c}$$

where

$$\left.\begin{aligned} K_0^A &= h^2|A_{j,k}| + 2(|\alpha^A_{j,k}| + |\beta^A_{j,k}|)\ , \\ K_1^A &= |\beta^A_{j,k}| - \beta^A_{j,k}\ , \qquad K_2^A = |\beta^A_{j,k}| + \beta^A_{j,k}\ , \\ K_3^A &= |\alpha^A_{j,k}| + \alpha^A_{j,k}\ , \qquad K_4^A = |\alpha^A_{j,k}| - \alpha^A_{j,k}\ , \end{aligned}\right\} \tag{6.35}$$

and the coefficients A, B, C and F in (6.32) are computed at appropriate grid points by central difference approximation. In particular, the mixed derivative $\partial^2\psi/\partial x \partial y$ may be computed everywhere in Ω_h in the form (6.28a) since ψ is not singular at re-entrant corners (see, however, the remarks in §6.2.4).

After solving (6.34a-c) for $\{T^{xx}_{j,k}\ ,\ T^{xy}_{j,k}\ ,\ T^{yy}_{j,k}\}$, the transformation (6.4) may be performed using the values of F already calculated.

Method 2   Using the transformation (6.4), equations (6.3) may be written as

$$\left.\begin{aligned} A_1 S^{xx} + W\,L\,S^{xx} &= 2BS^{xy} + G_1 \,, \\ A_2 S^{xy} + W\,L\,S^{xy} &= CS^{xx} + BS^{yy} + G_2 \,, \\ A_3 S^{yy} + W\,L\,S^{yy} &= 2CS^{xy} + G_3 \,, \end{aligned}\right\} \tag{6.36}$$

where A_m, B and C are given in (6.32) and

$$\left.\begin{aligned} G_1 &= 2W\left[\,2\left(\frac{\partial^2\psi}{\partial x\partial y}\right)^2 - \frac{\partial\psi}{\partial y}\frac{\partial^3\psi}{\partial x^2\partial y} + \frac{\partial\psi}{\partial x}\frac{\partial^3\psi}{\partial x\partial y^2} + \frac{\partial^2\psi}{\partial y^2}\left(\frac{\partial^2\psi}{\partial y^2} - \frac{\partial^2\psi}{\partial x^2}\right)\right] , \\ G_2 &= -W\left[\,2\frac{\partial^2\psi}{\partial x\partial y}\left(\frac{\partial^2\psi}{\partial x^2} + \frac{\partial^2\psi}{\partial y^2}\right) + \left(\frac{\partial\psi}{\partial y}\frac{\partial}{\partial x} - \frac{\partial\psi}{\partial x}\frac{\partial}{\partial y}\right)\left(\frac{\partial^2\psi}{\partial y^2} - \frac{\partial^2\psi}{\partial x^2}\right)\right] , \\ G_3 &= 2W\left[\,2\left(\frac{\partial^2\psi}{\partial x\partial y}\right)^2 + \frac{\partial\psi}{\partial y}\frac{\partial^3\psi}{\partial x^2\partial y} - \frac{\partial\psi}{\partial x}\frac{\partial^3\psi}{\partial x\partial y^2} - \frac{\partial^2\psi}{\partial x^2}\left(\frac{\partial^2\psi}{\partial y^2} - \frac{\partial^2\psi}{\partial x^2}\right)\right] . \end{aligned}\right\} \tag{6.37}$$

The discretized form of (6.36) is then

$$\begin{aligned} &K_0^{A_1} S^{xx}_{j,k} - K_1^{A_1} S^{xx}_{j+1,k} - K_2^{A_1} S^{xx}_{j-1,k} - K_3^{A_1} S^{xx}_{j,k+1} - K_4^{A_1} S^{xx}_{j,k-1} \\ &\qquad = h^2[2B_{j,k}\,S^{xy}_{j,k} + G_{1,j,k}]\ \mathrm{sign}\,(A_1)\,, \end{aligned} \tag{6.38a}$$

$$\begin{aligned} &K_0^{A_2} S^{xy}_{j,k} - K_1^{A_2} S^{xy}_{j+1,k} - K_2^{A_2} S^{xy}_{j-1,k} - K_3^{A_2} S^{xy}_{j,k+1} - K_4^{A_2} S^{xy}_{j,k-1} \\ &\qquad = h^2[C_{j,k}\,S^{xx}_{j,k} + B_{j,k}\,S^{yy}_{j,k} + G_{2,j,k}]\,, \end{aligned} \tag{6.38b}$$

$$\begin{aligned} &K_0^{A_3} S^{yy}_{j,k} - K_1^{A_3} S^{yy}_{j+1,k} - K_2^{A_3} S^{yy}_{j-1,k} - K_3^{A_3} S^{yy}_{j,k+1} - K_4^{A_3} S^{yy}_{j,k-1} \\ &\qquad = h^2[2C_{j,k}\,S^{xy}_{j,k} + G_{3,j,k}]\ \mathrm{sign}\,(A_3)\,, \end{aligned} \tag{6.38c}$$

where the coefficients K are defined in (6.35).

The derivatives in (6.37) are again approximated to $O(h^2)$ by central differences. In particular, the third derivatives of ψ may be calculated from

$$\frac{\partial^3\psi}{\partial x^3}(x_j,y_k) \simeq \frac{\psi_{j+2,k} - 2\psi_{j+1,k} + 2\psi_{j-1,k} - \psi_{j-2,k}}{2h^3}\,, \tag{6.39a}$$

$$\frac{\partial^3\psi}{\partial y^3}(x_j,y_k) \simeq \frac{\psi_{j,k+2} - 2\psi_{j,k+1} + 2\psi_{j,k-1} - \psi_{j,k-2}}{2h^3}\,, \tag{6.39b}$$

$$\frac{\partial^3\psi}{\partial x^2\partial y}(x_j,y_k) \simeq \frac{\psi_{j+1,k+1} - 2\psi_{j,k+1} + \psi_{j-1,k+1} - \psi_{j+1,k-1} + 2\psi_{j,k-1} - \psi_{j-1,k-1}}{2h^3}, \quad (6.39c)$$

$$\frac{\partial^3\psi}{\partial x\partial y^2}(x_j,y_k) \simeq \frac{\psi_{j+1,k+1} - 2\psi_{j+1,k} + \psi_{j+1,k-1} - \psi_{j-1,k+1} + 2\psi_{j-1,k} - \psi_{j-1,k-1}}{2h^3}, \quad (6.39d)$$

with corresponding computational molecules shown in Fig. 6.3(a-d).

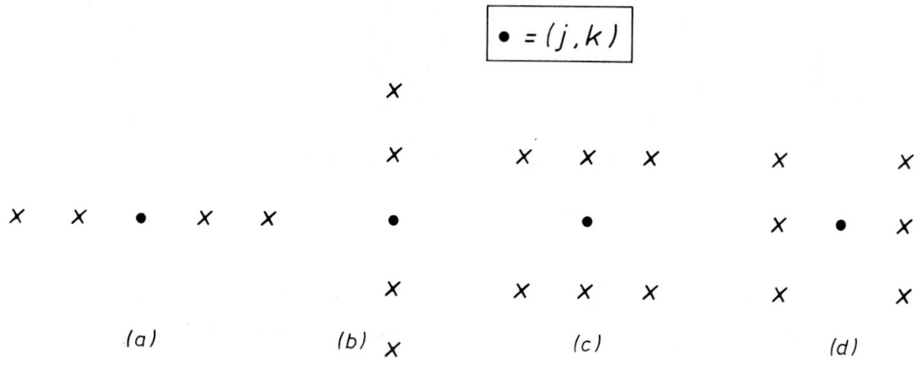

Fig. 6.3 Computational molecules for third derivatives.

At grid points adjacent to a solid boundary, formulae (6.39a,b) will involve ψ-values external to the flow geometry. These may be eliminated in the usual manner (§5.4.2) using a central difference approximation to the Neumann condition in (6.6). For example, in Fig. 6.4, (x_j,y_k), (x_{j+1},y_k) and (x_{j+2},y_k) are internal, boundary and external grid points, respectively. Using

$$\frac{\psi_{j+2,k} - \psi_{j,k}}{2h} = c + O(h^2) ,$$

we have, to $O(h^2)$,

$$\frac{\partial^3\psi}{\partial x^3}(x_m,y_n) \simeq \frac{-2\psi_{j+1,k} + \psi_{j,k} + 2\psi_{j-1,k} - \psi_{j-2,k}}{2h^3} + \frac{c}{h^2} ,$$

where $c = 0$ if the boundary is at rest. For the approximation of third derivatives near re-entrant corners, however, see the remarks at the end of §6.2.4 pertaining to (6.82).

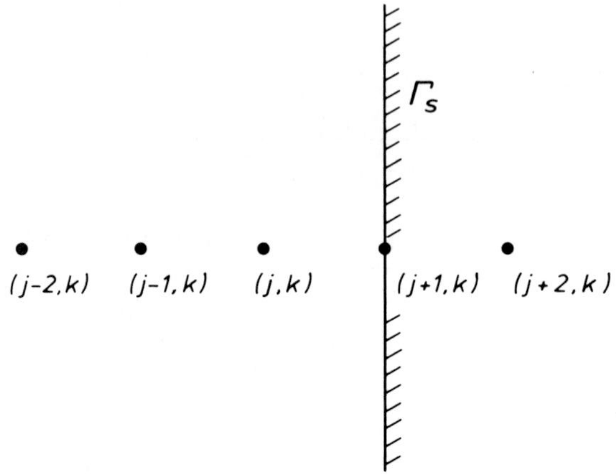

Fig. 6.4 External grid point (j+2,k).

The main advantage of Method 1 is that the right-hand sides of (6.34a-c) contain derivatives no higher than second. Its main disadvantage is that both sets of variables $\underline{T}$ and $\underline{S}$ must be kept throughout the computation and greater computer storage is needed. In Method 2, the transformation (6.4) need not be performed computationally, neither need $\underline{T}$ be stored; if required $\underline{T}$ may be evaluated once at the end of the computation. On the other hand, in Method 2 the cumbersome expressions (6.37) must be computed, with greater numerical error in approximating the third derivatives of ψ. A numerical comparison of both methods (Tiefenbruck and Leal 1982), however, indicates that there seems little to choose between them from the standpoints of numerical accuracy and time of computation. Since most finite difference workers have used Method 2 we shall restrict attention in what follows to this method; the modifications necessary for Method 1 are completely straightforward.

In (6.38a-c) the unknown stress variables appear on both the left and right-hand sides of the equations. Moving all unknown stresses over to the left-hand side would destroy the diagonal dominance of the system as a whole; it is therefore convenient to evaluate the right-hand sides using stress values from a previous iteration and to solve for the left-hand stresses as current iterates. Various methods for achieving this are discussed in §6.3. For the moment we shall assume that all right-hand sides in (6.38a-c) are known.

It is important to realize that although the block tridiagonal systems (6.38a-c) bear an algebraic resemblance to (6.17), the former are discretizations of hyperbolic equations, whereas the latter result from an elliptic equation.

The fundamental difference in type of the underlying partial differential equations is reflected in the corresponding discretized forms. Equations (6.17) must necessarily be solved implicitly as a matrix system over the whole of Ω_h, incorporating boundary conditions over the whole of Γ_h. This is not the case for (6.38a-c). In principle, these may be solved explicitly over any part of Ω_h in which the velocities u and v do not change sign, with stress boundary values required only on parts of Γ_h.

Consider, for example, the flow region in Fig. 6.5, which includes the entry Γ_{in}, and where $u,v \geq 0$. Assume for the present that $A_1 \geq 0$. Equation (6.38a) with $j = 1$ gives

$$K_0^{A_1} S^{xx}_{1,k} - K_2^{A_1} S^{xx}_{0,k} - K_4^{A_1} S^{xx}_{1,k-1} = h^2[2B_{1,k} S^{xy}_{1,k} + G_{1,1,k}] \,. \tag{6.40}$$

Thus $S^{xx}_{1,k}$ may be computed explicitly along the grid line $j = 1$ from the forward recurrence

$$S^{xx}_{1,k} = \left(\frac{K_4^{A_1}}{K_0^{A_1}}\right) S^{xx}_{1,k-1} + \left(\frac{K_2^{A_1} S^{xx}_{0,k} + h^2[2B_{1,k} S^{xy}_{1,k} + G_{1,1,k}]}{K_0^{A_1}}\right), \quad k = 1,\ldots,N, \tag{6.41}$$

using values of S^{xx} along the entry grid line $j = 0$ and at the boundary grid point $(x_1, 0)$ on $\Gamma_{s,1}$. Similar recurrences may be set up for successive grid lines $j = 2,\ldots,M$, using values of S^{xx} already calculated from the previous grid line and boundary stresses on $\Gamma_{s,1}$. We may therefore find S^{xx} everywhere in a simply connected subset of Γ_h bounded by Γ_{in} and $\Gamma_{s,1}$, where $u,v \geq 0$.

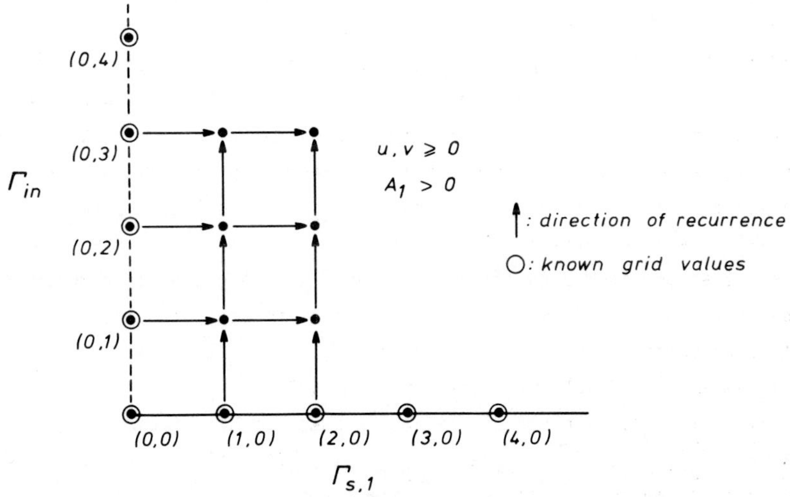

Fig. 6.5 Progress of an explicit method.

When there is a change of sign in u or v, (6.40) and (6.41) become invalid, and are replaced by similar recurrences which are backward in j if $u < 0$ and forward or backward in k depending on the sign of v. By choosing the appropriate recurrences and boundary stress values, S^{xx} may be found everywhere in Ω_h apart from recirculating vortex regions. Similarly for S^{xy} and S^{yy}. This reflects the fact that the solution of a hyperbolic equation reduces to that of integrating an ordinary differential equation along a characteristic curve, starting from an initial value. In the case of (6.38a-c) the characteristics are the flow streamlines; in regions where the streamlines are open, the appropriate recurrences in (6.38a-c) are solvable, but when the streamlines are closed the recurrences are not solvable since there will be no initial boundary stress value from which to start the integration. In practice the recurrences do not follow single streamlines but progress along grid lines, thereby following groups of streamlines, approximately, step by step.

If the coefficients A_1 or A_3 change sign, then the explicit recurrences above change direction (forward to backward and vice-versa). This avoids the onset of numerical instabilities but requires explicit boundary values from which to proceed in the reverse direction.

The computer programming of the explicit method just described becomes quite complicated, and we do not recommend this approach in general. Leal (1979) and Tiefenbruck and Leal (1982), however, have successfully employed an explicit method to simulate flows around spheres and bubbles, where the velocities do not change sign more than once along a grid line, and there are no recirculating vortex regions.

The approach used by most workers is to solve (6.38a-c) implicitly as matrix systems over the whole of Ω_h. No additional stress boundary conditions are needed in the implicit approach compared with the explicit approach since a coefficient K in (6.38a-c) multiplying a stress variable on Γ_h will vanish if u or v is zero. In particular, boundary stress values at the exit Γ_{out} never enter the implicit calculation since the corresponding coefficients K in (6.38a-c) multiplying stress variables on Γ_{out} will vanish. Moreover, the only essential stress boundary conditions are those at entry Γ_{in}; other boundary stress values are needed in the calculation, but these are not boundary conditions *per se*, since all solid and symmetry boundaries are characteristics. The calculation of boundary stresses is discussed in §6.2.5 and methods for solving the matrix systems (6.38a-c) are to be found in §6.3.

Finally, we note that the discretization used by Gatski and Lumley for the variables $\{T^{xx}_{j+\frac{1}{2},k+\frac{1}{2}}, T^{yy}_{j+\frac{1}{2},k+\frac{1}{2}}\}$ and $\{T^{xy}_{j,k}\}$ is similar to (6.34a-c). In (6.34a and c) all variables and coefficients are evaluated at the half-grid points, and to this end $T^{xy}_{j+\frac{1}{2},k+\frac{1}{2}}$ is approximated by $\frac{1}{4}(T^{xy}_{j+1,k+1} + T^{xy}_{j+1,k} + T^{xy}_{j,k+1} + T^{xy}_{j,k})$.

In (6.34b) all variables and coefficients are evaluated at the full-grid points, with $T^{xx}_{j,k}$ approximated by $\frac{1}{4}(T^{xx}_{j+\frac{1}{2},k+\frac{1}{2}} + T^{xx}_{j+\frac{1}{2},k-\frac{1}{2}} + T^{xx}_{j-\frac{1}{2},k+\frac{1}{2}} + T^{xx}_{j-\frac{1}{2},k-\frac{1}{2}})$, and similarly for $T^{yy}_{j,k}$. Gatski and Lumley solve the resulting matrix systems implicitly using Gauss-Seidel iteration.

As mentioned earlier, the basic discretization method described in this section is only first-order accurate; the false diffusion error associated with (6.38a) is

$$\pm \tfrac{1}{2}Wh\left\{\left|\frac{\partial\psi}{\partial y}\right|\frac{\partial^2 S^{xx}}{\partial x^2} + \left|\frac{\partial\psi}{\partial x}\right|\frac{\partial^2 S^{xx}}{\partial y^2}\right\} \qquad (6.42)$$

with similar expressions for (6.38b,c).

Very little effort has been expended so far on the development of stable second-order methods. In principle, the deferred correction methods described in the previous section are applicable, but these have not been tried for the constitutive equations discussed here.

6.2.3 Boundary vorticity approximation

We now describe how ω is calculated on solid boundaries to form part of the Dirichlet conditions for the vorticity equation (§6.1.1). The methods described in this section do not apply at singular corner points, which are discussed in the next section.

Fig. 6.6 shows part of a solid boundary parallel to the x-axis; (x_j,y_k) is a boundary grid point, whereas (x_j,y_{k+1}) is internal. A first-order approximation for $\omega(x_j,y_k)$ may be obtained from the Taylor expansion

$$\psi(x_j,y_{k+1}) = \psi(x_j,y_k) + hu(x_j,y_k) - \tfrac{1}{2}h^2\omega(x_j,y_k) + O(h^3),$$

whence

$$\omega(x_j,y_k) = -\frac{2}{h^2}[\psi(x_j,y_{k+1}) - \psi(x_j,y_k) - hu(x_j,y_k)] + O(h) ,$$

yielding the first-order formula

$$\omega_{j,k} = -\frac{2}{h^2}(\psi_{j,k+1} - \psi_{j,k} - hu_{j,k}) . \qquad (6.43)$$

Alternatively, if (x_j,y_{k-1}) is internal, then (6.43) is replaced by

$$\omega_{j,k} = -\frac{2}{h^2}(\psi_{j,k-1} - \psi_{j,k} + hu_{j,k}) . \qquad (6.44)$$

Similar formulae hold on solid boundaries parallel to the y-axis.

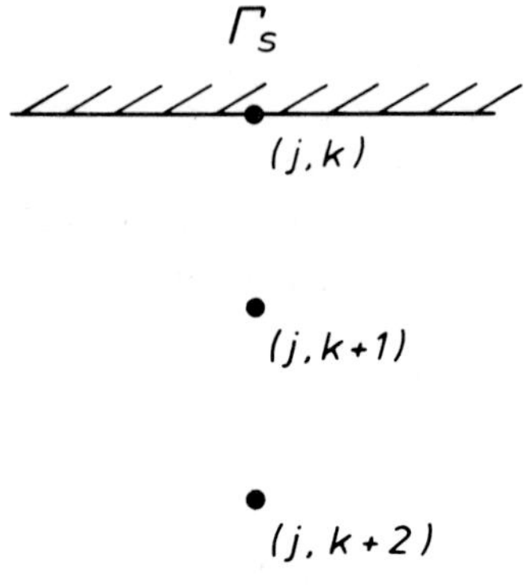

Fig. 6.6 Boundary vorticity approximation.

For Newtonian calculations, the first-order formulae (6.43), (6.44) prove quite stable in the iterative solution of the Navier-Stokes equations as discretized in (6.8), (6.17) and the UD1 scheme (6.20). For low R-values, when (6.20) is replaced by the CD scheme (6.18), the improvement in accuracy is often disappointing (see, for example, Gupta and Manohar 1979). This is because $O(h)$ accuracy of ω on Γ_h will often undermine the $O(h^2)$ discretization of (6.17) on Ω_h, yielding a *global* error in ω of $O(h)$. It is therefore natural to seek $O(h^2)$ approximations for ω on Γ_h.

Several second-order formulae are available (see, for example, Roache 1976; Gupta and Manohar 1979). Two of the more widely used of these formulae are those of Jensen (1959) and Woods (1954). In relation to Fig. 6.6, these are given respectively by

$$\omega_{j,k} = -\frac{1}{h^2}\left(-\frac{1}{2}\psi_{j,k+2} + 4\psi_{j,k+1} - \frac{7}{2}\psi_{j,k} - 3hu_{j,k}\right), \tag{6.45}$$

and

$$\omega_{j,k} = -\frac{3}{h^2}\left(\psi_{j,k+1} - \psi_{j,k} - hu_{j,k}\right) + \tfrac{1}{2}\omega_{j,k+1}. \tag{6.46}$$

Some care is needed in the use of second-order formulae, and we discuss the Newtonian and non-Newtonian cases separately.

(i) <u>The Newtonian case</u>. Several authors (see, for example, Roache 1976) claim to have found iterative instabilities when using Jensen's and Woods' formulae for boundary vorticity approximation. Webster (1979) and Court (1980) have found, however, that iterative convergence may be achieved with these formulae provided the iterates are smoothed (damped) according to

$$\omega^{[r+1]} \to (1 - \beta)\omega^{[r+1]} + \beta\omega^{[r]} , \qquad \beta \in [0,1] .$$

Unfortunately, this can engender quite slow iterative convergence, particularly when $\beta \approx 1$. These observations are confirmed by Gupta and Manohar (1979), who obtain near optimal values for the damping parameter β.

Second-order boundary vorticity approximation leads to improved accuracy on Ω_h even when the discretization of the vorticity equation here is $O(h)$. As a general rule, though, the more accuracy that is demanded on the boundary, the slower and more costly will be the iterative procedure. In the numerical simulation of flow generated in a square cavity by a moving plate (cf. Fig. 6.1c), Gupta and Manohar find Jensen's formula the most accurate of several second-order formulae but also the most costly to implement; Woods' formula is found to be significantly less accurate despite its $O(h^2)$ truncation error.

(ii) The non-Newtonian case. Here the situation is complicated by the emergence of the non-zero term, $h^2F_{j,k}$, on the right-hand side of (6.17), which is given, for example, by (6.29). Each stress term in (6.29) has a global discretization error of $O(h)$, at best, due to the first-order discretization of the constitutive equations. As W increases, this $O(h)$ error will increasingly undermine any improvement in accuracy obtainable from a second-order boundary vorticity approximation (and also from a second-order discretization of the vorticity equation). Webster (1979) argues that, until $O(h^2)$ accuracy is available for stress, there is little merit in using second-order formulae such as (6.45) and (6.46) in non-Newtonian simulations, except when W is very small.

The first-order formulae (6.43), (6.44) and their equivalents have proved quite successful in the finite-difference simulation of several non-Newtonian flow problems. Iterative convergence can usually be achieved without damping. With regard to second-order formulae, there is some numerical evidence (Webster 1979; Court 1980) to show that Jensen's formula can prove unstable in non-Newtonian calculations despite very heavy damping ($\beta \simeq 1$). On the other hand, Paddon (1979) and Holstein (1981) have used Woods' formula without incurring iterative instabilities, at least for small W. At present, however, there seems little to recommend the use of second-order boundary vorticity approximation, although the situation will need to be critically reviewed when second-order accuracy in stress is achievable.

6.2.4 Corner singularities

Moffatt (1964) has shown that ω is singular at a sharp corner. The discretization (6.17) therefore breaks down at grid points immediately adjacent to a corner, and so the finite difference schemes considered so far must be modified in some way to account for the singularity. A rigorous way of achieving this is to employ the local analytical form of the singularity; this is known for Newtonian fluids (Moffatt 1964) and certain Oldroyd fluids (Holstein 1981) but not for the Maxwell fluid. Moreover, we are referring to corners at the intersections of *stationary* solid boundaries; when the corner involves a moving boundary (e.g. Fig. 6.1c), the local analytical form of the singularity is not known completely, except for Stokesian fluids ($R = 0$) (Gupta et al 1981). The local forms constitute asymptotic expansions obtainable from separable-variable techniques.

Methods involving local singular forms are readily dovetailed into a global finite difference scheme. For Newtonian viscous flows techniques have been proposed by Ladevèze and Peyret (1974) and Holstein and Paddon (1981). For general elliptic partial differential equations with boundary singularities, techniques have been published by Fox and Sankar (1969) and Crank and Furzeland (1978). These last two methods can also be applied in principle to viscous flow problems.

Alternative methods for elliptic problems with boundary singularities are mesh refinement near the singularity; conformal transformation; modified integral equations; modified collocation; power series; dual series; and removal of the singularity. General references may be found in the paper by Crank and Furzeland (1978) and the books by Gladwell and Wait (1979, Chapters 3, 4) and Baker and Miller (1982). Unfortunately, only a few of the methods referred to have been applied directly to viscous flow problems.

In this section we first examine the local nature of the singularity for a Newtonian fluid at a re-entrant corner and then consider a few methods for modifying the finite difference scheme. Some remarks on the singular finite difference treatment of non-Newtonian fluids will conclude the section.

Re-entrant corners - Newtonian fluids

We consider the solution of the Navier-Stokes equations (6.1) and (6.5) with $F \equiv 0$ in the neighbourhood of the re-entrant corner (x_j, y_k) in Fig. 6.7. In such a neighbourhood, Moffatt has shown that the viscous forces dominate the inertial forces so that a viscous Newtonian fluid behaves as a Stokesian fluid.

Equations (6.1) and (6.5) may be written as

$$\nabla^4\psi = RJ(\psi,\omega) , \tag{6.47}$$

where

$$J(\psi,\omega) = \frac{\partial\psi}{\partial x}\frac{\partial\omega}{\partial y} - \frac{\partial\psi}{\partial y}\frac{\partial\omega}{\partial x} .$$

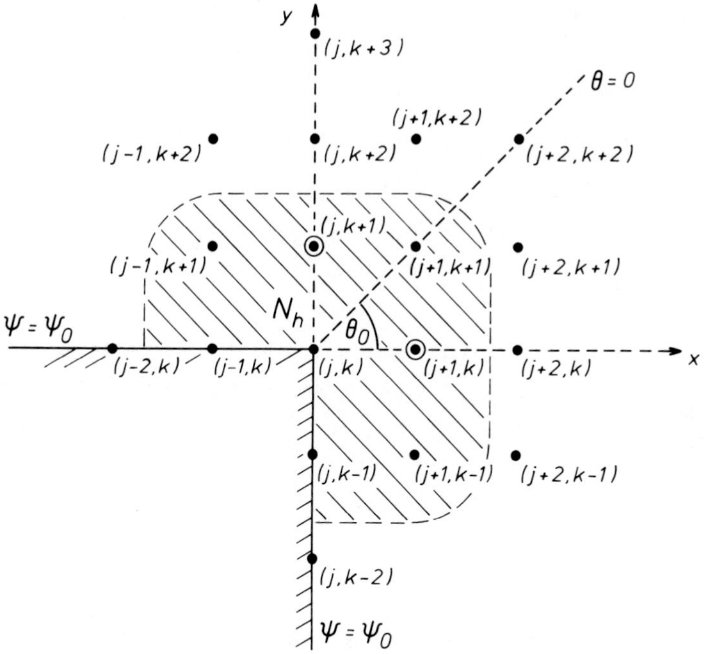

Fig. 6.7 A re-entrant corner.

We assume that the solution ψ of (6.47) can be written as a power series in R, at least in the vicinity of the corner, so that

$$\psi = \psi^{(0)} + R\psi^{(1)} + R^2\psi^{(2)} + \dots \quad . \tag{6.48}$$

Consequently the vorticity may be expressed as

$$\omega = \omega^{(0)} + R\omega^{(1)} + R^2\omega^{(2)} + \dots \quad , \tag{6.49}$$

where

$$\omega^{(n)} = - \nabla^2\psi^{(n)} , \qquad n = 0, 1, \dots . \tag{6.50}$$

Substituting (6.48) into (6.47) and comparing powers of R on both sides, we get

$$\left.\begin{aligned} \nabla^4\psi^{(0)} &= 0 , \\ \nabla^4\psi^{(n)} &= \sum_{m=0}^{n-1} J(\psi^{(n-m-1)}, \omega^{(m)}) , \quad n = 1, 2, \dots . \end{aligned}\right\} \tag{6.51}$$

The boundary conditions in the neighbourhood of (x_j, y_k) are

$$\left.\begin{aligned} \psi^{(0)} &= \text{constant } \psi_0 \text{ on } \Gamma \ , \\ \psi^{(n)} &= 0 \text{ on } \Gamma \ , \quad n = 1, 2, \ldots \ , \end{aligned}\right\} \tag{6.52}$$

and

$$\frac{\partial \psi^{(m)}}{\partial n} = 0 \text{ on } \Gamma \ , \quad m = 0, 1, \ldots \ . \tag{6.53}$$

It is convenient to introduce polar coordinates (r,θ) with origin at (x_j, y_k) and the line $\theta = 0$ along the internal bisector of the corner angle (Fig. 6.7). Then

$$x - x_j = r\cos(\theta + \theta_0) \ ,$$

$$y - y_k = r\sin(\theta + \theta_0) \ ,$$

where $\theta_0 = \pi/4$ is the angle between the line $\theta = 0$ and the x-axis. The Stokes equation $\nabla^4\psi^{(0)} = 0$ then has a solution of the form (Moffatt 1964; Holstein and Paddon 1981)

$$\psi^{(0)}(r,\theta) = \mathrm{Re}\{\sum_{m=0}^{\infty} a_m r^{\lambda_m} f_m(\theta)\} + \psi_0 \ , \tag{6.54}$$

where the exponents λ_m and functions f_m satisfy

$$f_m^{iv} + 2[(\lambda_m - 1)^2 + 1]f_m'' + \lambda_m^2(\lambda_m - 2)^2 f_m = 0 \tag{6.55}$$

and

$$f_m(\pm \tfrac{3}{4}\pi) = f_m'(\pm \tfrac{3}{4}\pi) = 0 \ , \quad m = 0, 1, \ldots \ , \tag{6.56}$$

where (6.56) comes from the boundary conditions (6.52), (6.53).

The case m = 0 corresponds to the trivial solution

$$\lambda_0 = 1 \ , \qquad f_0 \equiv 0 \ , \tag{6.57}$$

but for $m > 0$ we have

$$f_m(\theta) = \begin{cases} \cos[\tfrac{3}{4}\pi(\lambda_m - 2)]\cos(\lambda_m\theta) - \cos(\tfrac{3}{4}\pi\lambda_m)\cos[(\lambda_m - 2)\theta] & m = 1, 3, 5, \ldots, \\ \sin[\tfrac{3}{4}\pi(\lambda_m - 2)]\sin(\lambda_m\theta) - \sin(\tfrac{3}{4}\pi\lambda_m)\sin[(\lambda_m - 2)\theta] & m = 2, 4, 6, \ldots, \end{cases} \tag{6.58}$$

where the second equation in (6.56) constrains λ_m to satisfy

$$(\lambda_m - 1)(-1)^{m+1} = \sin[\tfrac{3}{2}\pi(\lambda_m - 1)] , \quad m = 1, 2, \ldots . \tag{6.59}$$

In addition we require that $\mathrm{Re}(\lambda_m) > 1$ to ensure that the fluid velocity tends to zero as the corner is approached. The values of λ_m, $m = 1, \ldots, 4$, ordered by increasing real part, are found to 4 decimal places to be

$$\lambda_1 = 1.5445, \quad \lambda_2 = 1.9085, \quad \mathrm{Re}(\lambda_3) = 2.6293, \quad \mathrm{Re}(\lambda_4) = 3.3013. \tag{6.60}$$

Since the Stokesian vorticity is given by

$$\omega^{(0)} = -\nabla^2\psi^{(0)} = O(r^{\lambda_1 - 2}) ,$$

it follows that ω becomes singular as $r \to 0$. Moreover, the expansion for $\omega^{(0)}$ may be written

$$\omega^{(0)}(r,\theta) = \mathrm{Re}\{\sum_{m=1}^{\infty} a_m r^{\lambda_m - 2} g_m(\theta)\} , \tag{6.61}$$

where

$$g_m(\theta) = \begin{cases} 4(\lambda_m - 1)\cos(\tfrac{3}{4}\pi\lambda_m)\cos[(\lambda_m - 2)\theta] , & m = 1, 3, 5, \ldots, \\ 4(\lambda_m - 1)\sin(\tfrac{3}{4}\pi\lambda_m)\sin[(\lambda_m - 2)\theta] , & m = 2, 4, 6, \ldots . \end{cases} \tag{6.62}$$

Substituting (6.54) and (6.61) into (6.51) then gives

$$\nabla^4\psi^{(1)} = O(r^{2\lambda_1 - 4}) ,$$

whence the second term in the power series (6.48) is

$$\psi^{(1)} = O(r^{2\lambda_1}) .$$

Since $\mathrm{Re}(\lambda_3) < 2\lambda_1 < \mathrm{Re}(\lambda_4)$, it follows that

$$\psi(r,\theta) = \text{Re}\{\sum_{m=1}^{3} a_m r^{\lambda_m} f_m(\theta)\} + \psi_0 + O(r^{2\lambda_1}) , \tag{6.63}$$

and

$$\omega(r,\theta) = \text{Re}\{\sum_{m=1}^{3} a_m r^{\lambda_m - 2} g_m(\theta)\} + O(r^{2\lambda_1 - 2}) , \tag{6.64}$$

which show that the viscous Newtonian and Stokesian corner flows share the first three expansion terms.

It is important to realize that if the corner involves one sliding and one stationary boundary (velocity discontinuity) then the boundary condition (6.53) for $m = 0$ is changed, rendering the trivial solution (6.57) nontrivial. The expansions (6.63), (6.64) are then no longer valid; moreover, the contribution of the term $\psi^{(1)}$ in the series (6.48) has not been fully worked out (Gupta et al 1981).

We now discuss how the finite difference approximation (6.17) may be modified at the points (x_{j+1},y_k), (x_j,y_{k+1}) in Fig. 6.7 to take account of the local singular form (6.64) in the neighbourhood N_h of the corner. Only the first three terms in the expansion will be considered. We assume that values of $\psi_{j,k}$ are known in N_h, usually as current iterates in an iterative procedure.

Several possibilities exist for estimating the constants a_1, a_2, a_3. For example, take three known grid values of ψ in N_h and use (6.63) to write down three equations which may be solved for a_1, a_2, a_3 (cf. Fox and Sankar 1969). We call this approach a *direct matching*. (We have assumed the constants to be real; when they are complex, more than three grid values are needed for the matching). Alternatively, a *least-squares matching* is possible, fitting several values of ψ in N_h to (6.63). Once the constants are known, the two vorticity values $\omega(x_{j+1},y_k)$ and $\omega(x_j,y_{k+1})$ could be estimated from (6.64) and the remaining values of ω found by solving (6.17) on $\Omega_h' = \Omega_h - \{(x_{j+1},y_k),(x_j,y_{k+1})\}$. A check on the feasibility of this approach could be made by comparing values of ω on $\Omega_h' \cap N_h$ determined from (6.17) with the corresponding values found from (6.64). The size of N_h should be such that the truncation error in (6.64) evaluated at the maximum value of r in N_h should not exceed the discretization error in $\omega_{j,k}$ outside N_h. In practice, only a few grid points in N_h should be required in a least-squares matching.

Using primitive variables (u,v,p) on interlocking grids, Ladevèze and Peyret (1974) determine the first few constants in the velocity expansion corresponding to (6.63) in an iterative manner. At each iteration the constants are redetermined by least-squares fit of the velocity expansions to current velocity grid values in a neighbourhood of the corner. This paper would appear to be the first successful attempt at incorporating the local analytical form of viscous corner flow in a finite difference scheme, but we omit the details here since the method is not immediately applicable to the Navier-Stokes equations in (ψ,ω) formulation.

In the matching schemes suggested above, no explicit use of difference approximations to singular variable derivatives at (x_{j+1},y_k) and (x_j,y_{k+1}) is made (cf. Crank and Furzeland 1978). A scheme which employs such approximations with optimal truncation errors for vorticity derivatives has been proposed by Holstein and Paddon (1981). Expressing (6.64) in the form

$$\omega(r,\theta) = \omega_n(r,\theta) + O(r^{\mu_{n+1}}) , \qquad n \leqslant 3 , \tag{6.65}$$

where

$$\omega_n(r,\theta) = \mathrm{Re}\{ \sum_{m=1}^{n} a_m r^{\lambda_m - 2} g_m(\theta)\}$$

and

$$\mu_{n+1} = \begin{cases} \mathrm{Re}(\lambda_{n+1} - 2) , & n = 1, 2, \\ 2\lambda_1 - 2 , & n = 3 , \end{cases}$$

Holstein and Paddon consider the function

$$\begin{aligned} W_n(\gamma,r,\theta) &\equiv r^{\gamma}[\omega(r,\theta) - \omega_n(r,\theta)] \\ &= b_{n+1} r^{\mu_{n+1}+\gamma} + O(r^{\mu_{n+2}+\gamma}) , \text{ say} , \end{aligned} \tag{6.66}$$

where $\mu_{n+2} > \mu_{n+1}$ and γ is a real constant to be chosen. Provided $\mu_{n+1} + \gamma > 0$, we have $W_n(\gamma,0,\theta) = 0$.

From conventional central difference approximations it follows that

$$\frac{\partial W_n}{\partial r}(\gamma,h,\theta) = \frac{1}{2h} W_n(\gamma,2h,\theta) + O(h^{\gamma+\varepsilon}) \tag{6.67}$$

and

$$\frac{\partial^2 W_n}{\partial r^2}(\gamma,h,\theta) = \frac{1}{h^2} [W_n(\gamma,2h,\theta) - 2W_n(\gamma,h,\theta)] + O(h^{\gamma+\varepsilon-1}) \tag{6.68}$$

where

$$\varepsilon = \begin{cases} \mu_{n+1} - 1 & \text{if } \mu_{n+1} + \gamma \neq 1, 2, \\ \mu_{n+2} - 1 & \text{if } \mu_{n+1} + \gamma = 1, 2. \end{cases}$$

Substituting (6.66) into (6.67) and (6.68) then gives

$$\left.\frac{\partial\omega}{\partial r}\right|_h = \left.\frac{\partial\omega_n}{\partial r}\right|_h + \frac{1}{2h}\left[-2\gamma(\omega-\omega_n)_h + 2^\gamma(\omega-\omega_n)_{2h}\right] + O(h^\varepsilon) , \tag{6.69}$$

and

$$\left.\frac{\partial^2\omega}{\partial r^2}\right|_h = \left.\frac{\partial^2\omega_n}{\partial r^2}\right|_h - \frac{1}{h^2}\left[\{\gamma(\gamma+1) - 2\}(\omega-\omega_n)_h + 2^\gamma(1-\gamma)(\omega-\omega_n)_{2h}\right] + O(h^{\varepsilon-1}). \tag{6.70}$$

These expressions make use of the values of ω and ω_n at the points (h,θ) and $(2h,\theta)$, and also the derivatives of ω_n which can be found analytically. The constants $\{a_m\}_{m=1}^n$ can be estimated as before by direct or least-squares matching of (6.63) to ψ-values in N_h.

The particular choices of γ given by

$$\gamma = 1 - \mu_{n+1} \quad \text{or} \quad 2 - \mu_{n+1} \tag{6.71}$$

both yield an optimal order of truncation error : $\varepsilon = \mu_{n+2} - 1$; direct differentiation of (6.65) yields only the order $\varepsilon = \mu_{n+1} - 1$. Using (6.71), for $n = 1$ and 2, respectively, we have $\varepsilon \simeq -0.37$ and $\varepsilon \simeq 0.09$, but the occurrence in (6.69) and (6.70) of negative orders for the absolute truncation errors does not appear to be a problem. It can be shown that the relative truncation errors in the vicinity of the singularity have positive orders (Holstein 1981).

In Fig. 6.7 the standard scheme (6.17) for the vorticity equation (6.5) breaks down at the grid points encircled. To modify the scheme at (x_{j+1},y_k) we may use (6.69) and (6.70) to give

$$\begin{aligned}\frac{\partial\omega}{\partial x}(x_{j+1},y_k) &= \frac{\partial\omega}{\partial r}(h,-\tfrac{\pi}{4})\\ &= \frac{\partial\omega_n}{\partial r}(h,-\tfrac{\pi}{4}) - \frac{\gamma}{h}[\omega(x_{j+1},y_k) - \omega_n(h,-\tfrac{\pi}{4})]\\ &\quad + \frac{2^{\gamma-1}}{h}[\omega(x_{j+2},y_k) - \omega_n(2h,-\tfrac{\pi}{4})] + O(h^\varepsilon) ,\end{aligned}$$

and

$$\begin{aligned}\frac{\partial^2\omega}{\partial x^2}(x_{j+1},y_k) &= \frac{\partial^2\omega_n}{\partial r^2}(h,-\tfrac{\pi}{4}) - \{\frac{\gamma(\gamma+1)-2}{h^2}\}[\omega(x_{j+1},y_k) - \omega_n(h,-\tfrac{\pi}{4})]\\ &\quad - \frac{2^\gamma(1-\gamma)}{h^2}[\omega(x_{j+2},y_k) - \omega_n(2h,-\tfrac{\pi}{4})] + O(h^{\varepsilon-1}) ,\end{aligned}$$

for the derivatives of ω with respect to x in (6.5). The derivatives with respect to y may be approximated by standard central differences since these do not involve the corner point :

$$\frac{\partial\omega}{\partial y}(x_{j+1},y_k) = \frac{1}{2h}[\omega(x_{j+1},y_{k+1}) - \omega(x_{j+1},y_{k-1})] + O(h^2) ,$$

$$\frac{\partial^2\omega}{\partial y^2}(x_{j+1},y_k) = \frac{1}{h^2}[\omega(x_{j+1},y_{k+1}) - 2\omega(x_{j+1},y_k) + \omega(x_{j+1},y_{k-1})] + O(h^2) .$$

After a little algebra it follows that at (x_{j+1},y_k), (6.5) may be approximated by the modified difference equation

$$K_0\omega_{j+1,k} - K_1\omega_{j+2,k} - K_3\omega_{j+1,k+1} - K_4\omega_{j+1,k-1} = - G_{j+1,k} , \qquad (6.72)$$

where

$$\left.\begin{aligned} &K_0 = \gamma(\gamma + 1 - 2\beta_{j+1,k}) , \qquad K_1 = 2^{\gamma}(\gamma - 1 - \beta_{j+1,k}) ,\\ &K_3 = 1 - \alpha_{j+1,k} , \qquad K_4 = 1 + \alpha_{j+1,k} ,\\ &G_{j+1,k} = \mathrm{Re}\{\sum_{m=1}^{n} \gamma_m a_m h^{\lambda_m-2} g_m(-\tfrac{\pi}{4})\} ,\\ &\text{and}\\ &\gamma_m = 2 - K_0 - 2^{\lambda_m-2} K_1 - (\lambda_m - 2)(\lambda_m - 3 - 2\beta_{j+1,k}) , \quad m = 1, 2, 3. \end{aligned}\right\} \qquad (6.73)$$

The formulae (6.13) and (6.16) have been used.

Similarly, at (x_j,y_{k+1}), (6.5) may be approximated by

$$K_0\omega_{j,k+1} - K_1\omega_{j+1,k+1} - K_2\omega_{j-1,k+1} - K_3\omega_{j,k+2} = - G_{j,k+1} , \qquad (6.74)$$

where

$$\left.\begin{aligned} &K_0 = \gamma(\gamma + 1 + 2\alpha_{j,k+1}) , \qquad K_3 = 2^{\gamma}(\gamma - 1 + \alpha_{j,k+1}) ,\\ &K_1 = 1 - \beta_{j,k+1} , \qquad K_2 = 1 + \beta_{j,k+1} ,\\ &G_{j,k+1} = \mathrm{Re}\{\sum_{m=1}^{n} \gamma_m a_m h^{\lambda_m-2} g_m(\tfrac{\pi}{4})\} ,\\ &\text{and}\\ &\gamma_m = 2 - K_0 + 2^{\lambda_m-2} K_3 - (\lambda_m - 2)(\lambda_m - 3 + 2\alpha_{j,k+1}) . \end{aligned}\right\} \qquad (6.75)$$

The above modified difference equations are based on central differences. This is not unreasonable since in the vicinity of the corner the local grid Reynolds number may be assumed small. A precise stability analysis of (6.17) in conjunction with (6.72) and (6.74) is, of course, difficult, but Holstein and Paddon argue that stability is better ensured by the choice of the first value of γ in (6.71), i.e., $\gamma = 1 - \mu_{n+1}$.

Roache (1976) describes nine different *ad hoc* methods for treating ω at sharp corners. Each method attaches a fictitious (sometimes multiple) finite value to the corner vorticity for purposes of incorporation in the standard finite difference scheme (6.17). There is little mathematical justification for most of these methods; the method of Kawaguti (1969) is, however, an exception. Referring to Fig. 6.7, this attaches a fictitious value

$$\omega^F_{j,k} = -\frac{2}{h^2}(\psi_{j+1,k} + \psi_{j,k+1} - 2\psi_{j,k}) \tag{6.76}$$

to the corner vorticity. Similar symmetric expressions are used at other sharp corners. Holstein and Paddon (1981, 1982) have justified Kawaguti's method by comparison with the asymptotic expansion method described above for the case $n = 1$.

When $n = 1$, the constant a_1 is estimated by

$$a_1 = \frac{\psi(h,\frac{\pi}{4}) + \psi(h,-\frac{\pi}{4}) - 2\psi_0}{2h^{\lambda_1} f_1(\frac{\pi}{4})} + O(h^{\mathrm{Re}\lambda_3 - \lambda_1}) .$$

With fictitious values $\omega^{F_j}(0,\theta)$, $j = 1, 2$, defined by

$$\frac{\partial\omega}{\partial r}(h,\theta) = \frac{1}{2h}[\omega(2h,\theta) - \omega^{F_1}(0,\theta)] ,$$

$$\frac{\partial^2\omega}{\partial r^2}(h,\theta) = \frac{1}{h^2}[\omega(2h,\theta) - 2\omega(h,\theta) + \omega^{F_2}(0,\theta)] ,$$

it may be shown that the use of these values at $\theta = \pm\frac{\pi}{4}$ are consistent with the asymptotic expansion provided

$$\omega^{F_j}(0,\pm\frac{\pi}{4}) \approx \frac{d_j}{h^2}[\psi(h,\frac{\pi}{4}) + \psi(h,-\frac{\pi}{4}) - 2\psi_0] , \quad j = 1, 2, \tag{6.77}$$

where

$$d_1 = [-2(\lambda_1 - 2) + 2^{\lambda_1 - 2}]\frac{g_1(\frac{\pi}{4})}{2f_1(\frac{\pi}{4})} \simeq -1.48 ,$$

and

$$d_2 = [(\lambda_1 - 2)(\lambda_1 - 3) - 2^{\lambda_1 - 2} + 2]\frac{g_1(\frac{\pi}{4})}{2f_1(\frac{\pi}{4})} \simeq -1.75 .$$

Equation (6.77) may be compared with (6.76), where $d_1 = d_2 = -2$.

For Newtonian problems, Holstein and Paddon (1982) go as far as to recommend the use of Kawaguti's formula (6.76) in preference to their asymptotic expansion method, at least for $n = 1$, because of the extreme simplicity of the former approach. Computations for low R-values using the two methods have yielded negligible differences in the ψ and ω fields both near and away from corners. (See also Cochrane et al 1982).

Re-entrant corners - non-Newtonian fluids

In Newtonian flow, the influence of small changes in corner conditions on flow characteristics away from the corner is weak. This has been observed experimentally and corroborated by numerical simulation (see, for example, Cochrane et al 1982; Walters and Webster 1982). In non-Newtonian flow, however, experiments indicate a strong interaction between corner conditions and fluid memory effects, which can extend to the whole of the flow field (Walters and Webster 1982). The correct treatment of re-entrant corners in non-Newtonian numerical simulation is therefore essential. Unfortunately, very little emphasis has been placed on this issue in the literature so far.

For non-Newtonian flow, (6.47) is replaced by

$$\nabla^4\psi = RJ(\psi,\omega) + F , \qquad (6.78)$$

where F depends on the stress derivatives (see (6.15)). The key feature in determining the local analytical form of the corner singularity, therefore, is the asymptotic behaviour of the stress components and their derivatives. Such behaviour can depend dramatically on the model structure; for example, when the time rates of change of the physical variables dominate the behaviour, as would be expected near corners, the asymptotic form of the stress is given (in dimensionalized variables) by

$$\underline{T} = \frac{2\eta_0}{\lambda_1}\underline{e} \qquad (6.79)$$

for the Maxwell model (2.72), where $\underline{e}$ is a strain tensor, but by

$$\underline{T} = 2\eta_0 \frac{\lambda_2}{\lambda_1} \underline{d} \tag{6.80}$$

for the Oldroyd B model (2.77). The asymptotic behaviour depicted by (6.79) is like that of an elastic solid, whereas the behaviour in (6.80) is like that of a viscous fluid (cf. Cochrane et al 1982).

The extension of Moffatt's analysis to the case of (6.78) where (6.79) pertains, i.e. the Maxwell fluid, has not yet been attempted. (See Holstein (1981) for a brief description of the Oldroyd B case). To make headway, therefore, workers have been forced to adopt simple heuristic strategies in the numerical treatment of corners. We have already mentioned some of these strategies in §§6.1 and 6.2; here only a few additional remarks are needed to complete the details.

At the points (x_{j+1},y_k) and (x_j,y_{k+1}) in Fig. 6.7 we consider :

(i) *The discrete vorticity equation (6.17).* Kawaguti's formula (6.77) has been widely used to provide fictitious values of $\omega_{j,k}$. The stress derivatives on the right-hand side of (6.17) may be approximated by formulae (6.27), (6.28b and c) without incurring singular grid values for the stress components. If (6.28a) is used for the mixed derivatives of normal stress components, then fictitious corner stress values are required (see §6.2.5).

(ii) *The discrete constitutive equations (6.38a-c).* Here, fictitious values of $S_{j,k}$ will again be required, unless the multiplying coefficient K vanishes. Also, in approximating the mixed derivative and certain third derivatives of stream function ψ to obtain values of the coefficients in (6.32) and (6.37), the no slip velocity condition at the corner should not be approximated explicitly by differences of ψ, since this can lead to ambiguities. Using $u = v = 0$ at (x_j,y_k), suppose that we infer

$$\psi_{j+1,k} - \psi_{j-1,k} = 0 \qquad \text{and} \qquad \psi_{j,k+1} - \psi_{j,k-1} = 0 \,. \tag{6.81}$$

Then, since

$$\psi_{j-\ell,k} = \psi_{j,k-\ell} = \psi_0 \,, \qquad \ell = 0, 1, \ldots , \tag{6.82}$$

we find $\psi_{j+1,k} = \psi_{j,k-1} = \psi_0$, which is usually not a property of the solution of the discrete Poisson equation (6.8) with part boundary conditions (6.82). To avoid the ambiguity, only (6.82) should be used in simplifying formulae (6.28a) for ψ and (6.39a-d) at (x_{j+1},y_k) and (x_j,y_{k+1}).

It should be emphasized that the heuristic strategies described above are likely to yield significant inaccuracies near corners, particularly as the elasticity parameter W increases. To assume a total discretization error of O(h) in vorticity and stress would be optimistic. The strategies should be viewed as temporary measures, and urgent attempts should be made to explore fully the local analytical form of singularities and to develop asymptotic expansion, mesh refinement, or other justifiable techniques for their treatment.

6.2.5 Boundary stress approximation

We complete our discussion of finite difference discretization for the flow of a Maxwell fluid by considering the calculation of stress values on solid and symmetry boundaries. Such values are needed in computing central differences for the right-hand side of (6.17) and in solving (6.38a-c) by explicit or implicit methods. We summarize the existing techniques below. Equations (6.36) are assumed to hold on solid boundaries, although we do not justify this assumption by limit arguments.

(i) *A stationary solid boundary parallel to the x-axis.* Substitution of the conditions

$$u = v = 0 , \qquad \frac{\partial u}{\partial x} = -\frac{\partial v}{\partial y} = 0 \tag{6.83}$$

into (6.36) yields a system of algebraic equations for the boundary values of S^{xx}, S^{xy} and S^{yy}, which may be solved to give

$$S^{xx} = 2W\left(\frac{\partial u}{\partial y}\right)^2 , \qquad S^{xy} = W\frac{\partial u}{\partial y} , \qquad S^{yy} = 0 . \tag{6.84}$$

Since $\partial u/\partial y = -\omega$ on the boundary, this derivative may be computed using the techniques of §6.2.3, in particular, formula (6.43).

(ii) *A stationary solid boundary parallel to the y-axis.* In a similar way we find

$$S^{xx} = 0 , \qquad S^{xy} = W\frac{\partial v}{\partial x} , \qquad S^{yy} = 2W\left(\frac{\partial v}{\partial x}\right)^2 , \tag{6.85}$$

where $\partial v/\partial x = \omega$.

(iii) *Re-entrant corners.* All velocity derivatives are undefined and some stress components are singular, but several authors have used fictitious corner stresses so that the standard difference equations (6.38a-c) are employable.

These fictitious values can be obtained from the conditions $u = v = 0$ only, by solving the resulting algebraic system

$$\begin{pmatrix} A_1 & -2B & 0 \\ -C & A_2 & -B \\ 0 & -2C & A_3 \end{pmatrix} \begin{pmatrix} S^{xx} \\ S^{xy} \\ S^{yy} \end{pmatrix} = \begin{pmatrix} G_1 \\ G_2 \\ G_3 \end{pmatrix}$$

obtained from (6.36), with fictitious values attached to the coefficients in (6.32) and (6.37). Referring to the re-entrant corner (x_j, y_k) in Fig. 6.7, the ψ-derivatives are treated using Kawaguti's method in the form

$$\left(\frac{\partial^2\psi}{\partial x^2}\right)^F_{j,k} = \frac{2}{h^2}(\psi_{j+1,k} - \psi_{j,k}) , \qquad \left(\frac{\partial^2\psi}{\partial y^2}\right)^F_{j,k} = \frac{2}{h^2}(\psi_{j,k+1} - \psi_{j,k}) , \tag{6.86}$$

and an extension of Kawaguti's method for the mixed derivative in the form

$$\left(\frac{\partial^2\psi}{\partial x \partial y}\right)^F_{j,k} = \frac{1}{4h^2}(2\psi_{j+1,k+1} - \psi_{j+1,k-1} - \psi_{j-1,k+1}) \tag{6.87}$$

(Cochrane et al 1982). Fortunately, the third derivatives do not enter in (6.37).

(iv) *A solid boundary parallel to the x-axis moving with constant velocity U.*
The conditions (6.83) are replaced by

$$u = U , \qquad v = 0 , \qquad \frac{\partial u}{\partial x} = -\frac{\partial v}{\partial y} = 0 , \tag{6.88}$$

so that (6.36) becomes a simplified differential system. If $U > 0$, the discrete equations (6.38a-c) reduce to

$$\left.\begin{aligned} (2U + h^2)S^{xx}_{j,k} - 2US^{xx}_{j-1,k} &= h^2(2B_{j,k}\, S^{xy}_{j,k} + G_{1,j,k}) , \\ (2U + h^2)S^{xy}_{j,k} - 2US^{xy}_{j-1,k} &= h^2(B_{j,k}\, S^{yy}_{j,k} + G_{2,j,k}) , \\ (2U + h^2)S^{yy}_{j,k} - 2US^{yy}_{j-1,k} &= h^2 G_{3,j,k} . \end{aligned}\right\} \tag{6.89}$$

The stresses on the moving boundary may be computed simultaneously with the internal stresses by solving (6.89) implicitly alongside (6.38a-c) (cf. §5.4.2). Alternatively, each equation in (6.89) may be solved explicitly by forward recurrence; the third equation should be solved first for $\{S^{yy}_{j,k}\}$, and then the second and first equations for $\{S^{xy}_{j,k}\}$ and $\{S^{xx}_{j,k}\}$ respectively.

Similarly for a moving boundary parallel to the y-axis.

(v) *An axis of symmetry parallel to the x-axis.* Here the conditions are $v = 0$, $\partial u/\partial y = \partial v/\partial x = 0$, again yielding a simplified form of (6.36) which may be discretized and solved as in (iv) above. It may be worth noting that an explicit forward recurrence can break down if the axial velocity u changes rapidly with x, thereby causing the coefficients A_1 or A_3 to change sign.

6.2.6 Matrix formulation

We summarize the discretization methods we have thus far developed by writing them in compact matrix form. This will permit us to study, in a fairly general setting, solution methods for the full discrete system (§§6.3, 6.4). First we introduce some notation. Let $\Gamma_h^i \subset \Gamma_h$ denote the set of boundary grid points at which the stress components are determined implicitly alongside the stresses on Ω_h, and let $\Gamma_h^e \subset \Gamma_h$ denote the set wherein they are determined explicitly by direct calculation or recurrence (§6.2.5(i)-(v)). We may then introduce the vectors of unknown variables

$$\underset{\sim}{S}^{xx} = \{ S^{xx}_{j,k} : (x_j,y_k) \in \Omega_h \cup \Gamma_h^i \} ,$$

$$\underset{\sim}{S}^{xy} = \{ S^{xy}_{j,k} : (x_j,y_k) \in \Omega_h \cup \Gamma_h^i \} ,$$

$$\underset{\sim}{S}^{yy} = \{ S^{yy}_{j,k} : (x_j,y_k) \in \Omega_h \cup \Gamma_h^i \} ,$$

$$\underset{\sim}{\omega} = \{ \omega_{j,k} : (x_j,y_k) \in \Omega_h \} ,$$

and

$$\underset{\sim}{\psi} = \{ \psi_{j,k} : (x_j,y_k) \in \Omega_h \} ,$$

in which the elements have the natural ordering of the grid points in $\Omega_h \cup \Gamma_h^i$ or Ω_h. It will also be convenient to write the three stress vectors as the partitioned vector

$$\underset{\sim}{S} = (\underset{\sim}{S}^{xx}, \underset{\sim}{S}^{xy}, \underset{\sim}{S}^{yy})^T .$$

On the boundary we have the vectors $\underset{\sim}{S}^e = (\underset{\sim}{S}^{xx,e}, \underset{\sim}{S}^{xy,e}, \underset{\sim}{S}^{yy,e})^T$ and $\underset{\sim}{\omega}^b$, where

$$\underset{\sim}{S}^{xx,e} = \{\underset{\sim}{S}^{xx}_{j,k} : (x_j,y_k) \in \Gamma_h^e\} \quad , \text{ etc.,}$$

and

$$\underset{\sim}{\omega}^b = \{\omega_{j,k} : (x_j,y_k) \in \Gamma_h\} .$$

These are to be determined, whereas the constant vector

$$\underset{\sim}{\psi}^b = \{\psi_{j,k} : (x_j, y_k) \in \Gamma_h\}$$

is assumed known.

We may then write the discrete constitutive equations (6.38a-c), the vorticity equation (6.17) and stream function equation (6.8), together with their computational boundary conditions, as the coupled matrix system

$$\left.\begin{aligned} \underline{A}\,\underset{\sim}{S} &= \underset{\sim}{b}\ , \\ \underline{B}\,\underset{\sim}{\omega} &= \underset{\sim}{c}\ , \\ \underline{C}\,\underset{\sim}{\psi} &= \underset{\sim}{d}\ , \end{aligned}\right\} \qquad \begin{aligned} &(6.90a) \\ &(6.90b) \\ &(6.90c) \end{aligned}$$

where $\underline{A}$ and $\underset{\sim}{b}$ have the partitioning

$$\underline{A} = \begin{bmatrix} \underline{A}^{xx} & & \bigcirc \\ & \underline{A}^{xy} & \\ \bigcirc & & \underline{A}^{yy} \end{bmatrix}, \qquad \underset{\sim}{b} = \begin{bmatrix} \underset{\sim}{b}^{xx} \\ \underset{\sim}{b}^{xy} \\ \underset{\sim}{b}^{yy} \end{bmatrix}, \qquad (6.91)$$

and the vectors $\underset{\sim}{b}^{xx}$, $\underset{\sim}{b}^{xy}$, $\underset{\sim}{b}^{yy}$, $\underset{\sim}{c}$ and $\underset{\sim}{d}$ again have their elements naturally ordered. The matrices $\underline{A}$, $\underline{B}$ and $\underline{C}$ are then block tridiagonal (cf. §5.4); $\underline{A}$ and $\underline{B}$ are in general asymmetric, whereas $\underline{C}$ is both symmetric and constant.

The system (6.90a-c) is nonlinearly coupled as a result of the functional dependences

$$\left.\begin{aligned} &\underline{A} = \underline{A}(\underset{\sim}{\psi};\underset{\sim}{\psi}^b)\ , \\ &\underline{B} = \underline{B}(\underset{\sim}{\psi};\underset{\sim}{\psi}^b)\ , \\ &\underset{\sim}{b} = \underset{\sim}{b}(\underset{\sim}{S},\ \underset{\sim}{S}^e,\ \underset{\sim}{\psi};\underset{\sim}{\psi}^b)\ , \\ &\underset{\sim}{c} = \underset{\sim}{c}(\underset{\sim}{S},\ \underset{\sim}{S}^e,\ \underset{\sim}{\omega}^b,\ \underset{\sim}{\psi};\underset{\sim}{\psi}^b), \\ &\text{and} \\ &\underset{\sim}{d} = \underset{\sim}{d}(\underset{\sim}{\omega};\underset{\sim}{\psi}^b)\ , \end{aligned}\right\} \qquad (6.92)$$

where $\underset{\sim}{\omega}^b = \underset{\sim}{\omega}^b(\underset{\sim}{\psi};\underset{\sim}{\psi}^b)$ and $\underset{\sim}{S}^e = \underset{\sim}{S}^e(\underset{\sim}{\omega}^b,\underset{\sim}{\psi};\underset{\sim}{\psi}^b)$.

The solution procedure must therefore be iterative, each iteration step involving the solution of one or more linearized systems. We thus begin the study of algorithms for solving (6.90a-c) by discussing methods for linear matrix equations in the next section.

6.3 SOLUTION OF LINEAR EQUATIONS

6.3.1 Introduction

The numerical solution of large sparse systems of linear equations is the object of much ongoing study. There are essentially two distinct groups of methods: *direct* and *iterative*. A direct method is an algorithm with a finite and predetermined number of steps at the end of which a solution is provided; in contrast, an iterative method requires an initial approximation and thereafter generates a sequence of vectors which, under favourable conditions, converges to the solution.

Direct methods are usually variations of basic Gauss elimination, making use of forward and backward substitutions. These correspond to the *decomposition* (or *factorization*) of the coefficient matrix into lower and upper triangular factors

$$\underline{A} = \underline{L}\,\underline{U}\,. \tag{6.93}$$

(Throughout this section, unless otherwise stated, $\underline{A}$ and $\underset{\sim}{b}$ will denote an arbitrary block tridiagonal matrix and a correspondingly partitioned vector, respectively.) If $\underline{A}$ has order $N \times N$, and typically only five non-zero entries in each row, then $\underline{L}$ and $\underline{U}$ have $O(N)$ non-zero entries in each row. The presence of these additional non-zeros is called *fill-in*, and the main problem in developing efficient direct methods for large sparse systems is to devise orderings of the equations so that fill-in is reduced as far as possible. For sparse matrices with band structure, some useful algorithms are those of Cuthill-Mckee and Reverse Cuthill-Mckee, and the nested dissection method of George. In finite element contexts, an important direct method which combines the assembly and solution stages of the algebraic equations, is the frontal elimination method of Irons (cf. §8.6). References to and descriptions of these and other direct methods are to be found in the books by Wait (1979), Gladwell and Wait (1979), and Meis and Marcowitz (1981).

To be efficient, direct methods must often have quite complex codes, and it would be foolish for the average programmer not to take advantage of computer library packages which are the fruits of many years development. Among the most widely used are the I.M.S.L. Library, the LINPACK routines of the Argonne Laboratory, the Harwell Subroutine Library, and the N.A.G. Library.

For the discrete Poisson equation, several important direct methods have emerged over the last decade or so which have become commonly known as Fast Poisson Solvers. Many of these algorithms fall into two distinct categories: those based on Fourier decomposition in one-dimension, using fast Fourier transform (FFT) techniques, and those based on block cyclic reduction (Buneman's algorithm). Both approaches are described in some detail by Buzbee et al (1970). In addition, a class of algorithms, abbreviated as FACR, combine Fourier and cyclic reduction techniques, and often prove faster than methods based on Fourier analysis or cyclic reduction alone. A useful review paper on Fast Poisson Solvers is that of Temperton (1979). Again, library routines are best used, and consequently we shall say no more about direct methods in this chapter.

Iterative methods have always proven very useful for solving the algebraic equations resulting from finite difference approximations of partial differential equations. Even today, they are the only feasible methods for solving very large systems. They make good use of sparsity and structure, no fill-in is involved, and often it is not necessary to store the matrix but simply to generate the non-zero elements when they are wanted. Furthermore, iterative methods are easy to code. We therefore devote the next two subsections to the description of the classical iterative methods of Gauss-Seidel (GS) and successive over-relaxation (SOR), and to the relatively recent preconditioned conjugate gradient methods.

6.3.2 Classical iterative methods

Classical GS and SOR are still perhaps the most widely used methods for solving the finite difference equations of Newtonian and non-Newtonian computational fluid dynamics. They have a well-established theoretical foundation expounded in the books by Varga (1963), Young (1971), Hageman and Young (1981), and others.

Consider a matrix system

$$\underline{A}\,\underset{\sim}{x} = \underset{\sim}{b} \; . \tag{6.94}$$

The basic idea is to replace (6.94) by an equivalent system

$$(\underline{I} - \underline{C})\underset{\sim}{x} = \underset{\sim}{d} \tag{6.95}$$

which has the same solution $\underset{\sim}{x}$. ($\underline{I}$ denotes the identity matrix). This is done by splitting $\underline{A}$ into a form

$$\underline{A} = \underline{E} - \underline{F} \quad , \tag{6.96}$$

where $\underline{E}$ is non-singular. Then (6.94) is equivalent to (6.95) with $\underline{C} = \underline{E}^{-1}\,\underline{F}$ and $\underset{\sim}{d} = \underline{E}^{-1}\,\underset{\sim}{b}$.

We then use the following fundamental result :

Theorem. The vector sequence defined by

$$\underset{\sim}{x}_{r+1} = \underline{C}\,\underset{\sim}{x}_r + \underset{\sim}{d} \; , \qquad r = 0, 1, \ldots , \tag{6.97}$$

$\mathbf{x}_0$ arbitrary, converges to the unique solution of (6.95) *if and only if* $\rho(\mathbf{C}) < 1$. (Here we use $\rho(.)$ to denote the *spectral radius* of a matrix, defined as the modulus of the largest eigenvalue of the matrix). For iterative convergence, therefore, the splitting (6.96) should satisfy $\rho(\mathbf{E}^{-1}\mathbf{F}) < 1$.

The *point GS* method has the splitting

$$\mathbf{A} = (\mathbf{D} - \mathbf{L}) - \mathbf{U} \quad , \tag{6.98}$$

where $\mathbf{D}$ is the diagonal part of $\mathbf{A}$, and $-\mathbf{L}$ and $-\mathbf{U}$ are the strictly lower and strictly upper triangular parts of $\mathbf{A}$, respectively. (The latter should not be confused with the triangular factors of $\mathbf{A}$ in (6.93). The iteration (6.97) then yields the point GS iterative process for (6.94), which may be written in the form

$$(\mathbf{D} - \mathbf{L})\mathbf{x}_{r+1} = \mathbf{U}\,\mathbf{x}_r + \mathbf{b} \; , \qquad r = 0, 1, \ldots \; . \tag{6.99}$$

The solution of (6.99) for one value of r is referred to as one sweep of GS iteration. Since $(\mathbf{D} - \mathbf{L})$ is lower triangular, a sweep is simply effected by forward substitution.

The *point SOR* method has the splitting (6.96) with

$$\mathbf{E} = \frac{1}{\alpha}\mathbf{D} - \mathbf{L} \qquad \text{and} \qquad \mathbf{F} = \frac{1}{\alpha}\mathbf{D} - (\mathbf{D} - \mathbf{U}) \; , \tag{6.100}$$

where $\alpha \neq 0$ is a constant relaxation parameter to be chosen. The point SOR iterative process is then

$$(\mathbf{D} - \alpha\mathbf{L})\mathbf{x}_{r+1} = [(1 - \alpha)\mathbf{D} + \alpha\mathbf{U}]\mathbf{x}_r + \alpha\mathbf{b} \; , \quad r = 0, 1, \ldots \; , \tag{6.101}$$

which may again be effected by forward substitution. It can be shown that a necessary (although by no means sufficient) condition for the convergence of SOR is that $0 < \alpha < 2$. When $0 < \alpha < 1$, the process is called under-relaxation, and when $1 < \alpha < 2$, over-relaxation; when $\alpha = 1$, SOR is simply GS. The advantage of SOR over GS is that, under favourable conditions, the relaxation parameter can be chosen to accelerate the convergence of the iterative process.

A sufficient condition for the convergence of SOR with $0 < \alpha \leq 1$ (i.e. under-relaxation and GS) is that the matrix $\mathbf{A}$ in (6.94) be irreducibly diagonally dominant. $\mathbf{A}$ is said to be *reducible* if, by a suitable permutation of its rows and corresponding columns, it may be written in the 2×2 block form

$$\begin{pmatrix} \mathbf{A}_{11} & \mathbf{A}_{12} \\ \mathbf{0} & \mathbf{A}_{22} \end{pmatrix} ,$$

where $\underset{\sim}{A}_{11}$ and $\underset{\sim}{A}_{22}$ are square diagonal blocks, and $\underset{\sim}{0}$ denotes a rectangular block of zeros; otherwise $\underset{\sim}{A}$ is *irreducible*. $\underset{\sim}{A}$ is said to be *irreducibly diagonally dominant* if it is both irreducible and diagonally dominant (see (6.10)) with the additional constraint that strict inequality in (6.10) pertains for at least one row. It is a simple matter to show that the matrices $\underset{\sim}{A}$, $\underset{\sim}{B}$ and $\underset{\sim}{C}$ in (6.90a-c) are all irreducibly diagonally dominant. (See Varga (1963) for a graph theoretic criterion for irreducibility). It can also be proved for these matrices that, when $0 < \alpha \leq 1$, the rate of convergence of SOR increases with α; thus GS converges faster than under-relaxation.

For the system (6.94), the theorems of Ostrowski and Reich infer that if $\underset{\sim}{A}$ is symmetric with positive diagonal elements, and $0 < \alpha < 2$, then SOR converges if and only if $\underset{\sim}{A}$ is positive-definite (i.e. all its eigenvalues are positive). This applies directly to the discrete Poisson equation (6.90c).

To determine an optimal relaxation factor α, we require that the matrix $\underset{\sim}{A}$ be two-cyclic and consistently ordered, and that the *associated point Jacobi matrix*, defined by $\underset{\sim}{B} = \underset{\sim}{D}^{-1}(\underset{\sim}{L} + \underset{\sim}{U})$, have real eigenvalues. (When $\underset{\sim}{B}$ has some complex eigenvalues the situation is much more complicated; see Young (1971)). $\underset{\sim}{A}$ is said to be *two-cyclic* if, by a suitable permutation of its rows and corresponding columns, it can be written in the 2×2 block form

$$\begin{pmatrix} \underset{\sim}{D}_1 & \underset{\sim}{A}_{12} \\ \underset{\sim}{A}_{21} & \underset{\sim}{D}_2 \end{pmatrix},$$

where $\underset{\sim}{D}_1$, $\underset{\sim}{D}_2$ are square diagonal matrices and $\underset{\sim}{A}_{12}$, $\underset{\sim}{A}_{21}$ are arbitrary rectangular matrices. If $\underset{\sim}{A}$ is two-cyclic, then it is *consistently ordered* if all the eigenvalues of the matrix $\beta\underset{\sim}{L} + \beta^{-1}\underset{\sim}{U}$ are independent of β, for all $\beta \neq 0$. It can then be shown that the optimal value of α is given by

$$\alpha_{opt} = \frac{2}{1 + \sqrt{1 - \rho^2(\underset{\sim}{B})}}, \tag{6.102}$$

where $\rho(\underset{\sim}{B})$ is the spectral radius of the associated point Jacobi matrix.

Again, it is fairly easy to show that the matrices in (6.90a-c) are two-cyclic and consistently ordered, but real eigenvalues of the associated point Jacobi matrices are guaranteed only for (6.90c), where the matrix is symmetric. The theory enables us to say with certainty, therefore, that an optimal over-relaxation parameter exists for (6.90c), but we can say nothing about (6.90a,b) other than GS is convergent for these systems. In practice we find that over-relaxation on (6.90a,b) invariably diverges (at least in conjunction with the functional iteration approach described in §6.4), whereas for (6.90c) the choice of a good over-relaxation parameter can significantly improve convergence.

Formula (6.102) is useful only when the spectral radius $\rho(\underline{B})$ can be estimated. Theoretical estimates are available for only the simplest cases, for example, if the flow region Ω is a rectangle of sides Mh and Nh, then the associated point Jacobi matrix $\underline{B}$ for the discrete Poisson matrix in (6.90c) has

$$\rho(\underline{B}) = \tfrac{1}{2}\left[\cos\left(\frac{\pi}{M}\right) + \cos\left(\frac{\pi}{N}\right)\right] .$$

In practice, the best value of α can be estimated by trial and error, or by more sophisticated means. Varga (1963) shows how to use a simple GS iteration to estimate $\rho^2(\underline{B})$. If $\underset{\sim}{x}_0$ is any vector with positive elements $x_m^{[0]}$, then the vectors $\underset{\sim}{x}_r$ generated by the GS iteration

$$(\underline{D} - \underline{L})\underset{\sim}{x}_{r+1} = \underline{U}\underset{\sim}{x}_r , \quad r = 0, 1, \ldots ,$$

also have positive elements $x_m^{[r]}$, and

$$\min_m \left(\frac{x_m^{[r+1]}}{x_m^{[r]}}\right) < \rho^2(\underline{B}) < \max_m \left(\frac{x_m^{[r+1]}}{x_m^{[r]}}\right) , \quad r = 0, 1, \ldots .$$

Both bounds converge to $\rho^2(\underline{B})$ as r increases, but it is better to overestimate $\rho(\underline{B})$ than to underestimate it. This result applies whenever $\underline{A}$ is irreducible, two-cyclic and consistently ordered, and its associated Jacobi matrix $\underline{B}$ is non-negative. Other methods of estimating optimal SOR parameters are discussed by Hageman and Young (1981).

We now consider in a little more detail the application of iterative methods to each system in (6.90a-c). One iteration of GS applied to the first subsystem of (6.90a) has the form (cf. (6.38a))

$$\left.\begin{aligned} S_{j,k}^{xx[r+1]} = \frac{1}{K_0^{A_1}} \Big[& K_1^{A_1} S_{j+1,k}^{xx[r]} + K_2^{A_1} S_{j-1,k}^{xx[r+1]} + K_3^{A_1} S_{j,k+1}^{xx[r]} \\ & + K_4^{A_1} S_{j,k-1}^{xx[r+1]} + b_{j,k}^{xx} \Big] , \quad (x_j,y_k) \in \Omega_h \cup \Gamma_h^i , \end{aligned}\right\} \qquad (6.103)$$

where $b_{j,k}^{xx} = h^2[2B_{j,k}\, S_{j,k}^{xy} + F_{1,j,k}]\, \mathrm{sign}\,(A_1)$.

In (6.103) each grid value of S^{xx} is updated by sweeping through the set $\Omega_h \cup \Gamma_h^i$ row by row, i.e. according to the natural ordering. All components of $\underset{\sim}{S}^{xx}$ on the right hand side of (6.103) are therefore the most recently available. (This may not be true of $\underset{\sim}{S}^{xy}$ on the right of (6.103); see §6.4). Explicitly determined boundary values (components of $\underset{\sim}{S}^{xx,e}$) which are present are not iterated upon, i.e. they are kept independent of r.

Similar remarks hold for the second and third subsystems of (6.90a).

One iteration of GS applied to (6.90b) is (cf. (6.17))

$$\omega_{j,k}^{[r+1]} = \frac{1}{K_0}[K_1\,\omega_{j+1,k}^{[r]} + K_2\,\omega_{j-1,k}^{[r+1]} + K_3\,\omega_{j,k+1}^{[r]} + K_4\,\omega_{j,k-1}^{[r+1]} - h^2F_{j,k}]\,,$$

$$(x_j,y_k) \in \Omega_h\,, \qquad (6.104)$$

where again boundary values in $\underset{\sim}{\omega}^b$ are kept independent of r.

Finally, one iteration of SOR applied to (6.90c) is (cf. (6.8))

$$\psi_{j,k}^{[r+1]} = (1-\alpha)\psi_{j,k}^{[r]} + \tfrac{1}{4}\alpha(\psi_{j+1,k}^{[r]} + \psi_{j-1,k}^{[r+1]} + \psi_{j,k+1}^{[r]} + \psi_{j,k-1}^{[r+1]} + h^2\omega_{j,k})$$

$$(x_j,y_k) \in \Omega_h\,, \qquad (6.105)$$

where the same remarks as above apply.

Deferred correction

In §6.1.2 we described the deferred correction method of Dennis and Chang in connection with the UD1 scheme for the vorticity equation. A contracted form of this technique can be implemented by means of the following modified GS iteration :

$$\omega_{j,k}^{[r+1]} = \left(1 - \frac{L_0}{M_0}\right)\omega_{j,k}^{[r]} + \frac{1}{M_0}[L_1\,\omega_{j+1,k}^{[r]} + L_2\,\omega_{j-1,k}^{[r+1]} + L_3\,\omega_{j,k+1}^{[r]}$$

$$+ L_4\,\omega_{j,k-1}^{[r+1]} + (L_2 - M_2)\omega_{j-1,k}^{[r]} + (L_4 - M_4)\omega_{j,k-1}^{[r]} - h^2F_{j,k}]\,,$$

$$(x_j,y_k) \in \Omega_h. \qquad (6.106)$$

Here the L's denote the coefficients (K) in the CD scheme (6.18) and the M's denote the UD1 coefficients (6.20). The scheme was proposed by Richards and Crane (1978) but is equivalent to that of Veldman (1973). We note that (6.106) is effectively an under-relaxation method with a parameter $\alpha = L_0/M_0 \leqslant 1$ which varies from grid-point to grid-point as a function of the two velocity components.

The modified iteration (6.106) can yield improved accuracy in Newtonian calculations, and it has also been generalized for second order fluid calculations (Pilate and Crochet 1977). For more general non-Newtonian calculations it remains to be seen whether O(h) errors in the non-Newtonian terms $F_{j,k}$ would dominate the overall accuracy (cf. §6.2.3).

The deferred correction in (6.106) can be used in conjunction with the SM scheme by simply choosing the coefficients M to be those in (6.22). The correction terms vanish for values of R below the critical CD limit. Richards and Crane (1979) also propose an alternative modification which they claim to be superior to both UD1 and SM corrections in Newtonian calculations.

Block iterative methods

We conclude this subsection by noting that for block tridiagonal systems it is usually possible to improve on the rates of convergence of point iterative methods by using block iteration. For (6.94) the *block GS* method has the splitting (6.98) where $\underline{D}$ becomes the block diagonal matrix consisting of the diagonal blocks of $\underline{A}$, and $-\underline{L}$ and $-\underline{U}$ are the strictly lower and strictly upper triangular matrices consisting of the remaining blocks of $\underline{A}$. Similarly, using the new definitions of $\underline{D}$, $\underline{L}$ and $\underline{U}$, the *block SOR* method has the splitting (6.96) with (6.100). The theory described above for point iterations is easily generalized to the block case.

Consider (6.94) with $\underline{A}$ written as $N \times N$ blocks

$$\underline{A} = \begin{pmatrix} \underline{A}_{11} & \underline{A}_{12} & & \bigcirc \\ \underline{A}_{21} & \underline{A}_{22} & \underline{A}_{23} & \\ & \cdot & \cdot & \cdot \\ & & \cdot & \cdot & \cdot \\ \bigcirc & & & \underline{A}_{N,N-1} & \underline{A}_{NN} \end{pmatrix} , \qquad (6.107)$$

where each diagonal submatrix $\underline{A}_{nn}$ is square, and the vectors $\tilde{x}$ and $\tilde{b}$ are each correspondingly partitioned into N vectors $\tilde{x}_n$ and $\tilde{b}_n$. Then a block SOR iteration is of the form

$$\underline{A}_{nn}\, \tilde{x}_n^{[r+1]} = (1-\alpha)\underline{A}_{nn}\, \tilde{x}_n^{[r]} + \alpha(\tilde{b}_n - \underline{A}_{n,n-1}\, \tilde{x}_{n-1}^{[r+1]} - \underline{A}_{n,n+1}\, \tilde{x}_{n+1}^{[r]}) , \quad n = 1, \ldots, N . \qquad (6.108)$$

Formula (6.108) consists of N sets of matrix equations, which, assuming that $\underline{A}_{nn}$ is non-singular for each n, can be solved by an efficient direct method. For each of the systems in (6.90a-c) the diagonal blocks $\underline{A}_{nn}$ are all diagonally dominant tridiagonal matrices. The individual systems in (6.108) may therefore be solved using Algorithm 5.1. (§5.2.2). This particular block iterative method is often called *successive line over-relaxation* (SLOR), using the idea that lines of grid points are treated as units.

When the flow region Ω is a rectangle then each coefficient matrix in (6.90a-c) has square blocks of equal order. When Ω is a union of rectangles then the coefficient matrices still retain square diagonal blocks, although not necessarily of the same order. The off-diagonal blocks are then rectangular, in general, but contain square submatrices which are diagonal. (We also note that, in principle, it is possible to solve a Dirichlet problem on a union of overlapping rectangles as a sequence of Dirichlet problems on the individual rectangles, (cf. the *Schwartz alternating procedure* described in Kantorovich and Krylov (1958)).

For an application of block iterative methods in non-Newtonian calculations see Townsend (1980b).

6.3.3 Preconditioned conjugate gradient methods

(i) *Symmetric positive definite matrices*

The basic conjugate gradient (CG) method was proposed by Hestenes and Stiefel (1952) to solve systems such as (6.94) when $\underline{A}$ is symmetric and positive definite. If $\underset{\sim}{y}$ is an approximate solution of (6.94) we consider the error norm defined by

$$E(\underset{\sim}{y}) \equiv \tfrac{1}{2}(\underset{\sim}{y} - \underset{\sim}{x})^T \underline{A}(\underset{\sim}{y} - \underset{\sim}{x}) = \tfrac{1}{2}\underset{\sim}{r}^T \underline{A}^{-1}\underset{\sim}{r} ,$$

where $\underset{\sim}{r} = \underset{\sim}{b} - \underline{A}\underset{\sim}{y}$ is called the *residual vector*. Because $\underline{A}$ is positive definite the norm has a unique minimum at $\underset{\sim}{y} = \underset{\sim}{x}$, and the CG algorithm attempts to find this minimum iteratively. Where appropriate we follow the account of Mitchell and Griffiths (1980).

At the m-th iteration let $\underset{\sim}{x}_m$ be the current approximation, $\underset{\sim}{r}_m = \underset{\sim}{b} - \underline{A}\underset{\sim}{x}_m$ the associated residual, and $\underset{\sim}{p}_m$ a search direction. The next approximation $\underset{\sim}{x}_{m+1}$ is determined by minimizing E in the direction $\underset{\sim}{p}_m$, i.e. by setting $\underset{\sim}{x}_{m+1} = \underset{\sim}{x}_m + \alpha_m \underset{\sim}{p}_m$, where α_m is a scalar chosen to minimize the one-dimensional function $E(\underset{\sim}{x}_m + \alpha \underset{\sim}{p}_m)$. By choosing the search direction $\underset{\sim}{p}_m$ to be the component of $\underset{\sim}{r}_m$ conjugate to each direction $\underset{\sim}{p}_0, \underset{\sim}{p}_1, \ldots, \underset{\sim}{p}_{m-1}$, i.e.

$$\underset{\sim}{p}_m^T \underline{A}\, \underset{\sim}{p}_n = 0 , \qquad n = 0, 1, \ldots, m-1,$$

then the solution $\underset{\sim}{x}$ can be attained, in principle, in at most N iterations, where N is the order of the system (6.94). The attraction of the method is that these search directions can be computed recursively as follows:

CG Algorithm

(0) Let m = 0. Choose an initial approximation $\underset{\sim}{x}_0$ and compute $\underset{\sim}{r}_0 = \underset{\sim}{b} - \underline{A}\underset{\sim}{x}_0$, $\underset{\sim}{p}_0 = \underset{\sim}{r}_0$.

(1) Compute: $\alpha_m = (\underset{\sim}{r}_m^T \underset{\sim}{r}_m)/(\underset{\sim}{p}_m^T \underline{A} \underset{\sim}{p}_m)$,

$$\underset{\sim}{x}_{m+1} = \underset{\sim}{x}_m + \alpha_m \underset{\sim}{p}_m ,$$

$$\underset{\sim}{r}_{m+1} = \underset{\sim}{r}_m - \alpha_m \underline{A} \underset{\sim}{p}_m ,$$

$$\beta_m = (\underset{\sim}{r}_{m+1}^T \underset{\sim}{r}_{m+1})/(\underset{\sim}{r}_m^T \underset{\sim}{r}_m) ,$$

$$\underset{\sim}{p}_{m+1} = \underset{\sim}{r}_{m+1} + \beta_m \underset{\sim}{p}_m .$$

(2) Let $m \rightarrow m+1$.

(3) Repeat steps (1) - (2) until convergence is reached.

If the calculations are performed with exact arithmetic then the exact solution $\underset{\sim}{x}$ is obtained in at most N iterations, and the method could then be classified as a direct method. In practice, however, rounding errors degrade the conjugacy of the sequence $\underset{\sim}{p}_0, \underset{\sim}{p}_1, \ldots$, and the finite termination property is lost. Fortunately, when N is large, it is often found that $\underset{\sim}{x}_m \simeq \underset{\sim}{x}$ for $m << N$.

The algorithm requires storage of only the non-zero elements of $\underline{A}$ together with the four vectors $\underset{\sim}{x}_m$, $\underset{\sim}{r}_m$, $\underset{\sim}{p}_m$ and $\underline{A}\, \underset{\sim}{p}_m$. This is greater than the storage requirements of SOR but may be offset by the fact that the parameters α_m, β_m are determined automatically, unlike SOR where the optimal relaxation factor α has to be estimated. Moreover Reid (1972) has shown that for structured matrices the CG algorithm may be more efficient than optimal SOR.

There is theoretical and numerical evidence to show that the CG algorithm converges very rapidly when a large proportion of the eigenvalues of $\underline{A}$ are clustered around a relatively small number of points on the real line. This contrasts with SOR where the optimum α is determined solely from an extreme eigenvalue. Unfortunately, when $\underline{A}$ results from the finite differencing of partial differential equations, it usually has a large spread of eigenvalues, and the basic CG method is not particularly successful. It is possible to restore the advantages of CG, however, by *preconditioning* the system (6.94).

Preconditioning involves the choice of a suitable positive definite matrix $\underline{M}$ so that (6.94) may be replaced by

$$\underline{M}^{-1}\underline{A}\, \underset{\sim}{x} = \underline{M}^{-1}\underset{\sim}{b} , \tag{6.109}$$

where $\underline{M}^{-1}\underline{A}$ should have a large number of its eigenvalues nearly equal or closely grouped. Several preconditioning methods in conjunction with CG are available (Duff 1979 serves as an excellent reference), but in this section we restrict attention to preconditioning by incomplete decomposition, which has been used successfully in non-Newtonian calculations by Court et al (1981) and Davies and Manero (1983).

The symmetric positive-definite matrix $\underline{A}$ may be factorized exactly as

$$\underline{A} = \underline{L}\,\underline{L}^T , \tag{6.110}$$

where $\underline{L}$ is a lower triangular matrix which may be computed efficiently by Cholesky decomposition (see, for example, Fox 1964). We have already mentioned that the exact $\underline{L}$ will be subject to fill-in. If we insist instead that $\underline{L}$ has non-zero entries only in those positions which correspond to non-zero elements in the lower triangle of $\underline{A}$, then we have the incomplete Cholesky decomposition

$$\underline{A} = \underline{L}\,\underline{L}^T + \underline{E} \quad ,$$

where $\underline{E}$ is an error matrix. If we now define $\underline{M} = \underline{L}\,\underline{L}^T$ and solve the resulting system (6.109) by CG, we have the Incomplete Cholesky Conjugate Gradient algorithm known as ICCG(0) (Meijerink and Van der Vorst 1977):

ICCG(O) Algorithm

(0) Let m = 0. Choose $\underset{\sim}{x}_0$ and compute $\underset{\sim}{r}_0 = \underset{\sim}{b} - \underline{A}\underset{\sim}{x}_0$, $\underset{\sim}{p}_0 = (\underline{L}\,\underline{L}^T)^{-1}\underset{\sim}{r}_0$.

(1) Compute:

$$\begin{aligned}
\alpha_m &= (\underset{\sim}{r}_m^T(\underline{L}\,\underline{L}^T)^{-1}\underset{\sim}{r}_m)/(\underset{\sim}{p}_m^T\,\underline{A}\,\underset{\sim}{p}_m), \\
\underset{\sim}{x}_{m+1} &= \underset{\sim}{x}_m + \alpha_m\underset{\sim}{p}_m , \\
\underset{\sim}{r}_{m+1} &= \underset{\sim}{r}_m - \alpha_m\,\underline{A}\,\underset{\sim}{p}_m , \\
\beta_m &= (\underset{\sim}{r}_{m+1}^T(\underline{L}\,\underline{L}^T)^{-1}\underset{\sim}{r}_{m+1})/(\underset{\sim}{r}_m^T(\underline{L}\,\underline{L}^T)^{-1}\underset{\sim}{r}_m), \\
\underset{\sim}{p}_{m+1} &= (\underline{L}\,\underline{L}^T)^{-1}\underset{\sim}{r}_{m+1} + \beta_m\underset{\sim}{p}_m .
\end{aligned}$$

(2) Let m → m+1.

(3) Repeat steps (1) - (2) until convergence is reached.

When $\underline{A}$ has the form (6.107) with $\underline{A}_{nn} = \underline{A}_{nn}^T$ and $\underline{A}_{n,n-1} = \underline{A}_{n-1,n}^T$, then $\underline{L}$ has the form

$$\underline{L} = \begin{pmatrix} \underline{L}_{11} & & & & \bigcirc \\ \underline{L}_{21} & \underline{L}_{22} & & & \\ & \cdot & \cdot & & \\ & & \cdot & \cdot & \\ & & & \cdot & \cdot \\ \bigcirc & & & \underline{L}_{N,N-1} & \underline{L}_{NN} \end{pmatrix} , \tag{6.111}$$

where the block matrices have the incomplete decompositions

$$\left.\begin{array}{ll} \underline{A}_{11} & \sim \underline{L}_{11}\,\underline{L}_{11}^{T}\,, \\ \underline{A}_{n,n-1} & \sim \underline{L}_{n,n-1}\,\underline{L}_{n-1,n-1}^{T}\,, \\ \underline{A}_{nn} & \sim \underline{L}_{n,n-1}\,\underline{L}_{n,n-1}^{T} + \underline{L}_{nn}\,\underline{L}_{nn}^{T}\,, \\ n = 2, \ldots, N. & \end{array}\right\} \tag{6.112}$$

Suppose now that the block matrices in (6.107) are square, of equal order $M \times M$, such that each diagonal block is a tridiagonal matrix and each adjacent codiagonal block is a diagonal matrix, i.e.

$$\underline{A}_{nn} = \begin{pmatrix} \beta_1^{(n)} & \alpha_2^{(n)} & & \bigcirc \\ \alpha_2^{(n)} & \beta_2^{(n)} & \alpha_3^{(n)} & \\ & \cdot & \cdot & \cdot \\ & \cdot & \cdot & \cdot \\ & & \cdot \quad \cdot & \cdot \\ \bigcirc & & \alpha_M^{(n)} & \beta_M^{(n)} \end{pmatrix}, \quad \underline{A}_{n,n-1} = \begin{pmatrix} \delta_1^{(n)} & & & \bigcirc \\ & \delta_2^{(n)} & & \\ & & \cdot & \\ & & & \cdot \\ & & & \cdot \\ \bigcirc & & & \delta_M^{(n)} \end{pmatrix}.$$

Then we may write

$$\underline{L}_{nn} = \begin{pmatrix} \mu_1^{(n)} & & & \bigcirc \\ \lambda_2^{(n)} & \mu_2^{(n)} & & \\ \cdot & \cdot & & \\ & \cdot & \cdot & \\ & & \cdot & \cdot \\ \bigcirc & & \lambda_M^{(n)} & \mu_M^{(n)} \end{pmatrix}, \quad \underline{L}_{n,n-1} = \begin{pmatrix} \sigma_1^{(n)} & & & \bigcirc \\ & \sigma_2^{(n)} & & \\ & & \cdot & \\ & & & \cdot \\ & & & \cdot \\ \bigcirc & & & \sigma_M^{(n)} \end{pmatrix}. \tag{6.113}$$

(When the diagonal blocks of $\underline{A}$ are of different orders, and the off-diagonal blocks regular, then the above forms are easily modified as are Algorithms 6.1 and 6.2 below). The entries of $\underline{L}$ may now be found by comparison of the left and right hand sides of (6.112) for each decomposition in turn. This leads to the following recursion:

Algorithm 6.1

(1) Let n = 1. Set $\sigma_m^{(1)} = \delta_m^{(N+1)} = 0$, m = 1, ..., M.

(2) Compute:
$$\mu_1^{(n)} = (\beta^{(n)} - [\sigma_1^{(n)}]^2)^{\frac{1}{2}} ,$$
$$\sigma_1^{(n+1)} = \delta_1^{(n+1)}/\mu_1^{(n)} .$$

(3) For m = 2, ..., M compute
$$\lambda_m^{(n)} = \alpha_m^{(n)}/\mu_{m-1}^{(n)} ,$$
$$\mu_m^{(n)} = (\beta_m^{(n)} - [\sigma_m^{(n)}]^2 - [\lambda_m^{(n)}]^2)^{\frac{1}{2}} ,$$
$$\sigma_m^{(n+1)} = \delta_m^{(n+1)}/\mu_m^{(n)} .$$

(4) Let n → n+1.

(5) If n ⩽ N repeat steps (2) - (5), otherwise stop.

It is essential that the diagonal elements μ of $\underline{L}$ be non-zero. Should a zero value occur (due to rounding error) then Kershaw (1978) reassigns a non-zero value in Algorithm 6.1. Moreover, square roots may be avoided if the incomplete decomposition $\underline{L}\,\underline{L}^T$ is replaced by $\underline{L}\,\underline{D}\,\underline{L}^T$, where $\underline{L}$ is a lower triangular matrix of the same structure as the lower part of $\underline{A}$ but with unit diagonal elements, and $\underline{D}$ is a diagonal matrix.

Throughout the ICCG(0) algorithm it is necessary to calculate scalar products of the form $\underset{\sim}{s}^T \underset{\sim}{s}$, where $\underset{\sim}{s} = \underline{L}^{-1}\underset{\sim}{r}$. Here $\underset{\sim}{s}$ is calculated efficiently by solving the lower triangular system

$$\underline{L}\,\underset{\sim}{s} = \underset{\sim}{r} \tag{6.114}$$

by forward substitution. Furthermore, evaluation of vectors of the form $\underset{\sim}{t} = (\underline{L}\,\underline{L}^T)^{-1}\underset{\sim}{r} = (\underline{L}^T)^{-1}\underset{\sim}{s}$ is best effected by solving the upper triangular system

$$\underline{L}^T\,\underset{\sim}{t} = \underset{\sim}{s} \tag{6.115}$$

by backward substitution.

The eigenvalues of $(\underline{L}\,\underline{L}^T)^{-1}\underline{A}$ have been computed by Meijerink and Van der Vorst (1977) and Kershaw (1978) for a wide variety of problems. It was found that a large proportion of eigenvalues were indeed closely grouped. This accounts for the rapid convergence rate observed in numerical experiments. The incomplete decomposition described above is closely related to the Strongly Implicit Method of Stone (1968) which is an important and powerful method for solving linear systems, unconnected with conjugate gradients. (See, for example, Gladwell and Wait 1979).

The ICCG(0) algorithm may be applied to the system (6.90c), on which it should perform considerably more efficiently than optimal SOR. (Kershaw 1978). Clearly, the incomplete decomposition of Algorithm 6.1 need be performed once only, and the non-zero elements of $\underline{L}$ then stored. The algorithm is not applicable to the asymmetric systems (6.90a.b) which we turn to next.

(ii) *Asymmetric non-singular matrices*

Again, several preconditioning methods are available (see, for example, Kershaw 1978; Duff 1979; Petravic and Kuo-Petravic 1979), but we shall consider only the generalization of ICCG suggested by Kershaw and used by Court et al and Davies and Manero.

The Cholesky decomposition (6.110) is replaced by the LU decomposition (6.93) where we take unit diagonal elements in the upper triangle $\underline{U}$. Insisting that $\underline{L}$ and $\underline{U}$ have the same structure as the lower and upper parts of $\underline{A}$ then gives rise to the Incomplete LU decomposition Conjugate Gradient algorithm (ILUCG) below :

ILUCG Algorithm

(0) Let m = 0. Choose $\underset{\sim}{x}_0$ and compute $\underset{\sim}{r}_0 = \underset{\sim}{b} - \underline{A}\underset{\sim}{x}_0$,

$$\underset{\sim}{p}_0 = (\underline{U}^T \underline{U})^{-1}\underline{A}^T(\underline{L}\ \underline{L}^T)^{-1}\underset{\sim}{r}_0 .$$

(1) Compute:

$$\alpha_m = (\underset{\sim}{r}_m^T(\underline{L}\ \underline{L}^T)^{-1}\underset{\sim}{r}_m)/(\underset{\sim}{p}_m^T\ \underline{U}^T\ \underline{U}\ \underset{\sim}{p}_m) ,$$

$$\underset{\sim}{x}_{m+1} = \underset{\sim}{x}_m + \alpha_m \underset{\sim}{p}_m ,$$

$$\underset{\sim}{r}_{m+1} = \underset{\sim}{r}_m - \alpha_m\ \underline{A}\ \underset{\sim}{p}_m ,$$

$$\beta_m = (\underset{\sim}{r}_{m+1}^T(\underline{L}\ \underline{L}^T)^{-1}\underset{\sim}{r}_{m+1})/(\underset{\sim}{r}_m^T(\underline{L}\ \underline{L}^T)^{-1}\underset{\sim}{r}_m) ,$$

$$\underset{\sim}{p}_{m+1} = (\underline{U}^T\ \underline{U})^{-1}\underline{A}^T(\underline{L}\ \underline{L}^T)^{-1}\underset{\sim}{r}_{m+1} + \beta_m \underset{\sim}{p}_m .$$

(2) Let m → m+1.

(3) Repeat steps (1) - (2) until convergence is reached.

When $\underline{A}$ is given by (6.107), $\underline{L}$ is again given by (6.111) whereas $\underline{U}$ has the form

$$\underline{U} = \begin{pmatrix} \underline{U}_{11} & \underline{U}_{12} & & \bigcirc \\ & \underline{U}_{22} & \underline{U}_{23} & \\ & \cdot & \cdot & \\ & & \cdot & \cdot \\ & & & \cdot \quad \cdot \\ \bigcirc & & & \underline{U}_{NN} \end{pmatrix},$$

where the block matrices have similar incomplete decompositions to (6.112) wherein the $\underline{L}^T$'s are replaced by $\underline{U}$'s.

If the block matrices of $\underline{A}$ have the tridiagonal and diagonal forms

$$\underline{A}_{nn} = \begin{pmatrix} \beta_1^{(n)} & \gamma_1^{(n)} & & & \bigcirc \\ \alpha_2^{(n)} & \beta_2^{(n)} & \gamma_2^{(n)} & & \\ \cdot & \cdot & \cdot & & \\ & \cdot & \cdot & \cdot & \\ & & \cdot & \cdot & \cdot \\ \bigcirc & & & \alpha_M^{(n)} & \beta_M^{(n)} \end{pmatrix},$$

$$\underline{A}_{n,n-1} = \begin{pmatrix} \delta_1^{(n)} & & & \bigcirc \\ & \delta_2^{(n)} & & \\ & & \cdot & \\ & & & \cdot \\ & & & \quad \cdot \\ \bigcirc & & & \delta_M^{(n)} \end{pmatrix}, \quad \underline{A}_{n-1,n} = \begin{pmatrix} \varepsilon_1^{(n-1)} & & & \bigcirc \\ & \varepsilon_2^{(n-1)} & & \\ & & \cdot & \\ & & & \cdot \\ & & & \quad \cdot \\ \bigcirc & & & \varepsilon_M^{(n-1)} \end{pmatrix},$$

then in addition to (6.113) we may write

$$\underline{U}_{nn} = \begin{pmatrix} 1 & \nu_1^{(n)} & & \bigcirc \\ & 1 & \nu_2^{(n)} & \\ & \cdot & \cdot & \\ & & \cdot & \cdot \\ & & & \cdot \quad \cdot \\ \bigcirc & & & 1 \end{pmatrix}, \quad \underline{U}_{n-1,n} = \begin{pmatrix} \tau_1^{(n-1)} & & & \bigcirc \\ & \tau_2^{(n-1)} & & \\ & & \cdot & \\ & & & \cdot \\ & & & \quad \cdot \\ \bigcirc & & & \tau_M^{(n-1)} \end{pmatrix},$$

where the entries of $\underline{L}$ and $\underline{U}$ may be found from the following recursion :

Algorithm 6.2

(1) Let n = 1. Set $\sigma_m^{(1)} = \tau_m^{(0)} = \delta_m^{(N+1)} = \varepsilon_m^{(N)} = 0$, m = 1, ..., M.

(2) Compute:

$$\mu_1^{(n)} = \beta_1^{(n)} - \sigma_1^{(n)}\tau_1^{(n-1)} ,$$

$$\nu_1^{(n)} = \gamma_1^{(n)}/\mu_1^{(n)} ,$$

$$\sigma_1^{(n+1)} = \delta_1^{(n+1)} ,$$

$$\tau_1^{(n)} = \varepsilon_1^{(n)}/\mu_1^{(n)} .$$

(3) For m = 2, ..., M-1 compute:

$$\lambda_m^{(n)} = \alpha_m^{(n)} ,$$

$$\mu_m^{(n)} = \beta_m^{(n)} - \lambda_m^{(n)}\nu_{m-1}^{(n)} - \sigma_m^{(n)}\tau_m^{(n-1)} ,$$

$$\nu_m^{(n)} = \gamma_m^{(n)}/\mu_m^{(n)} ,$$

$$\sigma_m^{(n+1)} = \delta_m^{(n+1)} ,$$

$$\tau_m^{(n)} = \varepsilon_m^{(n)}/\mu_m^{(n)} .$$

(4) Compute:

$$\lambda_M^{(n)} = \alpha_M^{(n)} ,$$

$$\mu_M^{(n)} = \beta_M^{(n)} - \lambda_M^{(n)}\nu_{M-1}^{(n)} - \sigma_M^{(n)}\tau_M^{(n-1)} ,$$

$$\sigma_M^{(n+1)} = \delta_M^{(n+1)} ,$$

$$\tau_M^{(n)} = \varepsilon_M^{(n)}/\mu_M^{(n)} .$$

(5) Let n → n+1.

(6) If n ⩽ N repeat steps (2) - (5), otherwise stop.

Again, it is important to ensure that the diagonal elements μ do not vanish. In the ILUCG algorithm, the techniques mentioned in conjunction with ICCG should be used to calculate scalar products and vectors of the form $(\underset{\sim}{U}^T \underset{\sim}{U})^{-1} \underset{\sim}{A}^T (\underset{\sim}{L}\, \underset{\sim}{L}^T)^{-1} \underset{\sim}{r}$. On various test problems Kershaw has found that the ILUCG algorithm requires about 1.5 to 2.5 as many iterations as ICCG(0) to converge to the same accuracy, together with about 50% more work per iteration. In comparing ILUCG on the systems (6.90a,b) with ICCG(0) on (6.90c), however, these ratios are likely to be greater because of the additional work in calculating the elements of the coefficient matrices in the former systems.

6.4 SOLUTION OF COUPLED SYSTEMS

Most functional iteration methods for solving the nonlinearly coupled systems (6.90a-c) are based on the following simple algorithm:

(0) Let $r = 0$. Guess an initial stream function vector $\underset{\sim}{\psi}_0$, and set up corresponding vectors $\underset{\sim}{\omega}_0$, $\underset{\sim}{\omega}_0^b$, $\underset{\sim}{S}_0$, $\underset{\sim}{S}_0^e$.

(1) Set up $\underline{A}_r \equiv \underline{A}(\underset{\sim}{\psi}_r;\underset{\sim}{\psi}^b)$ and $\underset{\sim}{b}_r \equiv \underset{\sim}{b}(\underset{\sim}{S}_r,\underset{\sim}{S}_r^e,\underset{\sim}{\psi}_r;\underset{\sim}{\psi}^b)$ in (6.90a) and *approximately* solve the linear equations

$$\underline{A}_r \, \underset{\sim}{S}_{r+1} = \underset{\sim}{b}_r \, . \tag{6.116a}$$

(2) Set up $\underline{B}_r \equiv \underline{B}(\underset{\sim}{\psi}_r;\underset{\sim}{\psi}^b)$ and $\underset{\sim}{c}_r \equiv \underset{\sim}{c}(\underset{\sim}{S}_{r+1},\underset{\sim}{S}_r^e, \underset{\sim}{\omega}_r^b, \underset{\sim}{\omega}_r;\underset{\sim}{\psi}^b)$ in (6.90b) and approximately solve

$$\underline{B}_r \, \underset{\sim}{\omega}_{r+1} = \underset{\sim}{c}_r \, . \tag{6.116b}$$

(3) Set up $\underset{\sim}{d}_r \equiv \underset{\sim}{d}(\underset{\sim}{\omega}_{r+1};\underset{\sim}{\psi}^b)$ in (6.90c) and approximately solve

$$\underline{C} \, \underset{\sim}{\psi}_{r+1} = \underset{\sim}{d}_r \, . \tag{6.116c}$$

(4) Compute the boundary vectors

$$\underset{\sim}{\omega}_{r+1}^b = \underset{\sim}{\omega}^b(\underset{\sim}{\psi}_{r+1};\underset{\sim}{\psi}^b) \, , \tag{6.117a}$$

$$\underset{\sim}{S}_{r+1}^e = \underset{\sim}{S}^e(\underset{\sim}{\omega}_{r+1}^b, \underset{\sim}{\psi}_{r+1};\underset{\sim}{\psi}^b) \, . \tag{6.117b}$$

(5) Let $r \to r+1$.

(6) Repeat steps (1) - (5) until all vectors have converged.

In this algorithm the linearization is effected simply by decoupling the systems at each step; the functional iteration may be classified rather broadly as of *Picard*-type. The rate of convergence is at best linear.

It is possible to solve approximately the three subsystems in (6.116a) separately, incorporating in the right hand vectors $\underset{\sim}{b}^{xy}$ and $\underset{\sim}{b}^{yy}$ of the second and third subsystems the most recent stress values $\underset{\sim}{S}_{r+1}^{xx}$ and $\underset{\sim}{S}_{r+1}^{xy}$ obtained from the first and second subsystems, respectively. The precise effect of this modification on the rate of convergence of the algorithm, however, does not appear to have been documented.

If the linear equations (6.116a-c) are themselves solved by iterative methods, then steps (1) - (3) will each involve a sequence of *inner* iterations, which for clarity we label by m (see below). For example, in solving (6.116a) one inner iteration might be a sweep of GS iteration, or alternatively one iteration of the ILUCG algorithm, whereas in solving (6.116c) one inner iteration might be a sweep of SOR or one iteration of ICCG(0). The iterations labelled by r in the above algorithm are then called *outer* iterations, and the rate of convergence of the algorithm is that of the outer iterations.

It is never necessary, nor indeed advisable, to continue a sequence of inner iterations to convergence, i.e. to solve a linear system exactly. The rate of convergence of the algorithm is often seriously affected by the number of inner iterations requested at each step. Let us express the approximate solution stages in steps (1) - (3) in the following general terms :

(1)' Define $\underset{\sim}{S}_{r+1}^{[0]} = \underset{\sim}{S}_r$ and perform m_1 inner iterations on (6.116a) such that

$$\| \underset{\sim}{S}_{r+1}^{[m_1]} - \underset{\sim}{S}_{r+1}^{[m_1-1]} \| \leqslant \varepsilon_1 \| \underset{\sim}{S}_{r+1}^{[m_1-1]} \|$$

for some prescribed tolerance ε_1. Define $\underset{\sim}{S}_{r+1} = \underset{\sim}{S}_{r+1}^{[m_1]}$.

(2)' Define $\underset{\sim}{\omega}_{r+1}^{[0]} = \underset{\sim}{\omega}_r$ and perform m_2 inner iterations on (6.116b) such that

$$\| \underset{\sim}{\omega}_{r+1}^{[m_2]} - \underset{\sim}{\omega}_{r+1}^{[m_2-1]} \| \leqslant \varepsilon_2 \| \underset{\sim}{\omega}_{r+1}^{[m_2-1]} \|$$

for some prescribed tolerance ε_2. Define $\underset{\sim}{\omega}_{r+1} = \underset{\sim}{\omega}_{r+1}^{[m_2]}$.

(3)' Define $\underset{\sim}{\psi}_{r+1}^{[0]} = \underset{\sim}{\psi}_r$ and perform m_3 inner iterations on (6.116c) such that

$$\| \underset{\sim}{\psi}_{r+1}^{[m_3]} - \underset{\sim}{\psi}_{r+1}^{[m_3-1]} \| \leqslant \varepsilon_3 \| \underset{\sim}{\psi}_{r+1}^{[m_3-1]} \|$$

for some prescribed tolerance ε_3. Define $\underset{\sim}{\psi}_{r+1} = \underset{\sim}{\psi}_{r+1}^{[m_3]}$.

Clearly, the numbers m_j, j = 1,2,3, are determined by the choice of tolerances ε_j, but it is difficult to give precise rules for choosing these tolerances so as to optimize the rate of convergence. For rough guidelines, we summarize below a few strategies used by various authors, but we do not claim that any of them are optimal.

Davies et al (1979) suggest that ε_j should be chosen in proportion to the discretization errors of the corresponding variables. Employing upwind-downwind differencing for the constitutive equations and the UD1 scheme for the vorticity equation, it was found that rough estimates of discretization errors led to the choice $(\varepsilon_1, \varepsilon_2, \varepsilon_3) = (10^{-3}, 10^{-1}, 10^{-5}) \times$ constant. (The constant was taken as unity.) Using under-relaxation in step (1)', GS in step (2)', and over-relaxation in step (3)', attempts to introduce these ε-values from the start of a computer run, however, caused the outer iterations to diverge. It was necessary to demand higher values of ε_j for the first few values of r and to reduce the ε_j to the required values as r increased. Within the functional iteration scheme it was also found more efficient to use under-relaxation (with an empirically chosen parameter) than GS in solving the discretized constitutive equations. The convergence criterion used in step (6) was that all successive outer iterates should satisfy the same norm inequalities as the successive inner iterates, i.e. $m_j = 1$, $j = 1,2,3$, for the final values of ε_j above. The maximum norm was used, which facilitated computation since max $|\psi_{j,k}|$ and max $|\omega_{j,k}|$ occurred on the boundary (cf. the maximum principle).

Holstein (1981) in calculations on axisymmetric flows keeps high ε-values throughout the algorithm so that only one inner iteration of GS is used on stresses, and no more than two inner iterations each of GS on vorticity and SOR on stream function. Holstein's convergence criterion in step (6) is a test on vorticity of the form

$$\| \underset{\sim}{\omega}_{r+1} - \underset{\sim}{\omega}_r \| \leqslant \varepsilon_\omega \| \underset{\sim}{\omega}_r \|$$

for a low tolerance ε_ω.

Manero (1980) has compared the effect of various choices of ε_j on the rapidity of convergence of functional iteration. In calculations restricted to planar flow through an L-shaped geometry, he used ILUCG in steps (1)$'$ and (2)$'$, and ICCG(0) in step (3)$'$. (It should be noted here that since the matrices $\underset{-}{A}_r$ and $\underset{-}{B}_r$ change with r, their incomplete LU decompositions must be recomputed at each step (1)$'$ and (2)$'$. This clearly affects the overall efficiency of the algorithm. The incomplete Cholesky decomposition of $\underset{-}{C}$ in step (3)$'$, on the other hand, does not need recomputing.) Manero found that rapid convergence was achieved by allowing more than one inner iteration in step (1)$'$, but only one inner iteration in each of steps (2)$'$ and (3)$'$. His final convergence criterion was the same as that of Davies et al with tolerances based on discretization error and taken to be $(\varepsilon_1, \varepsilon_2, \varepsilon_3) = (10^{-2}, 10^{-3}, 10^{-4})$. (A CD scheme was used for the vorticity equation). Restricting each step to one inner iteration only was comparatively inefficient.

Several authors have suggested that internal or boundary variables be smoothed after each outer iteration, in the form

$$\underset{\sim}{x}_{r+1} \to (1 - \beta_x)\underset{\sim}{x}_{r+1} + \beta_x \underset{\sim}{x}_r , \qquad \beta_x \in [0,1] ,$$

where $\underset{\sim}{x}$ denotes one of the vector variables $\underset{\sim}{S}$, $\underset{\sim}{S}^e$, $\underset{\sim}{\omega}$, $\underset{\sim}{\omega}^b$ or $\underset{\sim}{\psi}$, and β_x may depend on $\underset{\sim}{x}$. Apart from the smoothing of boundary vorticity in conjunction with second-order formulae (§6.2.3) it is not at all clear whether smoothing is actually necessary if a suitable choice of tolerances has been made in the above algorithm. (Crochet and Pilate 1976; Davies et al 1979; Manero 1980). As a general rule it is as well to avoid smoothing unless a gain in the overall rate of convergence is manifest, or, for comparatively large values of R or W, convergence is otherwise impossible to achieve (Townsend 1980b).

Finally, we mention that the choice of initial vectors in step (0) is of considerable importance. For low values of the elasticity parameter W, the quality of the initial guesses can affect quite markedly the number of outer iterations executed before convergence, whereas for higher values of W this quality also dictates whether the algorithm converges at all. *Continuation* with respect to the W parameter is advisable, and may be described briefly as follows.

Suppose that the Newtonian problem has been solved by omitting step (1) from the algorithm. Then for a problem with $W > 0$ we may use the Newtonian solution as an initial guess. If the algorithm fails to converge for this particular value of W, or alternatively the convergence is unacceptably slow, then the interval $[0,W]$ may be subdivided into $[0,W_1]$, $[W_1,W_2]$, ..., $[W_k,W]$, say, where $0 < W_1 < \ldots < W_k < W$, and the problem solved for successive values of W_j using the previous solution as an initial guess for the next problem. The continuation procedure often works provided the steps $W_j - W_{j-1}$ are not too large. A rigorous theory of continuation methods may be found in Ortega and Rheinboldt (1970) and Wacker (1978).

Walters and Webster (1982) use continuation with respect to both R and W parameters, which allows them to reach higher values of R and W than could be attained by continuation with respect to W with R fixed.

Unfortunately, for all practical viscoelastic flow problems there exist critical values of W above which the algorithm described fails to converge. As a critical value is approached, the steps in the continuation scheme become uneconomically small, and the radius of convergence of the nonlinear iteration tends to zero. This breakdown is not confined to the above algorithm; a limit on W is common to most published work on the numerical simulation of viscoelastic flows. It applies to finite difference and finite element methods, to Picard-type and Newton-type iteration schemes, to differential and integral constitutive equations, and to flow problems with and without abrupt changes in geometry. Some possible causes of the breakdown are discussed in Chapter 11.

A Newton algorithm

Let $\underset{\sim}{X}$ and $\underset{\sim}{F}(\underset{\sim}{X})$ denote the partitioned vectors

$$\underset{\sim}{X} = (\underset{\sim}{S}, \underset{\sim}{\omega}, \underset{\sim}{\psi})^T \quad ,$$

$$\underset{\sim}{F}(\underset{\sim}{X}) = (\underline{\underset{\sim}{A}}\underset{\sim}{S} - \underset{\sim}{b}, \underline{\underset{\sim}{B}}\underset{\sim}{\omega} - \underset{\sim}{c}, \underline{\underset{\sim}{C}}\underset{\sim}{\psi} - \underset{\sim}{d})^T \; . \qquad (6.118)$$

Then the Jacobian matrix of $\underset{\sim}{F}$ has the block structure

$$\nabla\underset{\sim}{F} \equiv \underline{J} = \begin{pmatrix} \underline{A} - \dfrac{\partial \underset{\sim}{b}}{\partial \underset{\sim}{S}}, & \underline{0}, & \dfrac{\partial \underline{A}}{\partial \underset{\sim}{\psi}} \underset{\sim}{S} - \dfrac{\partial \underset{\sim}{b}}{\partial \underset{\sim}{\psi}} \\ -\dfrac{\partial \underset{\sim}{c}}{\partial \underset{\sim}{S}}, & \underline{B}, & \dfrac{\partial \underline{B}}{\partial \underset{\sim}{\psi}} \underset{\sim}{\omega} - \dfrac{\partial \underset{\sim}{c}}{\partial \underset{\sim}{\psi}} \\ \underline{0}, & -\dfrac{\partial \underset{\sim}{d}}{\partial \underset{\sim}{\omega}}, & \underline{C} \end{pmatrix}, \tag{6.119}$$

in which the matrices $\dfrac{\partial \underset{\sim}{b}}{\partial \underset{\sim}{S}}$ and $\dfrac{\partial \underline{A}}{\partial \underset{\sim}{\psi}} \underset{\sim}{S}$, for example, have elements

$$\left(\frac{\partial \underset{\sim}{b}}{\partial \underset{\sim}{S}}\right)_{mn} = \frac{\partial b_m}{\partial S_n}, \qquad \left(\frac{\partial \underline{A}}{\partial \underset{\sim}{\psi}} \underset{\sim}{S}\right)_{mn} = \sum_{\ell} \left(\frac{\partial A_{m\ell}}{\partial \psi_n}\right) S_{\ell},$$

where b_m denotes the mth element of $\underset{\sim}{b}$, etc.

A Newton algorithm for solving the system $\underset{\sim}{F}(\underset{\sim}{X}) = \underset{\sim}{0}$ is as follows:

(0) Let $r = 0$. Set up initial vectors $\underset{\sim}{X}_0$, $\underset{\sim}{\omega}_0^b$ and $\underset{\sim}{S}_0^e$.

(1) Set up $\underline{J}_r \equiv \underline{J}(\underset{\sim}{X}_r)$ and $\underset{\sim}{F}_r \equiv \underset{\sim}{F}(\underset{\sim}{X}_r)$. Solve

$$\underline{J}_r \, \delta\underset{\sim}{X}_r = -\underset{\sim}{F}_r . \tag{6.120}$$

(2) Compute

$$\begin{aligned} \underset{\sim}{X}_{r+1} &= \underset{\sim}{X}_r + \delta\underset{\sim}{X}_r , \\ \underset{\sim}{\omega}^b_{r+1} &= \underset{\sim}{\omega}^b(\underset{\sim}{\psi}_{r+1};\underset{\sim}{\psi}^b) , \\ \underset{\sim}{S}^e_{r+1} &= \underset{\sim}{S}^e(\underset{\sim}{\omega}_{r+1},\underset{\sim}{\psi}_{r+1};\underset{\sim}{\psi}^b). \end{aligned}$$

(3) Let $r \to r+1$.

(4) Repeat steps (1) - (3) until convergence is reached.

Each block in the Jacobian (6.119) is itself a block tridiagonal matrix, so that (6.120) may be solved by a direct method for banded systems, or alternatively by a preconditioned CG method for non-symmetric systems.

A continuation scheme may efficiently be incorporated within a Newton algorithm as follows. Consider an increment $\delta W_j = W_{j+1} - W_j$. Then if $\underset{\sim}{X}(W)$ denotes a solution of the system $\underset{\sim}{F}(\underset{\sim}{X};W) = \underset{\sim}{0}$, we have, approximately,

$$\underset{\sim}{X}(W_{j+1}) = \underset{\sim}{X}(W_j) + \frac{\partial \underset{\sim}{X}}{\partial W}(W_j)\,\delta W_j , \tag{6.121}$$

where

$$\underline{J}(\underset{\sim}{X};W)\frac{\partial \underset{\sim}{X}}{\partial W} = -\frac{\partial \underset{\sim}{F}}{\partial W} . \tag{6.122}$$

When the Newton sequence $\{\underset{\sim}{X}_r\}$ has almost converged with $W = W_j$, then the Jacobian in (6.120) is the same as that in (6.122) which may be solved to give $\frac{\partial \underset{\sim}{X}}{\partial W}(W_j)$. A good approximation to the solution of $\underset{\sim}{F}(\underset{\sim}{X};W_{j+1}) = \underset{\sim}{0}$ may thus be found from (6.121), used in step (0), and the process repeated.

6.5 EXAMPLES

To illustrate what may be achieved by the methods described in this chapter, we look at the numerical simulation of two flows which have been studied experimentally in the laboratory. The first example will be that of mixing-and-separating flow in a rectangular channel, produced by the insertion of two barriers AD and BG of finite thickness AA' = BB', with a gap AB between them (Fig. 6.8(a)). The flow directions and flow rates in the inlet and outlet arms can be varied in numerous ways. A second example will be flow over a deep hole (Fig. 6.8(b)). Both flows are planar, and we will accept that no three-dimensional effects are present in the laboratory experiments.

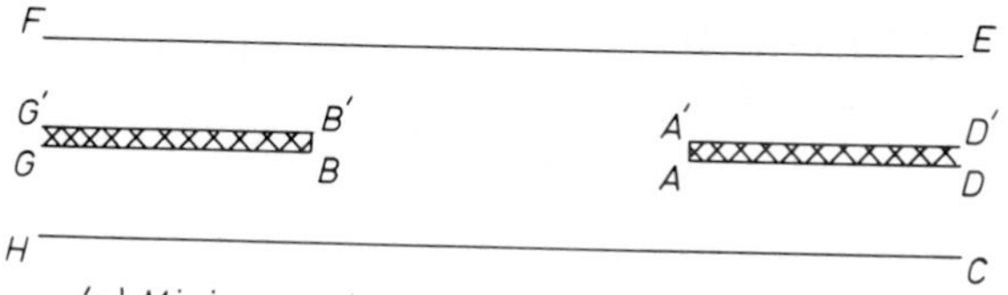

(a) Mixing and separating flow geometry.

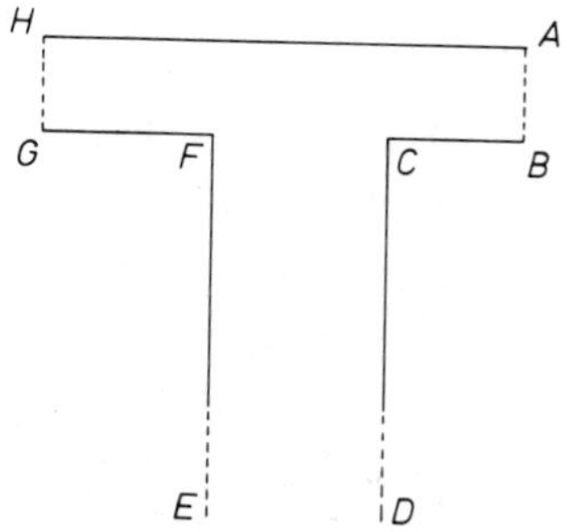

(b) Deep hole geometry.

Fig. 6.8

6.5.1 Mixing-and-separating flow

We shall use a Maxwell model to simulate the behaviour of a Boger test fluid, details of which may be found in the paper of Cochrane et al. (1981). In non-dimensionalizing the variables (see (3.18)) and estimating the parameters R and W ((3.14), (3.15)), the characteristic length L is taken as the inlet channel width CD (= D'E) while the characteristic velocity U is the flow rate into this inlet arm divided by its cross-sectional area.

A square mesh is chosen with eleven grid-lengths across the wide channel CE; the barrier thickness AA' occupies one grid-length, allowing five across each of the narrow channels. This gives approximately 2500 grid-points over the whole geometry.

Fully developed Poiseuille flow conditions are imposed at the inlets CD and FG' and the outlets D'E and GH. Over CD, for example, we have the velocity profile

$$u = 6y(1 - y) , \quad v = 0 , \quad 0 \leqslant y \leqslant 1 ,$$

or equivalently, $\psi = y^2(3 - 2y) , \quad 0 \leqslant y \leqslant 1 .$

The mean flow rate across CD, as given by $\psi_D - \psi_C$, is unity as required. The vorticity on CD is

$$\omega = - \frac{\partial u}{\partial y} = 6(2y - 1) ,$$

while the non-Newtonian stress components are

$$S^{xx} = 2W\left(\frac{\partial u}{\partial y}\right)^2 , \quad S^{xy} = S^{yy} = 0 .$$

Over all solid boundaries, the no-slip condition u = v = 0 is used; the vorticity values are computed from the first-order formula (6.43) and the boundary stress values are found from (6.84) and (6.85). At the four re-entrant corners, Kawaguti's method is used to treat the singularities in vorticity (see (6.76)), while singular stress components are treated by the method discussed in §6.2.5(iii).

The vorticity equation is discretized using the CD scheme of (6.17) and (6.18), with a switch to UD1 (6.20) when the grid Reynolds number condition (6.19) is violated. The constitutive equations are discretized in the form (6.38(a-c)) and the coupled matrix systems are solved by the functional iteration method of §6.4, using inner GS iterations for systems (6.116(a) and (b)) with inner SOR iterations for (6.116(c)). Prescribed tolerances for convergence of the vectors ($\underline{S}$, $\underline{\omega}$, $\underline{\psi}$) are estimated from the local discretization errors of the finite difference formulae used. The values (10^{-1}, 10^{-2}, 10^{-4}) are found to be appropriate, but it is necessary to demand higher values at early stages of the functional iteration, reducing these values to the required tolerances as the outer iterations progress. No explicit smoothing of the outer iterates is used

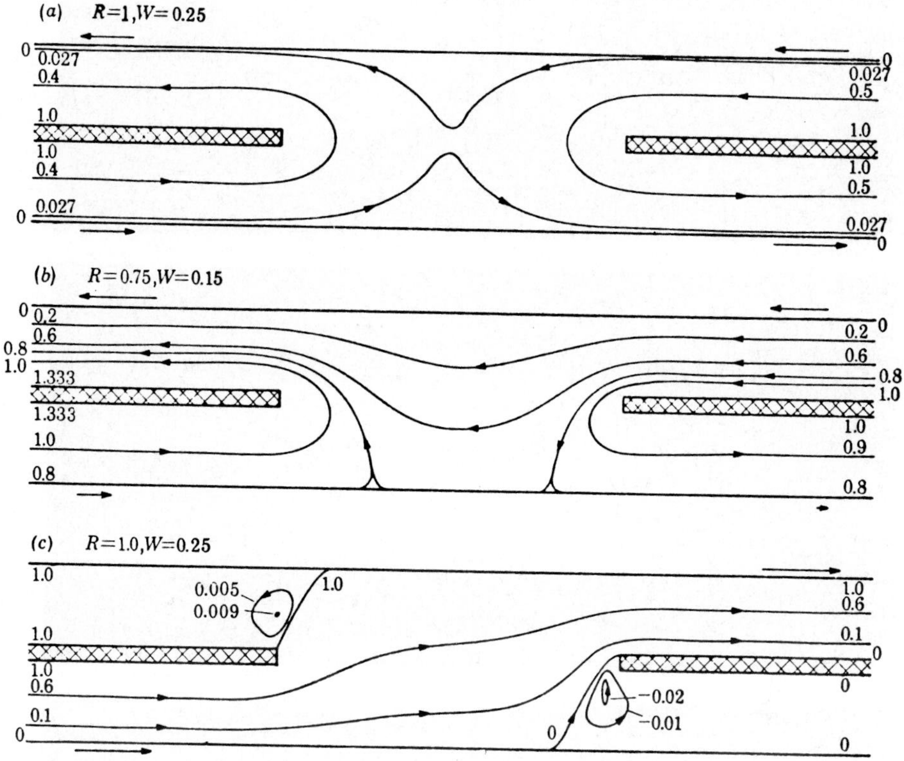

Fig. 6.9 Numerical simulation for combined mixing and separating flow with relative flow rate in each arm indicated by length of arrows.

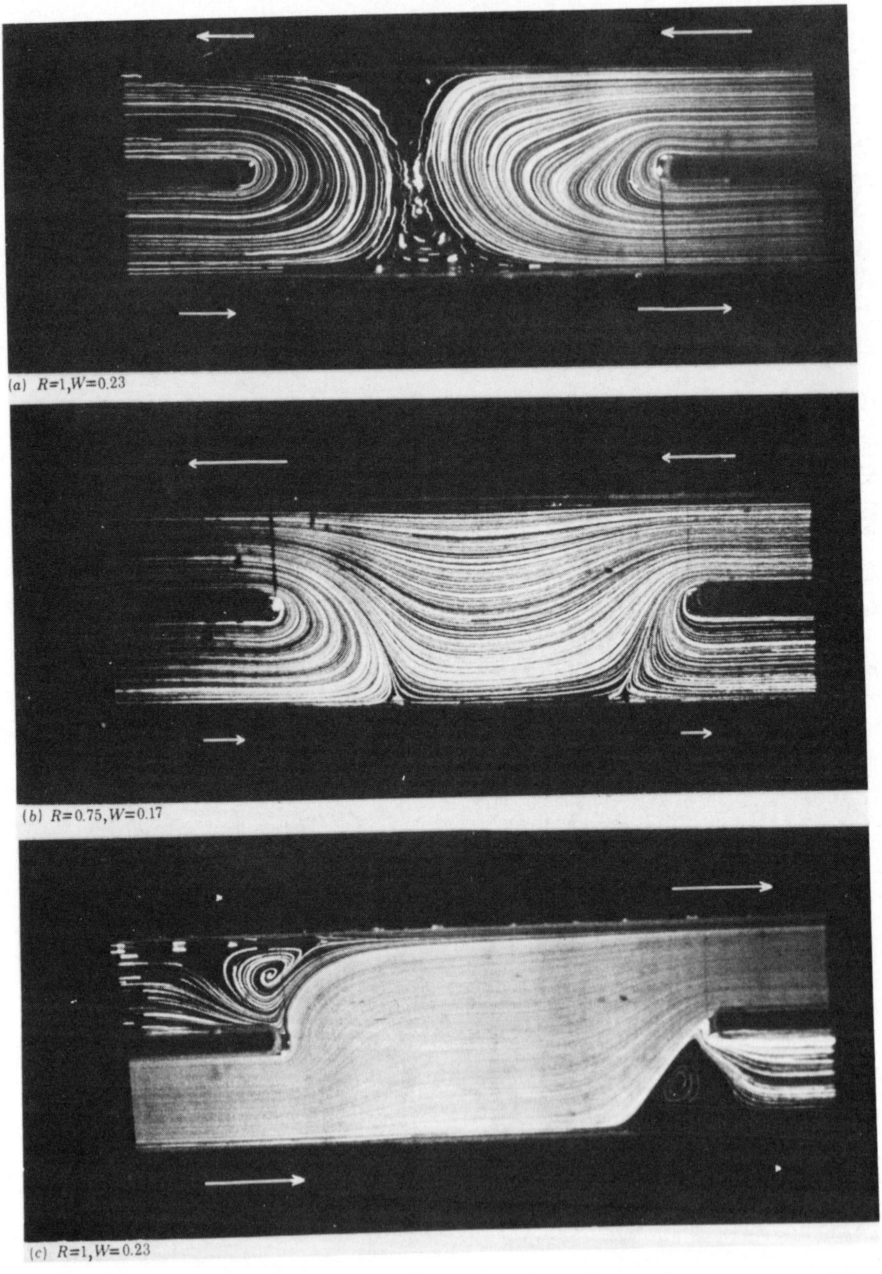

Fig. 6.10 Combined mixing and separating flow: Boger test fluid; gap AB = 25 mm; relative flow rates indicated by lengths of arrows; Reynolds number and Weissenberg number based on flow rate measurement in upper right arm.

in this example, but it is useful as W increases to under-relax the vorticity iterations, and in some cases the stress iterations, in order to facilitate overall convergence. This is an implicit form of smoothing.

Fig. 6.9(a-c) shows numerically simulated stream-function contours corresponding to three different flow patterns induced by different flow rates and directions. The flow rate in each arm is indicated by the length of the corresponding arrow. A comparision of Fig. 6.9 with Fig. 6.10, wherein flow visualization photographs of a Boger test fluid are shown (Cochrane et al. 1981), indicates very good agreement between numerical simulation and experiment in all three cases, with even the finer details of the flow described.

The numerical algorithm fails to converge in this example for an elasticity parameter above $W \simeq 0.3$. There is, however, nothing to indicate any breakdown of flow pattern for $W > 0.3$ in the experiments with the Boger fluid.

6.5.2 Flow over a deep hole

In this example the characteristic length L is taken to be the width of the channel AB (Fig. 6.8(b)) while U is the mean velocity across AB. Fully developed Poiseuille flow conditions are imposed over AB and GH. As before, a square mesh is used, with ten grid-lengths across AB.

Experimental data are available for W-values in the range $0 \leq W \leq 0.75$ (Cochrane et al. 1981) and four cases are shown in Fig. 6.11. The initial symmetry which is known to be present when $R = W = 0$ is destroyed by the opposing influences of inertia and elasticity, and there is a tendency for the vortex in elastic liquids to retreat deeper into the hole.

Numerical simulations for the Newtonian cases $R = 3$ and $R = 6$ are shown in Fig. 6.12(a) and (c), where agreement with experiment is excellent. The numerical algorithm of the previous example, however, fails to yield a converged solution for $W = 0.75$. Even for $W = 0.38$ the Maxwell model fails to give a solution, but convergence is attainable at this value of W for an Oldroyd 4-constant model (2.78). For this model, W is given by $W = (\lambda_1 - \lambda_2)U/L$, and the time constants λ_2 and μ_0 are normalized by multiplication by U/L. The convergence of the numerical algorithm for the Oldroyd model at $W = 0.38$ may be attributed to its variable viscosity behaviour.

The numerical results are shown in Fig. 6.12(b) and (d). There is no quantitative agreement with experiment, only a trend in the right direction. This would suggest inadequacy of the model.

We have not examined the effects of mesh refinement in this example, due to the limitations of computer storage. In general, such effects should be carefully considered before any great weight is attached to the finer details of numerical simulation. In particular, care should be taken to distinguish between

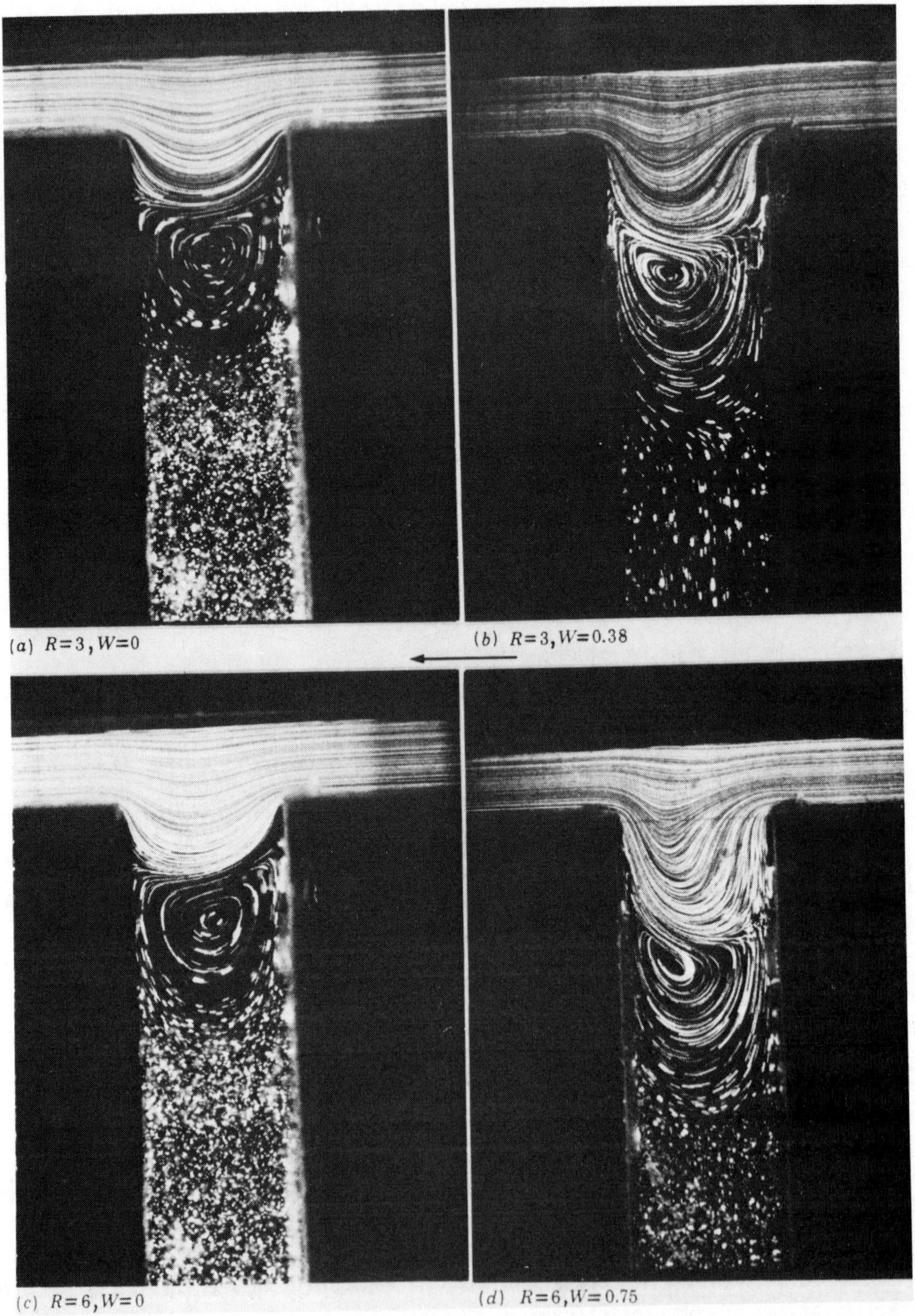

Fig. 6.11 Flow over a deep hole: (a) and (c) Newtonian, (b) and (d) Boger test fluid.

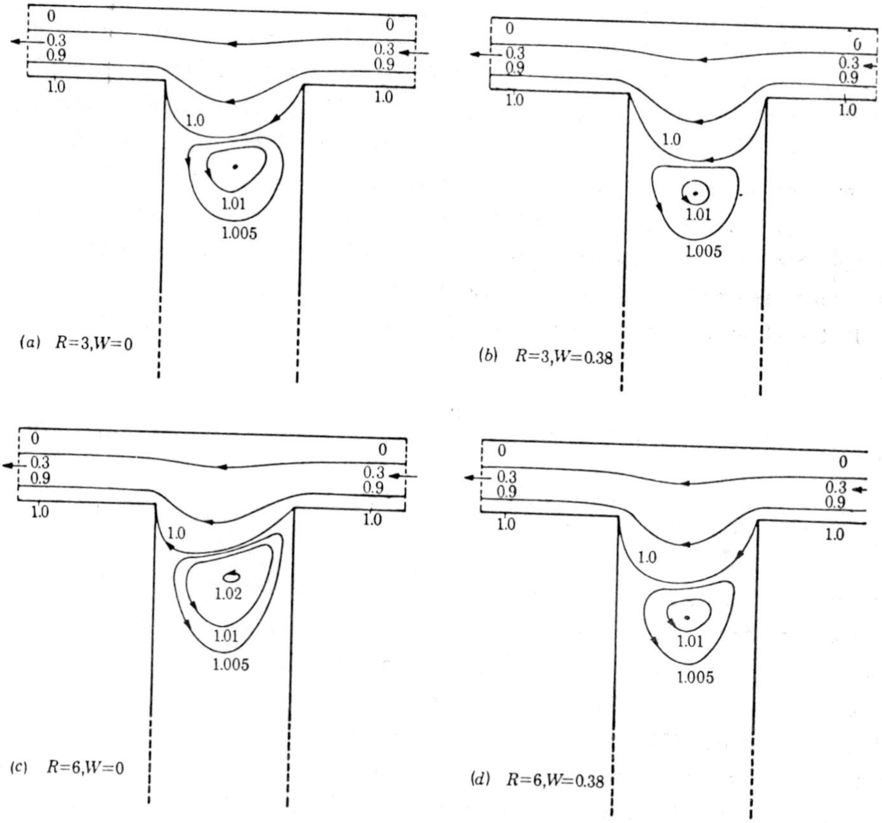

Fig. 6.12 Numerical simulation for flow over a deep hole, model (2·78) being used in (b) and (d).
Non-dimensional values of λ_1, λ_2 and μ_0 are $\lambda_1 = 0{\cdot}45$, $\lambda_2 = 0{\cdot}075$, $\mu_0 = 0{\cdot}0007$.

artifacts resulting from discretization error in non-Newtonian simulation and true non-Newtonian behaviour. (See, for example, Mendelson et al. 1982; Davies et al. 1983).

6.6 MISCELLANEOUS TOPICS

6.6.1 Pressure recovery

One of the advantages of working with the (ψ, ω, $\underline{S}$)-formulation of the governing equations is that the number of variables is reduced. Pressure was eliminated from the momentum equations (3.20) and (3.21) to give (3.28), which in turn was solved in the form (6.5) using the decomposition (6.4). We now consider the problem of recovering the pressure from our finite difference solution in terms of ψ, ω and $\underline{S}$. Even in the Newtonian case the recovery process can be susceptible to large numerical errors; the problem is more acute in the non-Newtonian case and is at present only partially resolved.

In the absence of body forces we may express the pressure gradients in (3.20) and (3.21) as

$$\frac{\partial p}{\partial x} = \nabla^2 u - R\left(u\frac{\partial u}{\partial x} + v\frac{\partial u}{\partial y}\right) + \frac{\partial S^{xx}}{\partial x} + \frac{\partial S^{xy}}{\partial y} \quad , \tag{6.123}$$

$$\frac{\partial p}{\partial y} = \nabla^2 v - R\left(u\frac{\partial v}{\partial x} + v\frac{\partial v}{\partial y}\right) + \frac{\partial S^{xy}}{\partial x} + \frac{\partial S^{yy}}{\partial y} \quad . \tag{6.124}$$

To obtain a pressure solution we may be tempted to start at an arbitrary point in the geometry at which we have assigned an arbitrary pressure value, and numerically integrate the right hand sides of (6.123) and (6.124) along grid-lines parallel to the x and y axes respectively. In the Newtonian case ($\underline{S} = \underline{0}$), that this method will give different answers when different paths are used to get to the same point is well documented (Roache 1976). The reason is in part due to numerical differentiation errors in computing the higher derivatives of ψ in (6.123) and (6.124), and in part due to the quadrature error associated with the rule chosen for numerical integration. Both these errors are path dependent. In the non-Newtonian case, numerical differentiation errors in computing the extra stress gradients magnify the difficulties involved. In all cases, direct numerical integration of (6.123) and (6.124) is especially susceptible to error when the path of integration is close to a sharp corner.

Existing work on non-Newtonian pressure calculation using finite differences contains some examples of estimating pressure variation along lines in the flow (Townsend 1980a; Tiefenbruck and Leal 1982), essentially using a direct integration approach. Townsend uses an averaging process to minimize numerical error.

In general, a more accurate solution can be determined from the Poisson form of the pressure equation, obtained from (6.123), (6.124) and the continuity equation (3.19) :

$$\nabla^2 p = 2R\left[\frac{\partial u}{\partial x}\frac{\partial v}{\partial y} - \frac{\partial v}{\partial x}\frac{\partial u}{\partial y}\right] + \frac{\partial^2 S^{xx}}{\partial x^2} + 2\,\frac{\partial^2 S^{xy}}{\partial x \partial y} + \frac{\partial^2 S^{yy}}{\partial y^2}$$

$$= G \text{ , say .} \tag{6.125}$$

This equation is subject to the Neumann boundary condition

$$\frac{\partial p}{\partial n} = g \ , \tag{6.126}$$

where n is normal to the wall and g is given by either (6.123) or (6.124), whichever is appropriate. For a no-slip wall, for example, where s is tangential to the wall, we have

$$\frac{\partial p}{\partial n} = -\frac{\partial \omega}{\partial s} + \frac{\partial S^{nn}}{\partial n} + \frac{\partial S^{ns}}{\partial s} \ . \tag{6.127}$$

Equations (6.125) and (6.126) may be treated numerically as a special case of the mixed boundary value problem discussed in §5.4.2 with the Dirichlet condition absent. The resulting system of linear equations is then singular, which reflects the fact that the pressure is undetermined up to an arbitrary constant. The system may be rendered non-singular by fixing the pressure at a convenient grid-point, thereby reducing the dimension of the system by one. Even so, the resulting matrix system will be much less well-conditioned than that obtained from truly mixed boundary conditions or a pure Dirichlet problem.

This means that classical iterative methods such as SOR often have difficulty in converging. The problem is further exacerbated by the breakdown, in its discrete form, of the compatibility condition

$$\int_{\Gamma} g \, d\Gamma = \int_{\Omega} G \, d\Omega \ , \tag{6.128}$$

which is necessary for a Poisson equation with pure Neumann conditions to possess a solution. It is not unusual for SOR iterations to drift slowly but endlessly. In the Newtonian case, remedies based on satisfying the constraint (6.128) have been proposed by Briley (1974) and Ghia et al. (1977). Webster (1979) has demonstrated, however, that none of these remedies is particularly successful in non-Newtonian calculations. On the other hand, Manero (1981) has shown that the problem of drift does not arise if preconditioned conjugate gradient methods are used to solve the discrete systems.

The problem of numerical error, however, still remains. Manero has compared ICCG(0) solutions of non-Newtonian pressure fields computed from (6.125) and

(6.126) with those computed from (6.125) subject to Dirichlet conditions. These latter conditions were obtained by direct numerical integration of (6.123) or (6.124) along a computational boundary to provide pressure values there. The computational boundary was parallel to the actual boundary but one grid-length interior to it. This avoids integration over sharp corners, but of course does not escape the influence of such corners on numerical variables in its vicinity. Manero found that the numerical pressure fields calculated from different boundary conditions deviated significantly from each other. This is not particularly surprising since the source term G in (6.125) is highly susceptible to errors of numerical differentiation, as are the boundary pressures computed by direct integration.

Further research on the recovery of pressure fields from non-Newtonian finite difference solutions in terms of ψ, ω and $\underset{\sim}{S}$ should therefore be directed towards reducing the noise levels inherent in the source terms G and g of (6.125) and (6.126).

6.6.2 Axisymmetric flow

Finite difference approximation in terms of cylindrical polar coordinates requires special mention, and in this section we briefly describe the treatment of axisymmetric flows. Let (r,θ,z) denote cylindrical polar coordinates, and let u and w denote radial and axial velocity components, respectively. There are four components of stress, namely, T^{zz}, T^{rz}, T^{rr} and $T^{\theta\theta}$.† Adopting the decomposition (6.4), the governing equations for the flow of a Maxwell fluid in terms of the six variables ψ, ω, S^{zz}, S^{rz}, S^{rr} and $S^{\theta\theta}$ are then :

the stream function equation

$$\frac{1}{r}\left(\frac{\partial^2\psi}{\partial r^2} + \frac{\partial^2\psi}{\partial z^2} - \frac{1}{r}\frac{\partial\psi}{\partial r}\right) = -\omega \;; \tag{6.129}$$

the vorticity equation

$$\frac{\partial^2\omega}{\partial r^2} + \frac{\partial^2\omega}{\partial z^2} + \frac{1}{r}\frac{\partial\omega}{\partial r} - \frac{\omega}{r^2} - R\left(u\frac{\partial\omega}{\partial r} + w\frac{\partial\omega}{\partial z} - \frac{u}{r}\omega\right)$$
$$= \left(\frac{\partial^2}{\partial r^2} - \frac{\partial^2}{\partial z^2} + \frac{1}{r}\frac{\partial}{\partial r} - \frac{1}{r^2}\right)S^{rz} + \frac{\partial^2}{\partial r\partial z}(S^{zz} - S^{rr}) + \frac{1}{r}\frac{\partial}{\partial z}(S^{\theta\theta} - S^{rr}); \tag{6.130}$$

and the constitutive equations

---

† As in previous sections, upper index notation here does not imply contravariant tensors. In this section stress components are physical.

$$\left.\begin{aligned}
A_1 S^{zz} + WLS^{zz} &= 2BS^{rz} + G_1 \\
A_2 S^{rz} + WLS^{rz} &= CS^{zz} + BS^{rr} + G_2 \\
A_3 S^{rr} + WLS^{rr} &= 2CS^{rz} + G_3 \\
A_4 S^{\theta\theta} + WLS^{\theta\theta} &= + G_4
\end{aligned}\right\} , \tag{6.131}$$

where

$$\left.\begin{aligned}
&L \equiv u\frac{\partial}{\partial r} + w\frac{\partial}{\partial z} , \\
&A_1 = 1 - 2W\frac{\partial w}{\partial z} , \qquad A_2 = 1 + W\frac{u}{r} , \\
&A_3 = 1 - 2W\frac{\partial u}{\partial r} , \qquad A_4 = 1 - 2W\frac{u}{r} , \\
&B = W\frac{\partial w}{\partial r} , \qquad C = W\frac{\partial u}{\partial z} , \\
&G_1 = -2W\left[L\left(\frac{\partial w}{\partial z}\right) - 2\left(\frac{\partial w}{\partial z}\right)^2 - \left(\frac{\partial u}{\partial z} + \frac{\partial w}{\partial r}\right)\frac{\partial w}{\partial r}\right] , \\
&G_2 = -W\left[\left(L + \frac{u}{r}\right)\left(\frac{\partial u}{\partial z} + \frac{\partial w}{\partial r}\right) - 2\left(\frac{\partial w}{\partial r}\frac{\partial u}{\partial r} + \frac{\partial w}{\partial z}\frac{\partial u}{\partial z}\right)\right] , \\
&G_3 = -2W\left[L\left(\frac{\partial u}{\partial r}\right) - 2\left(\frac{\partial u}{\partial r}\right)^2 - \left(\frac{\partial u}{\partial z} + \frac{\partial w}{\partial r}\right)\frac{\partial u}{\partial z}\right] , \\
&G_4 = -2W\left[L\left(\frac{u}{r}\right) - 2\left(\frac{u}{r}\right)^2\right] .
\end{aligned}\right\} \tag{6.132}$$

The velocities and stream-function are related by

$$u = -\frac{1}{r}\frac{\partial\psi}{\partial z} , \qquad w = \frac{1}{r}\frac{\partial\psi}{\partial r} , \tag{6.133}$$

and the continuity equation (3.19) has the form

$$\frac{\partial u}{\partial r} + \frac{\partial w}{\partial z} + \frac{u}{r} = 0 . \tag{6.134}$$

The relevant boundary conditions may be set up using the same principles as described in §6.1.1 and §§6.2.3 - 6.2.5, but the central axis of symmetry requires special consideration. At $r = 0$ we have the symmetry conditions

$$\psi = \text{constant}, \quad u = 0 , \quad \frac{\partial u}{\partial z} = \frac{\partial w}{\partial r} = \omega = 0 . \tag{6.135}$$

From L'Hôpital's rule we may also deduce

$$\lim_{r\to 0} \frac{u}{r} = \left.\frac{\partial u}{\partial r}\right|_{r=0} . \tag{6.136}$$

It then follows from (6.131) and (6.132) that

$$S^{rz} = 0 , \quad \text{at} \quad r = 0 .$$

It may be argued on physical grounds that this condition is unrestricted.

To find S^{zz} and S^{rr}, we can solve implicitly the discretized forms of the appropriate equations in (6.131), using (6.135) and the additional condition

$$\left.\frac{\partial u}{\partial r}\right|_{r=0} = -\frac{1}{2}\left.\frac{\partial w}{\partial z}\right|_{r=0}$$

deduced from (6.134) and (6.136). Alternatively we may use the derivative boundary conditions

$$\frac{\partial S^{rr}}{\partial r} = \frac{\partial S^{\theta\theta}}{\partial r} = \frac{\partial S^{zz}}{\partial r} = 0 , \quad r = 0 .$$

Using a square mesh in the (r,z)-plane, central differences may be used to approximate both first and second ψ-derivatives in (6.129). With natural ordering of the grid-points the resulting block tridiagonal scheme will be diagonally dominant as in the Cartesian case. Problems arise, however, in discretizing the left-hand-side of the vorticity equation (6.130), since even when upwind differencing is employed the resulting system will not in general be diagonally dominant. This is due to the presence of the coefficient of ω given by

$$\frac{1}{r}\left(Ru - \frac{1}{r}\right) ,$$

which may have any sign. To remedy this situation, we may transform to the new variable

$$\zeta = \frac{\omega}{r} .$$

The left-hand-side of (6.130) is then

$$r\left(\frac{\partial^2\zeta}{\partial r^2} + \frac{\partial^2\zeta}{\partial z^2}\right) + 3\frac{\partial\zeta}{\partial r} - Rr\left(u\frac{\partial\zeta}{\partial r} + w\frac{\partial\zeta}{\partial z}\right) ,$$

which yields an unconditionally diagonally dominant system when upwind differences are used to approximate $\partial\zeta/\partial r$ and $\partial\zeta/\partial z$. Torrance (1968) points out that at $r = 0$, ζ, although bounded, is not zero.

6.6.3 Non-rectangular geometries

Apart from the previous section, we have so far assumed that the flow region Ω is planar and is either a rectangle or a union of parallel-sided rectangles. In this section we briefly mention the treatment of more general regions.

If we are interested in planar flow around non-rectangular bodies which define a "natural" coordinate system, such as circles, ellipses, Jackowski airfoils, etc., then we may make a conformal transformation from the Cartesian plane into a new plane, wherein we may employ a rectangular finite difference mesh. One of the advantages of a conformal mapping, as opposed to any other coordinate transformation, is that the form of the Laplacian operator is essentially invariant. Thus, if (ξ,η) denote orthogonal curvilinear coordinates related to Cartesian coordinates (x,y) through a conformal mapping

$$z = f(\zeta) \ , \qquad z = x + iy \ , \qquad \zeta = \xi + i\eta \ ,$$

and if J denotes the Jacobian determinant of the transformation

$$J = \left|\frac{df}{d\zeta}\right|^2 = \frac{\partial(x,y)}{\partial(\xi,\eta)} \ ,$$

then the velocity coordinates in (ξ,η) coordinates are related to the stream function by

$$v^{\xi} = \frac{1}{J}\frac{\partial\psi}{\partial\eta} \ , \qquad v^{\eta} = -\frac{1}{J}\frac{\partial\psi}{\partial\xi} \ ,$$

while the vorticity ω satisfies

$$\frac{\partial^2\psi}{\partial\xi^2} + \frac{\partial^2\psi}{\partial\eta^2} = -J\omega \ .$$

Pilate and Crochet (1977), for example, have studied the flow of a second order fluid around an elliptical cylindrical surface of semi-axis a along the x-axis and semi-axis b along the y-axis. The appropriate conformal transformation is

$$z = \tfrac{1}{2}(a + b)\exp(\zeta) + \tfrac{1}{2}(a - b)\exp(-\zeta) \ .$$

Details and examples of other conformal transformations may be found in the books of Ahlfors (1953), Churchill (1960) and Henrici (1974).

Another use of a coordinate transformation has been made by Gatski and Lumley (1978a) to calculate vorticity boundary values along walls in the shape of rectangular hyperbolae.

The finite difference treatment of arbitrarily shaped boundaries which possess no natural coordinate system proves more difficult. Several authors have used interpolation between grid-points to represent boundary conditions on a curved boundary passing through a rectangular grid (see, for example, Bramble and Hubbard, 1965, and Roache, 1976). This results in the modification of finite difference schemes in ways which are not always easy to handle.

Recently a numerical scheme has been developed which automatically constructs a general curvilinear coordinate system with coordinate lines coincident with arbitrarily shaped boundaries. This scheme, the so called "boundary-fitted coordinates scheme", has been used successfully in the finite difference solution of the incompressible Navier-Stokes equations for flow about arbitrarily shaped two-dimensional bodies (Thames et al, 1977). A code (TOMCAT) is also available for dealing with multiply connected regions (Thompson et al., 1977).

The curvilinear coordinates are obtained by numerical solution of an elliptic differential system in the physical plane, but all numerical computations, both to generate the coordinate system and to solve the governing equations of flow on the coordinate system, are done on a rectangle with a square mesh. It is possible for the coordinate system to change with time as desired, while still carrying out all computations on the square mesh. This allows the treatment of deforming bodies and also free surfaces. An up-to-date discussion of boundary-fitted coordinate systems for the numerical solution of partial differential equations may be found in Thompson (1982) and Thompson et al. (1982).

Other ideas for the finite difference treatment of free boundary problems may be found in Fox and Sankar (1973), Baiocchi et al. (1974), Cryer (1977), Furzeland (1977) and Ryan and Dutta (1981). The last paper would appear to be the first to report the successful finite difference simulation of Newtonian extrudate swell, incorporating the effects of both surface tension and gravitational forces.

6.6.4 Mesh refinement and nonuniform grids

The importance of mesh refinement in both obtaining and assessing the accuracy of finite difference solutions to flow problems cannot be overstressed. Limits of computer store and cost, however, prohibit the adoption of a very fine mesh on the whole flow region. It is therefore necessary to concentrate fine meshes in areas of high flow activity while relying on coarser mesh representation elsewhere. Unfortunately, sudden changes in mesh size can induce large first-order truncation errors in finite difference approximations which are second-order on uniform grids. Considerable care is therefore needed.

A brief guide to the development of nonuniform grids is to be found in Roache (1976, Chapter VI); a recent and potentially important paper which describes the use of coordinate transformations to derive accurate difference approximations for nonuniform grids is that of Jones and Thompson (1980).

An elegant and efficient way of implementing local mesh refinement is to use the ideas of multigrid methods (see, for example, Brandt, 1973, 1977, 1982, and Nicolaides, 1975). Wesseling (1977) has used multigrid methods for solving the Navier-Stokes equations, and considerable developments are taking place in this area . A useful introduction to multigrids may be found in the book by Hackbusch and Trottenberg (1982), which also contains an extensive bibliography.

Chapter 7

Finite Difference Simulation: Time-Dependence

7.1 INTRODUCTION

In Newtonian fluid mechanics, time-dependent flows are synonymous with unsteady flows. Two large groups of problems are readily identified: first, the response of a flow to dynamic disturbances introduced extraneously, such as a periodic change in the driving pressure gradient or a change in the shape or orientation of rigid bodies within the flow; secondly, the evolution of unsteady flows which are self-generated and self-sustained, such as vortex-shedding or turbulence in cases where the boundary conditions are time-independent. The numerical simulation of unsteady viscous flows embodies a large and expanding literature; the books of Roache (1976), Noye (1978), Temam (1979), Telionis (1981) and Belytschko and Hughes (1983), for example, serve between them to illustrate most aspects of modern developments in this area.

Unsteady flows are of equal interest in non-Newtonian fluid mechanics. For memory fluids, moreover, time-integration is also relevant in the study of steady flows, as can be seen for the integral models introduced in Chapter 2. We accordingly divide the present chapter into two parts. In §7.2 we give a general discussion of the principles involved in choosing finite difference schemes for unsteady non-Newtonian flows governed by implicit differential constitutive equations. Unfortunately our discussion cannot fail to be anything but superficial since current experience among researchers in this field is limited. Nevertheless, sufficient material will be presented to enable readers to carry out their own time-dependent simulations. As in the previous chapter we take the Maxwell fluid as a model. In §7.3 we describe finite difference methods for treating steady flow governed by integral constitutive equations; we use the integral form of the Maxwell model as a means of illustration.

7.2 UNSTEADY FLOWS

The earliest work on finite difference simulation of unsteady non-Newtonian flows was confined to generalized Newtonian fluids and to the second-order fluid. An influential paper in the former category is that of Duda and Vrentas (1973) on entrance flows, whereas for the second-order fluid Gilligan and Jones (1970) considered transient flow past a circular cylinder while Baudier and Avenas (1973) studied flow in a square cavity. Their solutions were restricted to very low W-values due to the onset of numerical instability. Later, Crochet and Pilate (1975) were able to reach much higher W-values for flow of a second-order fluid in a square cavity using a fully implicit numerical method for time-integration. Crochet and Pilate's paper is also noteworthy since it gives a Fourier stability analysis of the implicit scheme used and finds a different stability criterion from that encountered for purely viscous transient flow.

Numerical work utilizing differential constitutive models was until very recently confined to time-dependent problems in one space dimension. Townsend (1973) provided a method for an Oldroyd model in this context, and this method was later used by Akay (1979) and Manero and Walters (1980). The only paper to appear on transient flow in two space dimensions at the time of writing is that of Townsend (1983), again for an Oldroyd fluid.

7.2.1 The governing equations

In terms of non-dimensionalized variables ψ, ω and $\underline{T}$, the governing equations for unsteady Maxwell flow are

$$\nabla^2\psi = -\omega \ , \tag{7.1}$$

$$L\omega - \frac{1}{R}\nabla^2\omega = -\frac{1}{R}\left[\frac{\partial^2}{\partial x \partial y}(S^{xx} - S^{yy}) + \left(\frac{\partial^2}{\partial y^2} - \frac{\partial^2}{\partial x^2}\right)S^{xy}\right] \ , \tag{7.2}$$

and

$$\left.\begin{aligned} A_1T^{xx} + WLT^{xx} &= 2BT^{xy} + F_1 \ , \\ A_2T^{xy} + WLT^{xy} &= CT^{xx} + BT^{yy} + F_2 \ , \\ A_3T^{yy} + WLT^{yy} &= 2CT^{xy} + F_3 \ , \end{aligned}\right\} \tag{7.3}$$

where the coefficients A_k, B, C and F_k are given in (6.32) as in steady flow, but the operator L is now given by

$$L = \frac{\partial}{\partial t} + \frac{\partial\psi}{\partial y}\frac{\partial}{\partial x} - \frac{\partial\psi}{\partial x}\frac{\partial}{\partial y} \ . \tag{7.4}$$

The stress tensors $\underline{S}$ and $\underline{T}$ are related as before by transformation (6.4).

Equations (7.1), (7.2) and (7.3) are elliptic, parabolic and hyperbolic in ψ, ω and $\underline{T}$, respectively. The type of the vorticity equation (7.2) is therefore different from its steady flow counterpart (the elliptic equation (6.5)), whereas the types of (7.1) and (7.3) are the same as for steady flow ((6.1) and (6.31)).

Equations (7.1) - (7.3) must be solved at each time step. To specify the solution at time $t > t_0$, an initial state ψ^0, ω^0 and $\underline{T}^0$ must be given at time t_0, with boundary conditions on ψ, ω and $\underline{T}$ at all times $t > t_0$. The boundary condition on $\underline{T}$ is required at entry only. Boundary values of ω may be computed at any time $t > t_0$ using the methods of §6.2.3.

7.2.2 The vorticity equation

Consider a square mesh of mesh-length h for the space variables x, y as before, and let k denote the step-length in time t. Without loss of generality we shall assume $t_0 = 0$. Let $\omega^n_{\ell,m}$ denote a finite difference approximation for the vorticity ω at $(x_\ell, y_m, t_n) = (\ell h, mh, nk)$, with a similar notation for other variables. Writing (7.2) in the form

$$\frac{\partial\omega}{\partial t} = \frac{1}{R}\nabla^2\omega - \left(\frac{\partial\psi}{\partial y}\frac{\partial\omega}{\partial x} - \frac{\partial\psi}{\partial x}\frac{\partial\omega}{\partial y}\right) - \frac{F}{R} \quad , \tag{7.5}$$

where F is the non-Newtonian source term defined in (6.15), we see that a simple explicit finite difference approximation to (7.5) is given by

$$\frac{\omega^{n+1}_{\ell,m} - \omega^n_{\ell,m}}{k} = \frac{1}{Rh^2}(- K^n_0\omega^n_{\ell,m} + K^n_1\omega^n_{\ell+1,m} + K^n_2\omega^n_{\ell-1,m} + K^n_3\omega^n_{\ell,m+1} + K^n_4\omega^n_{\ell,m-1} - h^2F^n_{\ell,m}) \quad , \tag{7.6}$$

where, if a CD-scheme is used for space-derivatives, the coefficients K^n_j are given by (6.18) with time-dependent variables

$$\alpha^n_{\ell,m} = \tfrac{1}{4}R(\psi^n_{\ell+1,m} - \psi^n_{\ell-1,m}) \; , \qquad \beta^n_{\ell,m} = \tfrac{1}{4}R(\psi^n_{\ell,m+1} - \psi^n_{\ell,m-1}) \; . \tag{7.7}$$

This scheme is often called FTCS (Roache 1976) - Forward Time differencing, Central Space differencing. The local truncation error is easily shown to be $O(k) + O(h^2)$.

It is convenient to express (7.6) in the form

$$\omega^{n+1}_{\ell,m} = (1 - \frac{r}{R}K^n_0)\omega^n_{\ell,m} + \frac{r}{R}(K^n_1\omega^n_{\ell+1,m} + K^n_2\omega^n_{\ell-1,m} + K^n_3\omega^n_{\ell,m+1} + K^n_4\omega^n_{\ell,m-1} - h^2F^n_{\ell,m}) \quad , \tag{7.8}$$

where $r = k/h^2$. This equation may also represent a UD1-scheme if the coefficients K^n_j are determined from (6.20) and (7.7). This latter scheme is of course less accurate than FTCS, with a local truncation error $O(k) + O(h)$.

The explicit formula (7.8) enables the computation of ω at an advanced time-level n+1 in terms of $(\psi, \omega, \underset{\sim}{S})$ at neighbouring spatial grid-points at the current time-level n. This process is often referred to as "time-marching". We note that the use of an explicit formula such as (7.8) does not demand the solution of a system of matrix equations at each time step; its implementation is therefore very fast. This advantage, however, is counteracted by a severe limit on the size of the ratio $r = k/h^2$ (and consequently the size of k) which is necessary to maintain numerical stability of the marching process.

In the context of time-integration, we define a finite difference formula such as (7.8) to be *stable* if small perturbations $\delta\omega^n_{\ell,m}$ in the exact solution of the difference equation remain bounded as n tends to infinity. Such perturbations arise from rounding errors in machine computation. A wide range of methods for investigating stability are available, for example, the von Neumann method, matrix method, and energy method (see Mitchell and Griffiths, 1980), the discrete perturbation method (Thom and Apelt, 1961), and Hirt's method (Hirt, 1965). A useful evaluation of stability criteria is also given by Roache. In what follows we restrict attention to von Neumann's method which is perhaps the most widely used. It is essentially a Fourier method.

The perturbation at a grid-point at a given time-level is expressed as a Fourier sum

$$\delta\omega^n_{\ell,m} = \sum_{\lambda,\mu} a^n_{\lambda,\mu}\, e^{i(\rho_\lambda x_\ell + \sigma_\mu y_m)} , \quad i = \sqrt{-1} , \tag{7.9}$$

where in general the frequencies ρ_λ and σ_μ are arbitrary. To examine the stability of (7.8) we first assume that the coefficients K_j and F are constants (independent of ℓ, m and n). Perturbations in ω then satisfy

$$\delta\omega^{n+1}_{\ell,m} = (1 - \frac{r}{R}K_0)\delta\omega^n_{\ell,m} + \frac{r}{R}(K_1\delta\omega^n_{\ell+1,m} + K_2\delta\omega^n_{\ell-1,m} + K_3\delta\omega^n_{\ell,m+1} + K_4\delta\omega^n_{\ell,m-1}) , \tag{7.10}$$

wherefrom, by direct substitution of (7.9), the Fourier coefficients are seen to obey the simple relation

$$a^{n+1}_{\lambda,\mu} = \xi\, a^n_{\lambda,\mu} , \tag{7.11}$$

where the number ξ, known as the *amplification factor*, is given by

$$\xi = (1 - \frac{r}{R}K_0) + \frac{r}{R}(K_1 e^{i\rho_\lambda h} + K_2 e^{-i\rho_\lambda h} + K_3 e^{i\sigma_\mu h} + K_4 e^{-i\sigma_\mu h}) . \tag{7.12}$$

Von Neumann's criterion for stability is that $|\xi| \leqslant 1$.

For FTCS, when α and β in (7.7) are assumed constant, the amplification factor simplifies to

$$\xi = \left[1 - \frac{4r}{R}(\sin^2\tfrac{1}{2}\rho_\lambda h + \sin^2\tfrac{1}{2}\sigma_\mu h)\right] + i\frac{2r}{R}(\alpha\sin\sigma_\mu h - \beta\sin\rho_\lambda h) .$$

The analysis of this expression proves intractable for general ρ_λ and σ_μ but consideration of the simplified case $\rho_\lambda = \sigma_\mu$ leads to the following necessary conditions for $|\xi| \leqslant 1$ (Mitchell and Griffiths 1980):

$$|\alpha| + |\beta| < 2 \quad \text{and} \quad r \leqslant \tfrac{1}{4}R \ . \tag{7.13}$$

The first inequality is clearly a condition on the grid-Reynolds number (cf. (6.19)), whereas the second inequality limits the size of the time step k consistent with stability.

For the UD1 scheme given by (6.20), the amplification factor is

$$\xi = \left[1 - \frac{4r}{R}\{(1 + |\beta|)\sin^2\tfrac{1}{2}\rho_\lambda h + (1 + |\alpha|)\sin^2\tfrac{1}{2}\sigma_\mu h\}\right]$$
$$- i\frac{2r}{R}(|\alpha| \sin \sigma_\mu h + |\beta| \sin \rho_\lambda h) \ ,$$

which, if $\rho_\lambda = \sigma_\mu$, leads to the single necessary condition

$$r \leqslant \frac{R}{4 + 2(|\alpha| + |\beta|)} \ . \tag{7.14}$$

This condition does not limit the grid-Reynolds number, only the time step. The price paid for this is the local truncation error of only $O(k) + O(h)$. (Cf. the CD and UD1 schemes of Chapter 6).

We make the following remarks concerning the von Neumann method of analyzing stability:

(i) The method applies only for linear difference equations with constant coefficients. When the coefficients are variable, and any inhomogeneous coefficient is not a function of the dependent variable, then the method can be applied locally. In this case it is often found that a difference scheme will be stable if the von Neumann condition, derived as though the coefficients were constant, is satisfied at every point of the field.

(ii) For difference approximations of the non-Newtonian vorticity equation, the inhomogeneous source term F is dependent on ω through the field equations. Perturbations $\delta\omega^n_{\ell,m}$ will therefore result in perturbations $\delta F^n_{\ell,m}$. In the stability analysis of (7.8), for example, a term of the form $-(k/R)\delta F^n_{\ell,m}$ must then be added to the right hand side of (7.10), which modifies the expression for the amplification factor in (7.12). The stability criteria (7.13) and (7.14) therefore represent best possible conditions; in practice more stringent criteria are likely, particularly for high W-values. Precise analyses of the term $\delta F^n_{\ell,m}$ will almost certainly prove intractable except in special cases. One example where a partial analysis has been made is for the second-order fluid (Crochet and Pilate 1975).

(iii) Boundary conditions are neglected by the von Neumann method which applies in theory to pure initial value problems with periodic initial data. It does, however, provide necessary conditions for stability of constant coefficient problems regardless of the type of boundary condition.

It is well-known that improved stability properties are associated with *implicit* finite difference schemes. Let us replace each of the terms $\omega^n_{\ell+p,m+q}$, $-1 \leq p,q \leq 1$, on the right of (7.6) by the corresponding weighted average

$$(1-\theta)\omega^n_{\ell+p,m+q} + \theta\,\omega^{n+1}_{\ell+p,m+q}$$

for a fixed θ, $0 < \theta \leq 1$. We then obtain the implicit scheme

$$\begin{aligned}(1 + \frac{r\theta}{R} K^n_0)\omega^{n+1}_{\ell,m} &- \frac{r\theta}{R}(K^n_1\omega^{n+1}_{\ell+1,m} + K^n_2\omega^{n+1}_{\ell-1,m} + K^n_3\omega^{n+1}_{\ell,m+1} + K^n_4\omega^{n+1}_{\ell,m-1}) \\ = \left[1 - \frac{r(1-\theta)}{R}K^n_0\right]\omega^n_{\ell,m} &+ \frac{r(1-\theta)}{R}(K^n_1\omega^n_{\ell+1,m} + K^n_2\omega^n_{\ell-1,m} + K^n_3\omega^n_{\ell,m+1} + K^n_4\omega^n_{\ell,m-1}) \\ &- \frac{k}{R}F^n_{\ell,m}\,. \end{aligned} \tag{7.15}$$

We shall assume in what follows that the coefficients K^n_j are given by the central space differencing formulae (6.18) and (7.7). When $\theta = 1$ we have a fully implicit scheme with local truncation error $O(k) + O(h^2)$, whereas for $\theta = \frac{1}{2}$ we have a scheme of Crank-Nicolson type (Crank and Nicolson 1947) with the improved truncation error $O(k^2) + O(h^2)$. Both these schemes are unconditionally stable for constant coefficients. For example, for $\theta = \frac{1}{2}$ the amplification factor associated with (7.15) is

$$\xi = \frac{1 - \frac{2r}{R}(\sin^2\frac{1}{2}\rho_\lambda h + \sin^2\frac{1}{2}\sigma_\mu h) + i\frac{r}{R}(\alpha\sin\sigma_\mu h - \beta\sin\rho_\lambda h)}{1 + \frac{2r}{R}(\sin^2\frac{1}{2}\rho_\lambda h + \sin^2\frac{1}{2}\sigma_\mu h) - i\frac{r}{R}(\alpha\sin\sigma_\mu h - \beta\sin\rho_\lambda h)}\,,$$

the modulus of which is easily shown to be less than unity for all ρ_λ and σ_μ provided $r > 0$.

The disadvantage of implicit schemes such as (7.15) is that they require the solution of a matrix system of equations at each time level. This problem is eased by using alternating direction implicit (ADI) methods which proceed by splitting the solution procedure into two steps each of which requires the solution of a tridiagonal system (cf. Algorithm 5.1). The first step involves solving for variables along grid-lines parallel to the x-axis while the second step involves grid-lines parallel to the y-axis. This splitting is achieved by introducing an intermediate set of variables, e.g. $\{\omega^{n+1*}_{\ell,m}\}$, which may or may not correspond to approximations at an intermediate time t^*_{n+1}, $t_n < t^*_{n+1} < t_{n+1}$.

For example, the split formula

$$\left.\begin{aligned}&(1+\tfrac{r}{R})\omega^{n+1*}_{\ell,m} - \tfrac{r}{2R}(K^n_1\,\omega^{n+1*}_{\ell+1,m} + K^n_2\omega^{n+1*}_{\ell-1,m})\\ &\quad = (1-\tfrac{r}{R})\omega^n_{\ell,m} + \tfrac{r}{2R}(K^n_3\omega^n_{\ell,m+1} + K^n_4\omega^n_{\ell,m-1}) \,, && (7.16a)\\ &(1+\tfrac{r}{R})\omega^{n+1}_{\ell,m} - \tfrac{r}{2R}(K^n_3\omega^{n+1}_{\ell,m+1} + K^n_4\omega^{n+1}_{\ell,m-1})\\ &\quad = (1-\tfrac{r}{R})\omega^{n+1*}_{\ell,m} + \tfrac{r}{2R}(K^n_1\omega^{n+1*}_{\ell+1,m} + K^n_2\omega^{n+1*}_{\ell-1,m}) + G^n_{\ell,m} \,, && (7.16b)\end{aligned}\right\}$$

with $\omega^{n+1*}_{\ell,m} = \omega^{n+\frac{1}{2}}_{\ell,m}$, is the ADI method of Peaceman and Rachford (1955) applied to the non-Newtonian vorticity equation. The inhomogeneous terms $G^n_{\ell,m}$ may be calculated from the source terms $F^n_{\ell,m}$ by solving the tridiagonal system

$$(1+\tfrac{r}{R})G^n_{\ell,m} - \tfrac{r}{2R}(K^n_1\,G^n_{\ell+1,m} + K^n_2\,G^n_{\ell-1,m}) = -\tfrac{k}{R}F^n_{\ell,m} \quad . \tag{7.17}$$

Very little additional work is needed for this step since the matrix in (7.17) is identical with that of (7.16a) for which the decomposition in Algorithm 5.1 is already available.

The Peaceman-Rachford formula has accuracy $O(k^2) + O(h^2)$ and is unconditionally stable for constant coefficients; it has been widely used in Newtonian calculations. Several other ADI methods abound in the literature.

Finally we consider the leapfrog method of Du Fort and Frankel (1953) which is an explicit three-time-level scheme with better stability properties than the simple two-time-level explicit schemes considered above. Townsend (1983) has used the Du Fort-Frankel method to discretize the vorticity equation for the unsteady flow of an Oldroyd fluid past a circular cylinder. Referring to (7.6), the forward time difference on the left hand side is replaced by a central time difference over 2k, while the centre node value $\omega^n_{\ell,m}$ on the right which comes from the diffusion term is replaced by its average at time-levels n+1 and n-1 :

$$\begin{aligned}\frac{\omega^{n+1}_{\ell,m} - \omega^{n-1}_{\ell,m}}{2k} = \frac{1}{Rh^2}&(-\tfrac{1}{2}K^n_0\omega^{n+1}_{\ell,m} - \tfrac{1}{2}K^n_0\omega^{n-1}_{\ell,m} + K^n_1\omega^n_{\ell+1,m} + K^n_2\omega^n_{\ell-1,m}\\ &+ K^n_3\omega^n_{\ell,m+1} + K^n_4\omega^n_{\ell,m-1} - h^2\bar{F}^n_{\ell,m}) \quad .\end{aligned} \tag{7.18}$$

The skipping of time-level n at the centre node accounts for the popular name "leapfrog". Townsend adopts the same averaging procedure for centre node values of non-Newtonian shear stress which arise from central difference approximations within the source term F (see (6.27)). We denote this change by $\bar{F}^n_{\ell,m}$ in (7.18). We note, however, that for Cartesian coordinates and a square mesh, the centre node values $S^{xy}_{\ell,m}$ are missing from (6.29) due to cancellation, so that in this case $\bar{F}^n_{\ell,m} = F^n_{\ell,m}$. This is not so for a rectangular mesh or for more general coordinate systems.

Equation (7.18) yields the explicit three-time-level formula

$$\omega^{n+1}_{\ell,m} = \left(\frac{2r}{R+4r}\right)(K^n_1\omega^n_{\ell+1,m} + K^n_2\omega^n_{\ell-1,m} + K^n_3\omega^n_{\ell,m+1} + K^n_4\omega^n_{\ell,m-1})$$
$$- \left(\frac{4r}{R+4r}\right)\omega^{n-1}_{\ell,m} - \left(\frac{2k}{R+4r}\right)\bar{F}^n_{\ell,m} \quad . \tag{7.19}$$

The local truncation error is $O(k^2) + O(h^2) + O(k^2/(Rh^2))$ which corresponds to second order accuracy in space and time if $k^2/(Rh^2)$ is very small. For the convection-diffusion equation in one space dimension the stability condition for Du Fort Frankel leapfrog is $C \leqslant 1$ where $C = uk/h$ is the Courant number, u being a constant velocity. Schumann (1975), however, has shown that in more than one dimension, large R-values may reduce this criterion by more than 50%.

When a three-level scheme is used, initial data are required at $t = -k$ and $t = 0$ (or alternatively at $t = 0$ and $t = k$) to start the calculation. If data at $t = 0$ only are available then data at $t = k$ may be calculated using a two-level difference scheme of comparable accuracy with that of the three-level scheme.

Other explicit and implicit three-level schemes for solving the unsteady vorticity equation may be found in the books cited at the beginning of this chapter. Several other methods are also available, including explicit locally one-dimensional schemes and the so-called hopscotch methods. The latter lie somewhere between explicit and implicit and are associated with the name of Gourlay.

7.2.3 The constitutive equations

We now turn our attention to the hyperbolic system (7.3) for which we confine our discussion to explicit difference schemes. For convenience we use the notation $(T^{xx}, T^{xy}, T^{yy}) = (T^1, T^2, T^3)$.

The first equation of (7.3) may be written

$$\frac{\partial T^1}{\partial t} = - \left(u\frac{\partial T^1}{\partial x} + v\frac{\partial T^1}{\partial y}\right) + \frac{1}{W}(- A_1T^1 + 2BT^2 + F_1) \ .$$

Using a forward difference for the time derivative and upwind differences for the space derivatives, where everything on the right hand side is evaluated at time-level n, leads to the explicit first-order formula

$$T^{1,n+1}_{\ell,m} = (1 - \frac{r}{W}K^{1,n}_0)T^{1,n}_{\ell,m} + K^n_1 T^{1,n}_{\ell+1,m} + K^n_2 T^{1,n}_{\ell-1,m} + K^n_3 T^{1,n}_{\ell,m+1} + K^n_4 T^{1,n}_{\ell,m-1}$$
$$+ \frac{k}{W}(2B^n_{\ell,m} T^{2,n}_{\ell,m} + F^n_{1,\ell,m}) \ , \qquad (7.20)$$

where

$$\left.\begin{aligned} K^{1,n}_0 &= h^2 A^n_{1,\ell,m} + 2(|\alpha^{0,n}_{\ell,m}| + |\beta^{0,n}_{\ell,m}|) \ , \\ K^n_1 &= |\beta^{0,n}_{\ell,m}| - \beta^{0,n}_{\ell,m} \ , \qquad K^n_2 = |\beta^{0,n}_{\ell,m}| + \beta^{0,n}_{\ell,m} \ , \\ K^n_3 &= |\alpha^{0,n}_{\ell,m}| + \alpha^{0,n}_{\ell,m} \ , \qquad K^n_4 = |\alpha^{0,n}_{\ell,m}| - \alpha^{0,n}_{\ell,m} \ , \end{aligned}\right\} \qquad (7.21)$$

and

$$\alpha^{0,n}_{\ell,m} = \tfrac{1}{4}W(\psi^n_{\ell+1,m} - \psi^n_{\ell-1,m}) \ , \qquad \beta^{0,n}_{\ell,m} = \tfrac{1}{4}W(\psi^n_{\ell,m+1} - \psi^n_{\ell,m-1}) \ . \qquad (7.22)$$

Similar formulae may be found for the second and third equations of (7.3).

A stability analysis of (7.20) and its sister equations is not available, but it is likely that severe restrictions are imposed on the size of the time step k. Moreover, first-order accuracy in space and time will not, in general, be adequate. A fully second-order formula with good stability properties may, in principle, be developed for the system (7.3) (cf. the Lax-Wendroff method of §7.3), but this is complicated by the inclusion of time-dependent coefficients. We describe instead, therefore, the leapfrog scheme proposed by Townsend (1983) for treating hyperbolic systems, which complements the Du Fort Frankel leapfrog scheme discussed for the parabolic vorticity equation. It is to be expected that Townsend's scheme has much better accuracy and stability properties than the first-order scheme (7.20), although exact analyses are not yet available.

The first equation of (7.3) is approximated by

$$A^n_{1,\ell,m}\left(\frac{T^{1,n+1}_{\ell,m} + T^{1,n-1}_{\ell,m}}{2}\right) + W\left(\frac{T^{1,n+1}_{\ell,m} - T^{1,n-1}_{\ell,m}}{2k}\right)$$
$$+ W\left[\left(\frac{\psi^n_{\ell,m+1} - \psi^n_{\ell,m-1}}{2h}\right)\left(\frac{T^{1,n}_{\ell+1,m} - T^{1,n}_{\ell-1,m}}{2h}\right) - \left(\frac{\psi^n_{\ell+1,m} - \psi^n_{\ell-1,m}}{2h}\right)\left(\frac{T^{1,n}_{\ell,m+1} - T^{1,n}_{\ell,m-1}}{2h}\right)\right]$$
$$- 2B^n_{\ell,m}\left(\frac{T^{2,n+1}_{\ell,m} + T^{2,n-1}_{\ell,m}}{2}\right) = F^n_{1,\ell,m} \ ,$$

with similar approximations for the second and third equations. A simple rearrangement then yields the 3 × 3 matrix system

$$\begin{pmatrix} \frac{1}{2}(h^2A^n_{1,\ell,m} + \frac{W}{r}) , & -h^2B^n_{\ell,m} , & 0 \\ -\frac{1}{2}h^2C^n_{\ell,m} , & \frac{1}{2}(h^2 + \frac{W}{r}) , & -\frac{1}{2}h^2B^n_{\ell,m} \\ 0 , & -h^2C^n_{\ell,m} , & \frac{1}{2}(h^2A^n_{3,\ell,m} + \frac{W}{r}) \end{pmatrix} \begin{pmatrix} T^{1,n+1}_{\ell,m} \\ T^{2,n+1}_{\ell,m} \\ T^{3,n+1}_{\ell,m} \end{pmatrix}$$

$$= \begin{pmatrix} -\frac{1}{2}(h^2A^n_{1,\ell,m} - \frac{W}{r}) , & h^2B^n_{\ell,m} , & 0 \\ \frac{1}{2}h^2C^n_{\ell,m} , & -\frac{1}{2}(h^2 - \frac{W}{r}) , & \frac{1}{2}h^2B^n_{\ell,m} \\ 0 , & h^2C^n_{\ell,m} , & -\frac{1}{2}(h^2A^n_{3,\ell,m} - \frac{W}{r}) \end{pmatrix} \begin{pmatrix} T^{1,n-1}_{\ell,m} \\ T^{2,n-1}_{\ell,m} \\ T^{3,n-1}_{\ell,m} \end{pmatrix} + \begin{pmatrix} E^n_{1,\ell,m} \\ E^n_{2,\ell,m} \\ E^n_{3,\ell,m} \end{pmatrix} , \tag{7.23}$$

where

$$E^n_{j,\ell,m} = -\beta^{0,n}_{\ell,m}(T^{j,n}_{\ell+1,m} - T^{j,n}_{\ell-1,m}) + \alpha^{0,n}_{\ell,m}(T^{j,n}_{\ell,m+1} - T^{j,n}_{\ell,m-1}) + h^2F^n_{j,\ell,m} , \quad j = 1, 2, 3. \tag{7.24}$$

This system may be solved directly at each spatial grid-point (x_ℓ, y_m) to give the three components of extra-stress at the advanced time-level n+1 in terms of known variables at levels n and n-1.

To end this brief section we mention that finite difference approximations to hyperbolic equations and systems cannot be convergent as $h \to 0$ unless they satisfy the celebrated Courant-Friedrichs-Lewy (C.F.L.) condition (Courant et al. 1928). For any explicit difference scheme it is possible to trace back from a given grid-point $P(x_\ell, y_m, t_n)$ at $t = t_n$, using only the structure of the difference scheme, to the grid-points at $t = t_0$ which influence the numerical solution at P. The three-dimensional set of grid-points emanating back from P is called the domain of dependence at P of the difference scheme. Similarly, the characteristic surfaces of the hyperbolic system which pass through P define a three-dimensional region which is the domain of dependence of the differential system. The C.F.L. condition states that the convex hull of the domain of dependence of the difference scheme must contain that of the differential system. In some simple cases, the C.F.L. criterion reduces to the von Neumann condition for stability, but this is not so generally. Discussions of the C.F.L. condition for hyperbolic systems in more than one space dimension can be found in Wilson (1972) and Telionis (1981).

7.2.4 Solution of the coupled equations

We have seen that the choice of explicit difference approximations to the vorticity and constitutive equations removes the need for solving large matrix systems at each time step. The small time steps which are required for stability, however, call for a large amount of computation, and the non-Newtonian time-integration problem cannot fail to be expensive by present-day computing standards. The need to concentrate research on developing stable and accurate schemes which can solve the coupled equations efficiently is therefore quite clear.

A general algorithm for the time-integration of the coupled constitutive equations, vorticity equation, and stream-function equation may be summarized as follows, if we restrict attention to explicit two-time-level schemes. In terms of known vectors $\underset{\sim}{T}^n$, $\underset{\sim}{\omega}^n$ and $\underset{\sim}{\psi}^n$ at time-level n we may write

$$\underset{\sim}{T}^{n+1} = \underset{\sim}{b}(\underset{\sim}{T}^n, \underset{\sim}{\omega}^n, \underset{\sim}{\psi}^n) , \tag{7.25a}$$

$$\underset{\sim}{\omega}^{n+1} = \underset{\sim}{c}(\underset{\sim}{S}^n, \underset{\sim}{\omega}^n, \underset{\sim}{\psi}^n) , \tag{7.25b}$$

$$\underset{\sim}{\psi}^{n+1} = \underset{\sim}{d}(\underset{\sim}{\omega}^{n+1}) , \tag{7.25c}$$

where $\underset{\sim}{b}$, $\underset{\sim}{c}$ and $\underset{\sim}{d}$ are vector functions defined by the difference schemes and boundary conditions. Equation (7.25c) implies the inversion of the usual discrete Poisson equation for stream-function, possibly using a Fast Poisson Solver (Townsend, 1983). Clearly, we may march in time by executing (7.25a-c) in order, but in addition at each time-level n+1 Townsend recommends an iterative refinement of the form

$$\left.\begin{aligned} \underset{\sim}{T}^{n+1}_{r+1} &= \underset{\sim}{b}(\underset{\sim}{T}^n, \underset{\sim}{\omega}^{n+1}_r, \underset{\sim}{\psi}^{n+1}_r) , \\ \underset{\sim}{\omega}^{n+1}_{r+1} &= \underset{\sim}{c}(\underset{\sim}{S}^{n+1}_r, \underset{\sim}{\omega}^n, \underset{\sim}{\psi}^{n+1}_r) , \\ \underset{\sim}{\psi}^{n+1}_{r+1} &= \underset{\sim}{d}(\underset{\sim}{\omega}^{n+1}_{r+1}) , \end{aligned}\right\} \quad r = 0, 1, \ldots,$$

where (7.25a-c) determine $\underset{\sim}{T}^{n+1}_0$, $\underset{\sim}{\omega}^{n+1}_0$ and $\underset{\sim}{\psi}^{n+1}_0$, respectively. Although this greatly increases the amount of work to be done at each time-level, the iterative refinement significantly improves the overall stability of the algorithm. The above description is easily extended to three-time-level schemes.

In common with the simulation of steady flow problems, Townsend finds that as W is increased, numerical solutions become more difficult to obtain. To extend the range of elasticity for which convergence is attainable he introduces a relaxation procedure, which, after each time step replaces the newly computed variables with a weighted average of their values at time levels n+1 and n. This implies, of course, that the time variation is not modelled correctly; for a two-time-level scheme, the relaxation is essentially equivalent to taking a fraction of the original time-step k, but for a three-time-level scheme the situation is more complex. Townsend remarks that if the algorithm converges to a unique steady state then relaxation does not affect the form of that solution. If, however, there are multiple solutions, then the relaxation could influence which branch is followed. It is found for the flow of an Oldroyd liquid past a cylinder that the relaxation process pushes the limiting W-value to an order of magnitude higher than is otherwise reached.

7.3 INTEGRAL CONSTITUTIVE MODELS

Existing work on finite difference simulation with integral constitutive models is limited to the approach of Court et al. (1981) which we shall discuss in this section. Several authors have pursued alternative finite element treatments, which are discussed in §10.10. Whichever approach is followed, the basic problem is that of "tracking" the position of particles, given a known velocity field. From this information the deformation tensor can be calculated as a function of time-lapse and hence the stress components evaluated by an appropriate choice of quadrature. The velocity field is then updated, the tracking adjusted, and new stress components found; the whole process is repeated iteratively until convergence is obtained.

Tracking and time-integration make up the major part of the computational cost of simulation using integral models. In the case of steady flow this cost can be significantly higher than for simulation using equivalent differential models, when they exist. Moreover, algorithms treating integral models are currently no more accurate nor more stable than can be found for a differential analogue. The main interest in working with integral forms, therefore, stems from the wider spectrum of models which become available.

7.3.1 Tracking

The integral form of the Maxwell model, in terms of non-dimensionalized variables is (cf. (3.6))

$$T^{ij} = \frac{1}{W^2} \int_0^\infty e^{-s/W} H^{ij}(s)\, ds \quad , \tag{7.26}$$

where the deformation tensor H^{ij} is given by

$$H^{ij} = (C^{-1})^{ij} - \delta^{ij} \tag{7.27}$$

and $(C^{-1})^{ij}$ is the Finger tensor defined in (3.8), which is also the inverse (in the matrix sense) of the Cauchy-Green tensor (cf.(2.25))

$$C^{ij} = \frac{\partial x'^m}{\partial x^i} \frac{\partial x'^m}{\partial x^j} \quad . \tag{7.28}$$

(Here we use upper indices to denote tensor components; lower indices are reserved for spatial grid points).

Recall that $x'^i = x'^i(x^j, t; t-s)$ is the position at time t-s, $s \geq 0$, of the particle that is instantaneously at the point x^j at time t. For steady flow, which is our concern here, x'^i is independent of t and may be written $x'^i = x'^i(x^j, s)$, where s is the time lapse.

In steady two-dimensional planar flow, the components of the displacement function x'^i are $x' = x'(x, y, s)$ and $y' = y'(x, y, s)$. Court et al. (1981) find $H^{ij}(s)$ in (7.27) by calculating x' and y' at all spatial grid-points (x_ℓ, y_m) at various times s. The first derivatives in (7.28) are then estimated by central space-differencing, and finally the 2×2 matrix C is inverted directly at each grid-point.

The displacement functions are found by solving numerically the hyperbolic equations (Oldroyd 1950)

$$\frac{\partial x'}{\partial s} + u \frac{\partial x'}{\partial x} + v \frac{\partial x'}{\partial y} = 0 \quad , \tag{7.29}$$

$$\frac{\partial y'}{\partial s} + u \frac{\partial y'}{\partial x} + v \frac{\partial y'}{\partial y} = 0 \quad . \tag{7.30}$$

It is important to choose a scheme which is at least second-order accurate in space since the numerical differentiation performed in (7.28) effectively reduces the spatial order of accuracy by one. There is a variety of methods available which are second-order in both space and time (see, for example, Mitchell and Griffiths (1980)), but here we describe only the explicit two-time-level scheme of Lax and Wendroff (1964) which has been widely used in various forms for treating hyperbolic equations and systems in two space dimensions.

If k denotes a time step, then by Taylor expansion correct to $O(k^2)$ we have, using (7.29),

$$x'(s + k) = x' - k\left(u \frac{\partial x'}{\partial x} + v \frac{\partial x'}{\partial y}\right)$$
$$+ \tfrac{1}{2}k^2\left[u \frac{\partial}{\partial x}\left(u \frac{\partial x'}{\partial x}\right) + u \frac{\partial}{\partial x}\left(v \frac{\partial x'}{\partial y}\right) + v \frac{\partial}{\partial y}\left(u \frac{\partial x'}{\partial x}\right) + v \frac{\partial}{\partial y}\left(v \frac{\partial x'}{\partial y}\right)\right] ,$$

where everything on the right hand side is evaluated at time s. The basic

Lax-Wendroff scheme approximates the first-order terms by central differences and the second-order terms by a mixture of forward and backward differences. In terms of the forward and backward operators defined by

$$\Delta_x w_{\ell,m} = w_{\ell+1,m} - w_{\ell,m} \ , \qquad \nabla_x w_{\ell,m} = w_{\ell,m} - w_{\ell-1,m} \ ,$$

$$\Delta_y w_{\ell,m} = w_{\ell,m+1} - w_{\ell,m} \ , \qquad \nabla_y w_{\ell,m} = w_{\ell,m} - w_{\ell,m-1} \ ,$$

the Lax-Wendroff scheme for (7.29), which has second-order accuracy in space and time, is

$$\begin{aligned} x'^{n+1}_{\ell,m} = [1 &- \tfrac{1}{2}pu_{\ell,m}(\Delta_x + \nabla_x) - \tfrac{1}{2}pv_{\ell,m}(\Delta_y + \nabla_y) \\ &+ \tfrac{1}{4}p^2(u_{\ell,m}\Delta_x u_{\ell,m}\nabla_x + u_{\ell,m}\nabla_x u_{\ell,m}\Delta_x + u_{\ell,m}\Delta_x v_{\ell,m}\nabla_y + u_{\ell,m}\nabla_x v_{\ell,m}\Delta_y) \\ &+ \tfrac{1}{4}p^2(v_{\ell,m}\Delta_y u_{\ell,m}\nabla_x + v_{\ell,m}\nabla_y u_{\ell,m}\Delta_x + v_{\ell,m}\Delta_y v_{\ell,m}\nabla_y + v_{\ell,m}\nabla_y v_{\ell,m}\Delta_y)]x'^{n}_{\ell,m} \ . \end{aligned} \tag{7.31}$$

We have used the fact that, for steady flow, the velocities u and v at (x_ℓ, y_m) are independent of s. In (7.31) $p = k/h$ is the ratio of time and space grid-lengths, and $x'^{n}_{\ell,m}$ denotes the approximation to $x'(x_\ell, y_m, s_n)$, $s_n = nk$. Similarly for equation (7.30).

Court et al. use a slightly less accurate modification of (7.31) based on a Lax-Wendroff scheme of Mitchell and Griffiths (1980), but scheme (7.31) is preferable. For constant coefficients u and v, (7.31) is stable provided (cf. Lax and Wendroff 1964)

$$p^2 \leqslant \frac{1}{8\max(u^2, v^2)} \ . \tag{7.32}$$

For variable coefficients this stability criterion can easily be checked locally at each grid-point.

Equation (7.31) and its sister equation are solved subject to the initial conditions

$$x'(x,y,0) = x \ , \qquad y'(x,y,0) = y \ . \tag{7.33}$$

Displacement function boundary conditions are required at entry for all $s \geqslant 0$, and these may be found from the fully-developed flow condition there by integrating back along a streamline. Values of x' and y' on solid boundaries are easily obtained from the no-slip condition, yielding

$$x'(x,y,s) = x \quad , \qquad y'(x,y,s) = y \quad , \quad s \geqslant 0 \ ,$$

on a stationary boundary, and

$$x'(x,y,s) = x - Us \ , \quad y'(x,y,s) = y \ , \quad s \geqslant 0 \ ,$$

for example, on a boundary moving parallel to the x-axis with constant speed U.

7.3.2 Computation of stress

Gaussian quadrature rules are available for the accurate estimation of several product-type integrals. For integrals of the form

$$\int_0^\infty e^{-z} f(z)\, dz \quad , \tag{7.34}$$

the appropriate set of rules is that of Gauss-Laguerre, which, given any integer $N \geqslant 1$, replaces (7.34) by a weighted sum

$$\sum_{i=0}^{N} w_i f(z_i) \quad , \tag{7.35}$$

where z_i are the zeros of the (N+1)th Laguerre polynomial, and the weights w_i are chosen to make the quadrature exact when f is a polynomial of degree 2N+1. The weights and zeros depend only on N, and are given in Table 10.3 for $N = 1$ and 2. For arbitrary functions f, which are sufficiently smooth, the accuracy of (7.35) increases with N.

For given N, we find from (7.26) that

$$T^{xx}_{\ell,m} \simeq \frac{1}{W} \sum_{i=0}^{N} w_i H^{xx}_{\ell,m}(s_i) \quad , \quad s_i = Wz_i \quad . \tag{7.36}$$

We note that the calculation of C in (7.28) and its inverse need only be carried out at the nodes s_i, i=0,...,N, which are unevenly distributed. The displacement functions, however, must be found at intermediate time-steps s_n, which are equally spaced. It is therefore convenient to divide each interval $[s_i, s_{i+1}]$ into an equal number of steps, M say, and vary the time-step $k_i = (s_{i+1} - s_i)/M$ for each interval for the purpose of the Lax-Wendroff integration. M must be chosen in accordance with the stability criterion (7.32). Similar expressions to (7.36) hold for the other stress components.

We observe that, for fixed N, the nodes s_i in (7.36) spread out as W increases thereby requiring longer time integration. For longer times, numerical errors in H^{ij} tend to dominate over pure quadrature errors and so there is little advantage in using high-order Gauss-Laguerre quadrature. Court et al. choose N in the range $1 \leqslant N \leqslant 5$.

Stresses at entry may of course be found from (7.26) - (7.28) by using analytical expressions for fully-developed flow. To find stress components at points where $u = v = 0$, as on stationary boundaries, Court et al. use the following procedure.

Differentiating (7.29) with respect to x and y yields the pair of equations

$$\frac{\partial}{\partial s}\left(\frac{\partial x'}{\partial x}\right) + \frac{\partial u}{\partial x}\frac{\partial x'}{\partial x} + \frac{\partial v}{\partial x}\frac{\partial x'}{\partial y} = 0 \quad ,$$

$$\frac{\partial}{\partial s}\left(\frac{\partial x'}{\partial y}\right) + \frac{\partial u}{\partial y}\frac{\partial x'}{\partial x} - \frac{\partial u}{\partial x}\frac{\partial x'}{\partial y} = 0 \quad ,$$

the solutions of which subject to (7.33) are

$$\frac{\partial x'}{\partial x} = \cosh(\alpha s) - \frac{1}{\alpha}\frac{\partial u}{\partial x}\sinh(\alpha s) \quad , \qquad \frac{\partial x'}{\partial y} = -\frac{1}{\alpha}\frac{\partial u}{\partial y}\sinh(\alpha s) \quad ,$$

where

$$\alpha^2 = \left[\left(\frac{\partial u}{\partial x}\right)^2 + \frac{\partial v}{\partial x}\frac{\partial u}{\partial y}\right] .$$

Similar expressions may be found for $\frac{\partial y'}{\partial x}$ and $\frac{\partial y'}{\partial y}$.

From (7.28) it is then found that

$$\left.\begin{aligned}
(C^{-1})^{xx} &= \frac{1}{\alpha^2}\left[\left(\frac{\partial u}{\partial x}\right)^2 + \left(\frac{\partial u}{\partial y}\right)^2\right]\sinh^2(\alpha s) + \cosh^2(\alpha s) + \frac{1}{\alpha}\frac{\partial u}{\partial x}\sinh(2\alpha s) \quad , \\
(C^{-1})^{xy} &= \frac{1}{2\alpha}\left(\frac{\partial v}{\partial x} + \frac{\partial u}{\partial y}\right)\sinh(2\alpha s) - \frac{1}{\alpha^2}\frac{\partial u}{\partial x}\left(\frac{\partial u}{\partial y} - \frac{\partial v}{\partial x}\right)\sinh^2(\alpha s) \quad , \\
(C^{-1})^{yy} &= \frac{1}{\alpha^2}\left[\left(\frac{\partial u}{\partial x}\right)^2 + \left(\frac{\partial v}{\partial x}\right)^2\right]\sinh^2(\alpha s) + \cosh^2(\alpha s) - \frac{1}{\alpha}\frac{\partial u}{\partial x}\sinh(2\alpha s) \quad .
\end{aligned}\right\} \quad (7.37)$$

These expressions are independent of the constitutive model, and are valid at any point at which $u = v = 0$. Slightly more complicated expressions may be found, for example, when u = constant U and $v = 0$. Substitution of (7.37) into the integral representation for stress can lead to analytic expressions. For example, in the case of the Maxwell model (7.26), we find

$$\left.\begin{aligned}
T^{xx} &= \frac{2}{1-4\alpha^2W^2}\left[\frac{\partial u}{\partial x} + W\left\{\frac{\partial u}{\partial y}\frac{\partial v}{\partial x} + 2\left(\frac{\partial u}{\partial x}\right)^2 + \left(\frac{\partial u}{\partial y}\right)^2\right\}\right] , \\
T^{xy} &= \frac{1}{1-4\alpha^2W^2}\left[\frac{\partial u}{\partial y} + \frac{\partial v}{\partial x} + 2W\left(\frac{\partial u}{\partial y}\frac{\partial v}{\partial y} + \frac{\partial u}{\partial x}\frac{\partial v}{\partial x}\right)\right] , \\
T^{yy} &= \frac{2}{1-4\alpha^2W^2}\left[-\frac{\partial u}{\partial x} + W\left\{\frac{\partial u}{\partial y}\frac{\partial v}{\partial x} + 2\left(\frac{\partial u}{\partial x}\right)^2 + \left(\frac{\partial v}{\partial x}\right)^2\right\}\right] ,
\end{aligned}\right\} \qquad (7.38)$$

which are consistent with (6.84) and (6.85).

Court et al. use (7.38) to calculate fictitious stress values at re-entrant corners for use in their finite difference scheme. This is consistent with the approach described in §6.2.5(iii).

7.3.3 Nonlinear coupling

The stress computations of §7.3.2 take the place of those of §6.2.2 in an iterative algorithm for solving the steady field equations (6.1) and (6.5) coupled with the constitutive equations. Using ICCG(0) for (6.1) and ILUCG for (6.5), Court et al. favour the following simple algorithm.

(0) Guess a $\underset{\sim}{\psi}$-field.

(1) Evaluate the internal stress field $\underset{\sim}{T}$ using Lax-Wendroff integration and Gaussian quadrature. Transform to $\underset{\sim}{S}$ using (6.4).

(2) Calculate boundary vorticity and stress.

(3) Perform one sweep of ILUCG on (6.5) followed by one sweep of ICCG(0) for (6.1).

(4) If $\|\underset{\sim}{\psi}_1 - \underset{\sim}{\psi}_0\| < \varepsilon \|\underset{\sim}{\psi}_0\|$, stop; otherwise let $\underset{\sim}{\psi}_0 = \underset{\sim}{\psi}_1$ and repeat (1) - (4).

Much more sophisticated algorithms can be envisaged, particularly in conjunction with a Newton algorithm (cf. §10.10).

Court et al. have compared simulation of the flow of a Maxwell fluid over an obstruction using both integral and differential models, for a range of W-values. The numerical results are in good agreement in both cases. It is found, however, that particles which pass near re-entrant corners are not tracked accurately along streamlines. Moreover the continuity condition det $C^{-1} = 1$ is not satisfied numerically near singularities.

It is of interest that breakdown of iterative convergence in the integral case takes place for approximately the same value of W as in the differential case.

Chapter 8

Introduction to Finite Elements

8.1 INTRODUCTION

The relative merits of finite difference and finite element techniques have been the object of long and often inconclusive debates. The use of finite elements for solving viscous flow is of a much more recent origin than finite differences; the first paper on finite elements for solving Navier-Stokes equations is only twelve years old at this time of writing (Oden 1970), and the search for the best finite element formulation is far from being ended. When finite differences are compared to finite elements, most researchers will agree on the following observations: i. finite difference techniques are relatively easy to understand and to implement for a newcomer to the field, while the development of a finite element code requires a non-negligible amount of programming; ii. when the same problem can be solved with the use of both techniques, the finite difference method will usually be cheaper on the computer than the finite element method; iii. the finite element method has a tremendous advantage over finite differences for solving flows in complex geometries, which, more often than not, cannot even be approached with the latter (cf. §4.4).

While the basic thought behind finite differences is the substitution of finite intervals Δx for infinitesimal intervals dx in the definition of differential operators, the philosophy of finite elements proceeds from another approach. Here, it is assumed at the outset that the *unknown functions* are approximated in some specific way in terms of a finite number of parameters; the selection of these parameters is based upon the satisfaction of the field and constitutive equations, and the boundary conditions, in a sense to be defined. The success of a finite element algorithm depends upon the appropriateness of the approximation and of the rule for satisfying the equations.

Although the present book is mainly concerned with the capabilities of numerical work for solving rheological problems, it is impossible to explain and compare several available methods without being specific on the type of element and the kind of criterion used for satisfying the equations. The reader who knows about finite elements may skip over to Chapter 9; for the reader who has not been exposed to finite elements, we will attempt in the present chapter to introduce and explain the basic concepts. Since our emphasis on the theoretical background will necessarily be limited, we will proceed by induction and start with simple one-dimensional problems which, in general, would not be solved by means of the finite element technique; their simplicity, however, will facilitate the introduction of new concepts. We will then generalize and briefly explain how to solve two-dimensional problems; the Poisson equation will provide an example for applying the theory. The concepts will then be used in Chapter 9 for calculating the flow of viscous fluids.

8.2 FINITE ELEMENT REPRESENTATION

We will consider the one-dimensional problems exposed in section 5.1, and seek a function u, defined over the real interval Ω, $x_0 \leq x \leq x_N$, which satisfies (5.1). We assume that there is no hope of obtaining the analytical form of u; we wish to calculate an approximate representation which will be close to the actual solution in a sense to be made precise in sections 8.3 and 8.4. Our present issue is how to define an approximation for u; we will not cover all the possibilities and will rather concentrate on the notions which will be useful in later chapters.

In order to clarify the problem, we will assume that u is a known function of x, say $u = \cos x$, $0 \leq x \leq \pi$; we want to approximate u by means of piecewise polynomials. In Fig. 8.1, we consider a very simple case; the domain of u (i.e. $0 \leq x \leq \pi$) is divided into two equal segments, called finite elements. On both elements, we want to substitute for u (i.e. $\cos x$) a polynomial function of x.

On Fig. 8.1a, we show the simplest case, where u is approximated by a zero-order polynomial, i.e. a constant, over each element; the constant is chosen here to be the average of u over the element. The approximation, which we denote by $\tilde{u}$, is piecewise continuous; we will say, for brevity, that the approximation is of the type P^0-C^{-1}, where we simultaneously refer to the order of the polynomial and the continuity properties of the approximation. In the sequel, we will say that a function is C^{-1} continuous when it is piecewise continuous; a function is C^k continuous when its kth derivative is continuous and its $(k+1)^{th}$ derivative is piecewise continuous.

On Fig. 8.1b, the approximation $\tilde{u}$ is a linear interpolation of u by means of first-order polynomials; $\tilde{u}$ is now continuous, and the type of the approximation is P^1-C^0. On Fig. 8.1c, $\tilde{u}$ is a second-order polynomial in each element; the coefficients of the polynomial are known through the imposition of the value of u at the ends of the element and at mid-distance from the ends.

We will find later that, in some applications, one may need an approximation $\tilde{u}$ which is C^1-continuous; Fig. 8.1d shows such an approximation. Here, $\tilde{u}$ is a third-order polynomial in each element; the four coefficients are identified by imposing the value of $\tilde{u}$ and its first order derivative at the ends of the element. The approximation $\tilde{u}$ is now continuous together with its first-order derivative, and its type is P^3-C^1.

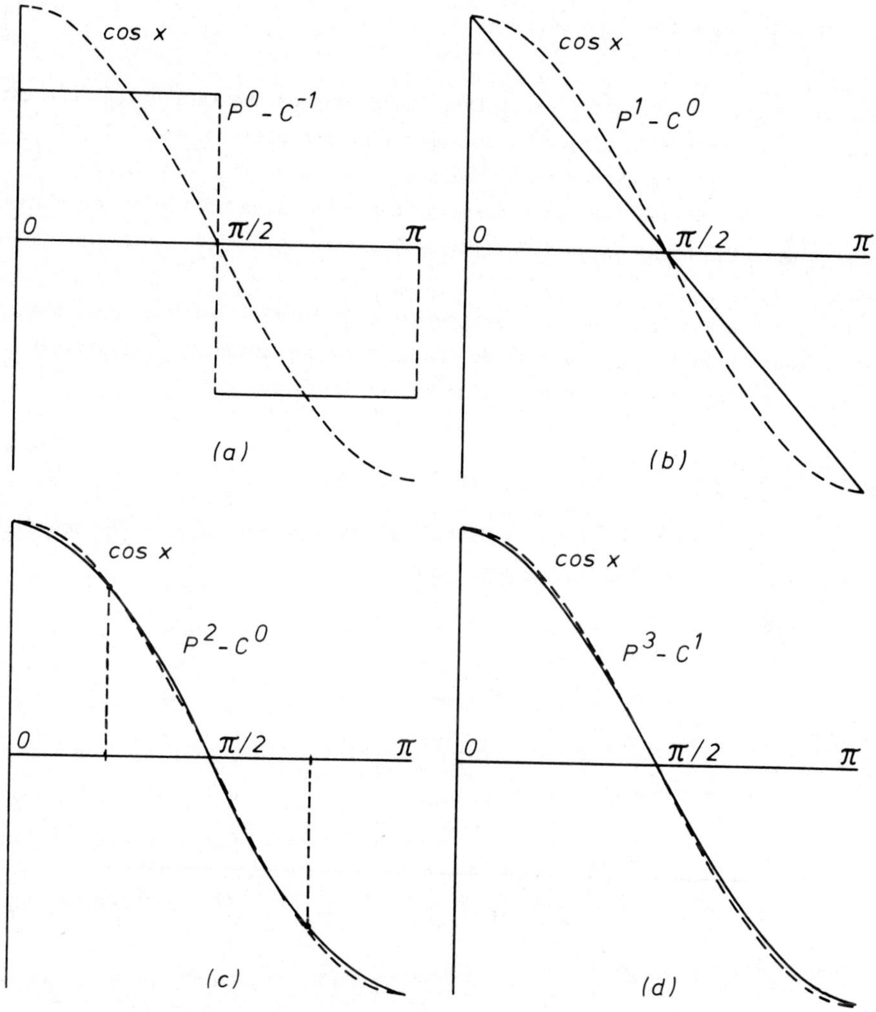

Fig. 8.1 Finite element approximation of the function $u = \cos x$.

Now that we have presented the problem by means of a simple example, we need to be systematic, and define in more general terms the process of approximating u over the domain Ω by means of piecewise polynomials. First, we divide the domain Ω into N non-overlapping elements Ω_i, such that

$$\bigcup_{i=1}^{N} (\Omega_i) = \Omega \ , \quad \Omega_i = \{x : x_{i-1} \leq x \leq x_i\} \ . \tag{8.1}$$

On the element Ω_i we identify *nodes*; these nodes may be the ends of the element, or the ends together with a node in the middle of the element, etc. The approximation $\tilde{u}$ will be expressed by means of *nodal values*; these may be the value of u or its derivatives at a node, or the mean value of u over the element (in which case the nodal value is not directly associated with a node of the element).

Over each element, we want to characterize a polynomial which assumes the selected nodal values. In order to do that, it is convenient to establish a 1:1 affine mapping of each element Ω_i onto the closed interval ω,

$$\omega = \{\xi : -1 \leqslant \xi \leqslant 1\} \ . \tag{8.2}$$

The mapping is shown on Fig. 8.2; ω is called the *parent element*. The coordinates $\xi \in \omega$ and $x \in \Omega_i$ are related by the equations

$$\xi = (2x - x_{i-1} - x_i)/(x_i - x_{i-1}) \quad ,$$

$$x = \xi(x_i - x_{i-1})/2 + (x_i + x_{i-1})/2 \ . \tag{8.3}$$

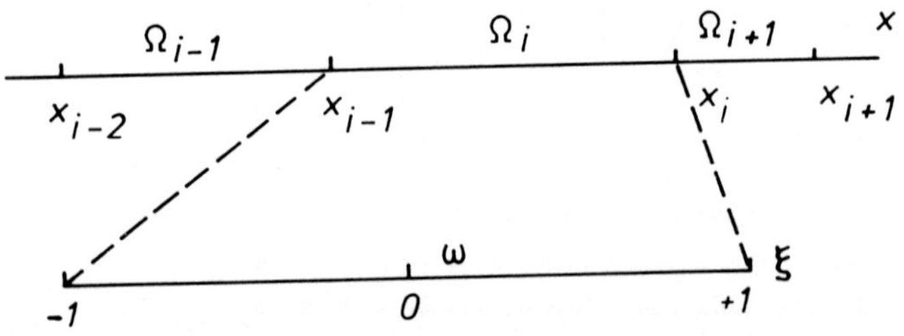

Fig.8.2 Mapping of the finite element Ω onto the parent element ω.

It is now much easier to define on ω a basis for polynomials of the required order; each element of the basis is associated with a nodal value, and is called a shape function.

For zero-order polynomials, we need a single shape function κ,

$$\kappa = 1 , \quad -1 \leq \xi \leq 1 . \tag{8.4}$$

For first-order polynomials, we identify two nodes at $\xi = -1$ and $\xi = 1$, and define shape functions which take the unit value at one node and vanish at the other, i.e.

$$\phi_1 = (1 - \xi)/2 , \quad \phi_2 = (1 + \xi)/2 , \quad -1 \leq \xi \leq 1 . \tag{8.5}$$

For second-order polynomials, we may identify three nodes located respectively at $\xi = -1$, $\xi = 0$ and $\xi = 1$, and use the same criterion for defining the shape functions, i.e.

$$\psi_1 = \xi(-1 + \xi)/2 , \quad \psi_2 = 1 - \xi^2 , \quad \psi_3 = \xi(1 + \xi)/2 , \quad -1 \leq \xi \leq 1 . \tag{8.6}$$

The choice of a basis for third-order polynomials depends upon the type of approximation which we want to accomplish. If the type of the approximation is P^3-C^1, the shape functions are associated with four nodal values, i.e. the value and the slope of the latter at the end-points of the parent element. For each shape function, one of the nodal values takes the unit value, and the others vanish; the shape functions are given by

$$\begin{aligned} \chi_1 &= (2 - 3\xi + \xi^3)/4 , & \chi_2 &= (2 + 3\xi - \xi^3)/4 , \\ \chi_3 &= (1 + \xi)(1 - \xi)^2/4 , & \chi_4 &= -(1 - \xi)(1 + \xi)^2/4 . \end{aligned} \tag{8.7}$$

The meaning of these functions is clearly seen on Fig. 8.3 where we show a graph of the shape functions defined in (8.4-7).

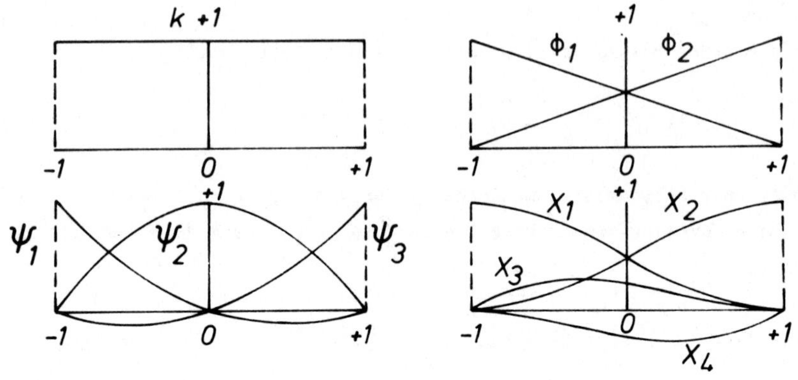

Fig.8.3 Shape functions on the parent element.

It is now an easy matter to define the approximation $\tilde{u}$ over the domain Ω. Let us consider an approximation of the type P^1-C^0, and let u_{i-1}, u_i be the values of u at the end-points of the element Ω_i; clearly, the approximation

$$\tilde{u} = u_{i-1}\ \phi_1[\xi(x)] + u_i\ \phi_2[\xi(x)]\ , \qquad x \in \Omega_i\ , \tag{8.8}$$

leads to a linear interpolation between u_{i-1} and u_i. Similarly, if $u_{i-1/2}$ denotes the value of u at the point of coordinate $(x_{i-1} + x_i)/2$, an approximation of the type P^2-C^0 is obtained by writing

$$\tilde{u} = u_{i-1}\ \psi_1[\xi(x)] + u_{i-1/2}\ \psi_2[\xi(x)] + u_i\ \psi_3[\xi(x)]\ , \quad x \in \Omega_i\ . \tag{8.9}$$

For obtaining an approximation of the type P^3-C^1, we select the nodal values u_{i-1}, u_i, $\dot{u}_{i-1}$, $\dot{u}_i$, where a dot denotes the derivative with respect to x; within the element Ω_i, the approximation is given by

$$\tilde{u} = u_{i-1}\ \chi_1[\xi(x)] + u_i\ \chi_2[\xi(x)]$$

$$+ \{\dot{u}_{i-1}\ \chi_3[\xi(x)] + \dot{u}_i\ \chi_4[\xi(x)]\}(x_i - x_{i-1})/2\ , \quad x \in \Omega_i\ . \tag{8.10}$$

At the present stage, we have shown how an interpolation may be constructed by associating nodal values and shape functions within each separate element. We will find it convenient in our further developments to write an expression for the approximation over the entire domain Ω rather than on individual elements.

Consider, as an example, the linear interpolation (8.8). We may define a function τ_i shown in Fig. 8.4 and such that

$$
\begin{aligned}
x \notin \Omega_i \cup \Omega_{i+1} &: \quad \tau_i = 0 , \\
x \in \Omega_i &: \quad \tau_i = \phi_2[\xi(x)] , \\
x \in \Omega_{i+1} &: \quad \tau_i = \phi_1[\xi(x)] .
\end{aligned}
\tag{8.11}
$$

The approximation $\tilde{u}$ may then be written as

$$
\tilde{u} = \Sigma\, U_i \tau_i , \tag{8.12}
$$

where the U_i's are nodal values, and the τ_i's are piecewise polynomials which vanish outside a small local support. It is easy to follow a similar argument for showing that the form (8.12) may also be constructed with approximations of the type (8.9) or (8.10). We will call the functions τ_i *global shape functions.*

Eq. (8.12) clearly shows that the approximation $\tilde{u}$ depends upon a finite set of nodal values U_i, which may include values of $\tilde{u}$ as well as its derivatives. Let $F(\Omega)$ denote the function space whose elements may serve as an approximation for the function u; (8.12) shows that (at least for conformal elements) the global shape functions τ_i form the basis of a finite dimensional subspace $\tilde{F}(\Omega)$ which is called an *approximating subspace.*

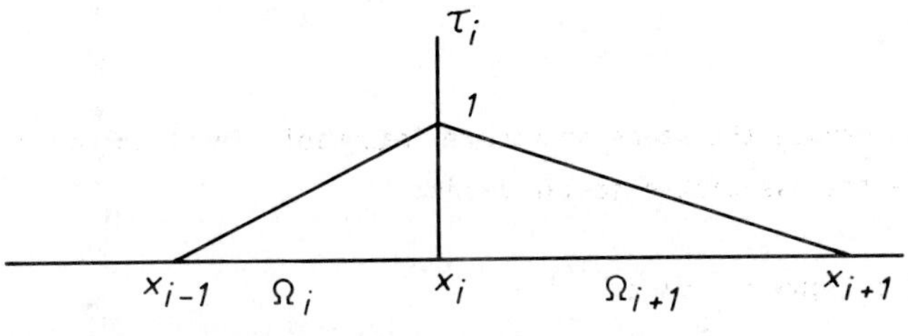

Fig. 8.4 Global shape function associated with the i-th node.

8.3 THE FINITE ELEMENT METHOD

In the previous section, we have described how a given function u may be approximated by some function $\tilde{u}$ which belongs to a finite-dimensional function space. In general, however, we do not know the function u at the outset, but we know that finding an approximate solution of the differential equation reduces to the specification of the nodal values. The central problems of the finite element method are the selection of the type of approximation which is compatible with the equations to be solved, and the determination of the nodal values; the two facets of the problem are actually closely related. To introduce the subject, we will first consider a limited class of problems for which the search for the best approximation to the actual solution is guided by the existence of a minimum principle; more precisely, the exact solution of a differential equation is such that some real-valued functional $\mathcal{J}(u)$ attains a minimum. When such a functional is available, one is led naturally to the Ritz method for defining a best approximation which may be summarized as follows: given a linear approximating subspace $\tilde{F}(\Omega)$, find $\tilde{u} \in \tilde{F}(\Omega)$ such that $\mathcal{J}(\tilde{u})$ is a minimum. Although we will not go into mathematical details, some degree of formalism is needed to understand the technique.

Let us recall the simple linear problem given in §5.1. We seek a function u such that

$$-u_{,xx} = f \,, \qquad 0 \leq x \leq \tfrac{1}{2}\pi \qquad , \tag{8.13}$$

with the boundary conditions

$$u(0) = u_0 \,, \qquad u(\tfrac{1}{2}\pi) = u_N \,, \tag{8.14}$$

and where the subscript x indicates differentiation with respect to x.

For convenience and without loss of generality, in view of the form of (8.13), we will now temporarily assume that the boundary conditions are homogeneous, i.e.

$$u_0 = u_N = 0 \,. \tag{8.15}$$

Let $L^2(\Omega)$ denote the space of square-integrable functions over Ω, and let $<\,;\,>$ denote the associated scalar product,

$$v,w \in L^2(\Omega) \; : \; <v;w> = \int_\Omega v\,w\,dx \,, \tag{8.16}$$

while the L^2-norm is given by

$$\| w \|_{0,\Omega} = <w;w>^{1/2} \,. \tag{8.17}$$

Here, we need to define a subspace of $L^2(\Omega)$ which is called a Sobolev space denoted by $H^1(\Omega)$; a function w belongs to $H^1(\Omega)$ if its H^1-norm, defined by

$$\| w \|_{1,\Omega} = \left[\int_\Omega (w^2 + w_{,x}{}^2) dx \right]^{1/2} \tag{8.18}$$

is finite.

It is clear, for example, that a piecewise polynomial of class C^0 belongs to $H^1(\Omega)$, while a function of class C^{-1} (and not C^0) does not. We further restrict the subspace by considering only those components of $H^1(\Omega)$ which vanish at the end-points of the interval; they form a subspace which is denoted by $H_0^1(\Omega)$.

For any $v \in H_0^1(\Omega)$, we may now define a real-valued functional

$$\mathcal{J}(v) = \frac{1}{2} < v_{,x} ; v_{,x} > - < f;v > \quad , \tag{8.19}$$

where $f \in L^2(\Omega)$ is the right-hand side of (8.13). It is an easy matter to show that solving (8.13) with the homogeneous boundary conditions amounts to finding $u \in H_0^1(\Omega)$ which minimizes the functional $\mathcal{J}(v)$.

At this stage, we have not introduced any approximation; our task now is to discretize the problem. For that purpose, we define a finite-dimensional approximating subspace, denoted by $\tilde{H}_0^1(\Omega)$, and obtained by selecting a finite basis among the components of $H_0^1(\Omega)$. Typically, if the global shape functions introduced in §8.2 belong to $H_0^1(\Omega)$, we may select them as the basis of $\tilde{H}_0^1(\Omega)$. For any $\tilde{v} \in \tilde{H}_0^1(\Omega)$, it is of course possible to calculate $\mathcal{J}(\tilde{v})$. The best approximation of the solution to (8.13) contained in $\tilde{H}_0^1(\Omega)$ is now $\tilde{u}$ such that

$$\mathcal{J}(\tilde{u}) = \inf \mathcal{J}(\tilde{v}) \quad , \qquad \tilde{v} \in \tilde{H}_0^1(\Omega) \ . \tag{8.20}$$

The procedure that we have just described is none other than the classical Ritz technique; the essence of the finite element method lies in the choice of the space $\tilde{H}_0^1(\Omega)$ containing the global shape functions introduced earlier. The reader may wonder why we have selected our approximating subspace in $H_0^1(\Omega)$; this is an important question which will require careful attention in our future developments. The present procedure is based on a minimum principle of the functional $\mathcal{J}$ given by (8.19); it is fairly clear that $v \in H_0^1(\Omega)$ is a sufficient condition for the existence of $\mathcal{J}(v)$. The choice of approximating subspace will of course depend upon the problem that we wish to solve. Consider for example the equation

$$u_{,xxxx} = f \ , \qquad 0 \leq x \leq 1 \ , \tag{8.21}$$

with the homogeneous boundary conditions

$$u(0) = u(1) = 0 \ , \qquad u_{,x}(0) = u_{,x}(1) = 0 \ ; \tag{8.22}$$

the functional to be minimized for obtaining the solution of (8.21) is now

$$\mathcal{J}'(v) = \frac{1}{2} < v_{,xx} \; ; \; v_{,xx} > - < f \; ; \; v > . \tag{8.23}$$

However, the existence of $\mathcal{J}'(v)$ now requires that the norm

$$\| v \|_{2,\Omega} = \left[\int_\Omega (v^2 + v_{,x}^2 + v_{,xx}^2) dx \right]^{1/2} \tag{8.24}$$

be finite, and v must belong to the Sobolev space $H_0^2(\Omega)$. In terms of global shape functions, it is clear that C^1-continuity is required from the elements of the approximating subspace.

We may further proceed with our understanding of the finite element method by introducing the approximation (8.12) in the functional $\mathcal{J}$ defined in (8.19). Since the boundary conditions are homogeneous, (8.12) is replaced by

$$\tilde{u} = \sum_{i=2}^{M-1} U_i \tau_i , \tag{8.25}$$

where we have assumed that U_1 and U_M are the values of u at the boundary points of Ω. Since $\tau_i \in H_0^1(\Omega)$, we can calculate $\mathcal{J}(\tilde{u})$ and obtain

$$\mathcal{J}(\tilde{u}) = \frac{1}{2} \sum_{i,j=2}^{M-1} < \tau_{i,x} ; \tau_{j,x} > U_i U_j - \sum_{i=2}^{M-1} < f \; ; \; \tau_i > U_i . \tag{8.26}$$

Defining the elements of the *stiffness matrix*

$$A_{ij} = < \tau_{i,x} \; ; \; \tau_{j,x} > \tag{8.27}$$

and the *nodal forces*

$$F_i = < f \; ; \; \tau_i > , \tag{8.28}$$

we find that $\mathcal{J}(\tilde{u})$ becomes a quadratic form,

$$\mathcal{J}(\tilde{u}) = \frac{1}{2} \sum_{i,j=2}^{M-1} A_{ij} U_i U_j - \sum_{i=2}^{M-1} F_i U_i . \tag{8.29}$$

Once we decide that the best approximation is the element of $\tilde{H}_0^1(\Omega)$ which minimizes $\mathcal{J}(\tilde{u})$, we find that the nodal values are the solution of the linear system

$$\sum_{j=2}^{M-1} A_{ij} U_j - F_i = 0 , \quad 2 \leqslant i \leqslant M-1 . \tag{8.30}$$

For completeness, let us examine two further points about the boundary conditions. First, we have assumed until now that the boundary conditions are homogeneous; let us now assume that u takes the non-vanishing value U_1 at $x = 0$. We write

$$\tilde{u} = U_1 \tau_1 + \sum_{j=2}^{M-1} U_j \tau_j \quad , \tag{8.31}$$

and by introducing (8.31) in (8.19) we obtain

$$\mathcal{J}(\tilde{u}) = \frac{1}{2} \sum_{i,j=2}^{M-1} < \tau_{i,x} ; \tau_{j,x} > U_i U_j + \frac{1}{2} < \tau_{1,x} ; \tau_{1,x} > U_1^2$$

$$- \sum_{i=2}^{M-1} [< f ; \tau_i > - U_1 < \tau_{1,x} ; \tau_{i,x} >] U_i - < f ; \tau_1 > U_1 \ . \tag{8.32}$$

With a new definition of the nodal force

$$F_i = < f ; \tau_i > - U_1 < \tau_{1,x} ; \tau_{i,x} > = < f ; \tau_i > - A_{1i} U_1 \ , \tag{8.33}$$

we find again that the problem of determining the best approximation reduces to (8.30). Secondly, let us assume that, instead of the boundary conditions (8.14), we impose

$$u(0) = 0 \ , \qquad u_{,x}(\pi/2) = 0 \ . \tag{8.34}$$

We introduce a new subspace $H^1_{0'}(\Omega)$ containing those components of $H^1(\Omega)$ which vanish at $x = 0$. It is now possible to show that solving (8.13) with the boundary conditions (8.34) amounts to finding $u \in H^1_{0'}(\Omega)$ which minimizes the functional $\mathcal{J}(v)$ in (8.19). With the use of a procedure entirely similar to what we have just described we find that the nodal values are the solution of the linear system

$$\sum_{j=2}^{M} A_{ij} U_j - F_i = 0 \ , \qquad 2 \leqslant i \leqslant M \ , \tag{8.35}$$

where U_M is now part of the unknowns. In (8.34), we say that the first equation is an *essential boundary condition*, while the second is a *natural boundary condition*.

Before closing this section, we wish to look at the extremum problem in a slightly different light. Let $\tilde{u}$ be the component of $\tilde{H}^1_0(\Omega)$ which minimizes $\mathcal{J}(u)$ (in the case of homogeneous boundary conditions). We must therefore have

$$\mathcal{J}(\tilde{u} + \lambda\tilde{v}) \geqslant \mathcal{J}(\tilde{u}) \ , \qquad \forall\ \tilde{v} \in \tilde{H}^1_0(\Omega) \ , \tag{8.36}$$

for any scalar λ, and thus

$$\left.\frac{d}{d\lambda}\right|_{\lambda=0} \mathcal{J}(\tilde{u} + \lambda\tilde{v}) = 0 \ . \tag{8.37}$$

In view of the definition (8.19) we have

$$\mathcal{J}(\tilde{u} + \lambda\tilde{v}) = \mathcal{J}(\tilde{u}) + \lambda[< \tilde{u}_{,x} \ ; \tilde{v}_{,x} > - < f \ ; \tilde{v} >] + \lambda^2 \frac{1}{2} < \tilde{v}_{,x} \ ; \tilde{v}_{,x} > \ , \tag{8.38}$$

and (8.37) becomes the variational equation

$$< \tilde{u}_{,x} \ ; \tilde{v}_{,x} > - < f \ ; \tilde{v} > = 0 \ , \qquad \forall\ \tilde{v} \in \tilde{H}^1_0(\Omega) \ . \tag{8.39}$$

Once we replace $\tilde{v}$ in (8.39) by the elements τ_i of the basis of $\tilde{H}^1_0(\Omega)$, we find again the system (8.30). The form (8.39) of the minimum principle will be useful in our further developments.

8.4 METHOD OF WEIGHTED RESIDUALS

Having explained the basic finite element concepts by means of a simple problem where the solution satisfies a minimum principle, we must recognize regretfully that in mathematical physics the existence of such a principle is the exception rather than the rule. In the present section, we make use of the same simple example (8.13) for introducing new concepts, and we proceed as if the solution to that equation did not satisfy a minimum principle. Once the concepts have been explained for that simple example, we will show how they may be generalized to more complex situations.

Our aim is to approximate the solution u of (8.13) by means of some function $\tilde{u}$ depending upon a finite number of parameters and given by (8.12); we must characterize the subspace to which $\tilde{u}$ should belong and find the necessary relations for calculating the U_i's.

The simplest method conceptually is the *collocation method*; we assume that $\tilde{u}$ is such that we may calculate the expression

$$R(\tilde{u}) = -\tilde{u}_{,xx} - f \ ; \tag{8.40}$$

clearly, $R(\tilde{u})$ is a remainder, which shows how well the approximation $\tilde{u}$ satisfies (8.13). Here, $\tilde{u}$ must be C^1-continuous and twice differentiable, since a second derivative appears on the right-hand side of (8.40). In order to calculate the nodal values, we force the differential equation to be satisfied at a number of points z_i equal to the number of unknown nodal values, i.e. (M-2) when $\tilde{u}$ is known at the boundaries of Ω, i.e.

$$R[\tilde{u}(z_i)] = 0 \; , \qquad 1 \leq i \leq M-2 \; . \tag{8.41}$$

Among the elements introduced in §8.2, only those of the type P^3-C^1 may be used for the collocation method applied to (8.13). Although one-dimensional problems are relatively simple, it is already clear that the collocation method requires rather complex elements because of the high degree of regularity of the global functions τ_i.

In order to explain a different approach, we must first reformulate the problem (8.13). Instead of seeking a solution $u \in H_0^2(\Omega)$ which satisfies (8.14), we may equivalently pose the following problem:

$$\text{find} \quad u \in H_0^2(\Omega) : \; < -u_{,xx} - f \; ; \; v > \; = 0 \; , \quad \forall \; v \in L^2(\Omega) \; . \tag{8.42}$$

Let us place a restriction on v, and assume that it belongs to $H_0^1(\Omega)$, i.e. v has a finite H^1-norm defined by (8.18) and vanishes at the boundary points where u has an imposed value; when such is the case we have after an integration by parts

$$< -u_{,xx} \; ; \; v > \; = \; < u_{,x} \; ; \; v_{,x} > \; . \tag{8.43}$$

Since only the first order derivative of u appears on the right-hand side of (8.43), it is meaningful to pose the following problem:

$$\text{find} \quad u \in H_0^1(\Omega) : \; < u_{,x} \; ; \; v_{,x} > - < f \; ; \; v > \; = 0 \; , \quad \forall \, v \in H_0^1(\Omega) \; ; \tag{8.44}$$

we have thus obtained a *weak formulation* of (8.13), which is the basis of our further developments. Note that, in view of (8.40), (8.42) may also be written as:

$$\text{find} \quad u \in H_0^2(\Omega) : \; < R(u) \; ; \; v > \; = 0 \; , \quad \forall \, v \in H_0^1(\Omega) \quad . \tag{8.45}$$

This equation justifies the expression of *method of weighted residuals* for the approach of the present section; more elaborate theoretical developments and numerous examples may be found in Finlayson (1972).

In order to return to the finite element concept, let us define for u an approximating subspace $\tilde{F}(\Omega)$, and a finite dimensional space $\tilde{V}(\Omega)$ of weighting functions $\tilde{v}$. Instead of (8.44) we may then write

$$\text{find } \tilde{u} \in \tilde{F}(\Omega) : < \tilde{u}_{,x} ; \tilde{v}_{,x} > - < f ; \tilde{v} > = 0 , \quad \forall \tilde{v} \in \tilde{V}(\Omega) ; \tag{8.46}$$

the elements of $\tilde{F}(\Omega)$ and $\tilde{V}(\Omega)$ should both belong to $H_0^1(\Omega)$, and the subspaces $\tilde{F}(\Omega)$ and $\tilde{V}(\Omega)$ have the same dimension since the number of equations must equal the number of unknowns. (Note that (8.46) is identical to (8.39)). The most widely used application of the discretized weak formulation (8.46) is the *Galerkin method* in which $\tilde{V}(\Omega)$ is identical to $\tilde{F}(\Omega)$, i.e. the weighting functions are the same as the global shape functions. With the use of (8.12), where the τ_i's must now belong to $H_0^1(\Omega)$, we obtain, from (8.46),

$$\sum_{j=2}^{M-1} < \tau_{j,x} ; \tau_{i,x} > U_j - < f ; \tau_i > = 0 , \quad 2 \leq i \leq M-1 . \tag{8.47}$$

These relations are identical to (8.30), with the definitions (8.27) and (8.28).

The collocation and Galerkin methods may be applied without difficulty to problems where a minimum principle does not exist, whether they are linear or non-linear. As a simple example, consider the problem

$$-u_{,xx} - (u_{,x}^2 + u^2)/2 = f , \quad 0 \leq x \leq 2\pi ,$$
$$u(0) = u(2\pi) = 0 . \tag{8.48}$$

The collocation technique is obvious, since it amounts to selecting an approximation in H_0^2, and to satisfying (8.48) at a finite number of collocation points. For applying the Galerkin method, we start with the weak form of (8.48), with the assumption that $v \in H_0^1$, i.e.

$$< u_{,x} ; v_{,x} > - \frac{1}{2} < u_{,x}^2 + u^2 ; v > - < f ; v > = 0 , \tag{8.49}$$

and next replace u by $\tilde{u}$, v by $\tilde{v}$, where $\tilde{u}$, $\tilde{v}$ belong to the same subspace $\tilde{H}_0^1$. Again, with the use of (8.12), we obtain

$$\sum_{j=2}^{M-1} A_{ij} U_j - \sum_{j,k=2}^{M-1} B_{ijk} U_j U_k = F_i , \quad 2 \leq i \leq M-1 , \tag{8.50}$$

where A_{ij} and F_i are given by (8.27) and (8.28), and

$$B_{ijk} = \frac{1}{2} < \tau_{j,x} \tau_{k,x} ; \tau_i > + \frac{1}{2} < \tau_j \tau_k ; \tau_i > . \tag{8.51}$$

Once the coefficients A_{ij}, B_{ijk} and the nodal forces F_i have been calculated, finding an approximate solution to the continuous problem (8.50) reduces to the solution of a finite non-linear algebraic system of equations.

8.5 CONSTRUCTION OF THE ALGEBRAIC SYSTEM

Now that we have explained the basic ideas behind the finite element method, it is desirable to explain briefly the main steps which must be accomplished for calculating the coefficients of the algebraic system which must be solved for generating the solution. Again, we will take (8.13) and its discretized form (8.30) as a typical example.

With the knowledge of the global shape functions τ_i and the function f, we must calculate the components A_{ij} of the stiffness matrix and the components F_i of the nodal force vector, given respectively by (8.27) and (8.28). We have seen earlier that a shape function τ_i is associated with node i, and vanishes outside the element(s) to which node i belongs. Let Ω_e and Ω_{e+1} be two such elements. Instead of calculating A_{ij} over the whole domain Ω, it is sufficient to perform the integration over these two elements only, and thus,

$$\begin{aligned} A_{ij} &= \int_{\Omega_e} \tau_{i,x}\,\tau_{j,x}\,dx + \int_{\Omega_{e+1}} \tau_{i,x}\,\tau_{j,x}\,dx\,, \\ F_i &= \int_{\Omega_e} f\,\tau_i\,dx + \int_{\Omega_{e+1}} f\,\tau_i\,dx\,. \end{aligned} \tag{8.52}$$

It is clear that most coefficients A_{ij} will vanish identically since τ_j vanishes everywhere in Ω_e and Ω_{e+1} unless node j belongs to one of these elements. In most finite element problems, the stiffness matrix A_{ij} will be sparse, and some special techniques are needed for avoiding the storage of numerous zeros. On Fig. 8.5, we show a shape function τ_i of the type P^1-C^0; node i belongs to Ω_i and Ω_{i+1}. Here $\tau_i = \phi_2$ in Ω_i, and $\tau_i = \phi_1$ in Ω_{i+1}. When node i is not a boundary node we have,

$$\begin{aligned} A_{i,i-1} &= \int_{\Omega_i} \phi_{1,x}\,\phi_{2,x}\,dx\,, \qquad A_{i,i} = \int_{\Omega_i} \phi_{2,x}^2\,dx + \int_{\Omega_{i+1}} \phi_{1,x}^2\,dx\,, \\ A_{i,i+1} &= \int_{\Omega_{i+1}} \phi_{1,x}\,\phi_{2,x}\,dx\,, \\ A_{ij} &= 0 \quad \text{for} \quad j \neq i-1\,,\ i\,,\ i+1\,. \end{aligned} \tag{8.53}$$

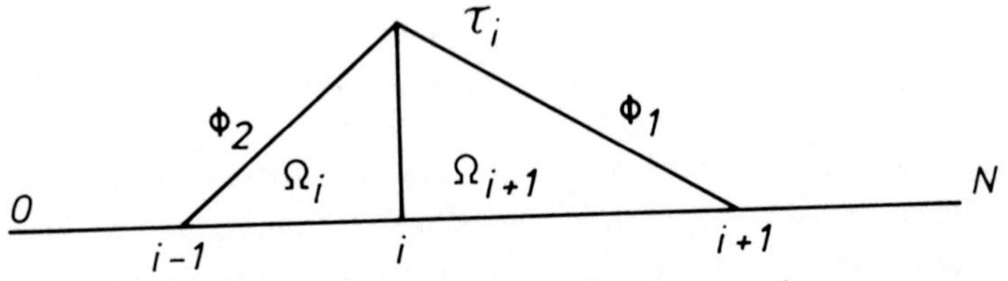

Fig. 8.5 Shape function τ_i of the P^1-C° type.

Similarly, we have

$$F_i = \int_{\Omega_i} f\,\phi_2\,dx + \int_{\Omega_{i+1}} f\,\phi_1\,dx\,. \tag{8.54}$$

Equations (8.53) and (8.54) suggest an easy way of calculating the components A_{ij} and F_i. Instead of calculating them over the whole domain Ω, we may consider separately every individual element, and calculate a *local stiffness matrix* and a *local nodal force vector* for that element. For example, consider element Ω_i on Fig. 8.5; we establish the relationship

$$\begin{aligned} &\text{node 1 of } \Omega_i = \text{global node } (i-1)\,, \\ &\text{node 2 of } \Omega_i = \text{global node } i\,, \end{aligned} \tag{8.55}$$

and calculate the local components,

$$\begin{aligned} &A^i_{11} = \int_{\Omega_i} \phi^2_{1,x}\,dx\,, \qquad A^i_{22} = \int_{\Omega_i} \phi^2_{2,x}\,dx\,, \\ &A^i_{12} = A^i_{21} = \int_{\Omega_i} \phi_{1,x}\,\phi_{2,x}\,dx\,. \end{aligned} \tag{8.56}$$

From the relationship (8.55) we know that A^i_{11} is part of $A_{i-1,i-1}$, A^i_{22} is part of A_{ii}, etc. Thus, whenever a local stiffness matrix has been constructed for an element, the global stiffness matrix is *assembled* on the basis of (8.55). Eq. (8.53_2) shows for example that

$$A_{i,i} = A^i_{22} + A^{i+1}_{11}\,. \tag{8.57}$$

Apart from the book-keeping operations involved in assembling the global matrices, the calculation reduces to the evaluation of scalar products over a single element, e.g.

$$A_{ij}^{e} = \int_{\Omega_e} \tau_{i,x}\, \tau_{j,x}\, dx \quad . \tag{8.58}$$

A common procedure is to reduce the domain of integration to the parent element ω shown on Fig. 8.2; using the transformation (8.3) we may write

$$\int_{\Omega_e} \tau_{i,x}\, \tau_{j,x}\, dx = \int_{\omega} \tau_{i,\xi}\, \tau_{j,\xi} \frac{d\xi}{dx}\, d\xi \quad . \tag{8.59}$$

For achieving the integration, one may use numerical quadrature. Consider an integral of the form

$$I = \int_{-1}^{1} k(\xi)\, d\xi \quad , \tag{8.60}$$

of the same form as the right-hand side of (8.59). The Gauss-Legendre quadrature rules specify a set of n *sampling points* ξ_i and the associated *weights* w_i such that, when $k(\xi)$ is a polynomial of order (2n-1) or less, I is given exactly by

$$I = \sum_{i=1}^{n} k(\xi_i)\, w_i \quad . \tag{8.61}$$

Sampling points and weights for values of n up to 3 are given in Table 8.1. Since ω extends from $\xi = -1$ to $\xi = 1$, it is clear that the right-hand side of (8.59) may be evaluated quite easily by numerical quadrature, which may be programmed without difficulty. Further details on numerical integration may be found, for example, in Zienkiewicz (1977) or Strang and Fix (1973).

TABLE 8.1

Gauss-Legendre quadrature points and weights

n	ξ_i	w_i
1	0	2
2	$-1/\sqrt{3}$	1
	$1/\sqrt{3}$	1
3	$-\sqrt{0.6}$	5/9
	0	8/9
	$\sqrt{0.6}$	5/9

8.6 SOLUTION OF THE ALGEBRAIC SYSTEM

In most problems which will be discussed throughout the next two chapters the final system consists of non-linear algebraic equations which have the general form,

$$Z_i(\underset{\sim}{U}) = F_i \ , \quad 1 \leq i \leq M \ , \tag{8.62}$$

where Z_i is a non-linear function of the U_j's, which we write for brevity as the vector $\underset{\sim}{U}$. A typical example is given by the system (8.50), to which we add the first and last equations

$$U_1 = 0 \ , \quad U_M = 0 \ . \tag{8.63}$$

The system (8.62) cannot be handled unless one introduces some kind of linearization; typically, one calculates a sequence

$$U_j^1 \ , \ U_j^2 \ , \ \ldots \ , \ U_j^{n-1} \ , \ U_j^n \ , \ U_j^{n+1} \ , \ \ldots \tag{8.64}$$

starting from an initial guess by means of an iterative technique, which is stopped when a pre-defined convergence criterion is satisfied.

The most widely used iterative procedure is Newton's method which has been discussed in §5.3. Briefly, the (n+1)-th iteration is written as follows,

$$U_j^{n+1} = U_j^n + \delta U_j \ , \tag{8.65}$$

and it is assumed at the outset that δU_j is small. By substituting U_j^{n+1} for U_j in (8.62), expanding Z_i in terms of δU_j, and neglecting terms of order higher than the first, we obtain

$$\sum_{j=1}^{M} \frac{\partial Z_i}{\partial U_j}(\underset{\sim}{U}^n)\, \delta U_j = F_i - Z_i(\underset{\sim}{U}^n) \ . \tag{8.66}$$

As a general rule the iterative procedure converges to a solution when the right-hand side of (8.66) tends to zero in some specified sense; indeed, by comparing (8.62) and the right-hand side of (8.66), we find that the latter should vanish for the exact solution. In practice we claim that a solution is found when

$$|F_i - Z_i(\underset{\sim}{U}^n)| < \varepsilon \ , \quad 1 \leq i \leq M \ , \tag{8.67}$$

where ε is a fixed positive constant. Another verification is the relative convergence test; let $\|\underset{\sim}{U}^n\|$ be a norm of the solution after the nth iteration, e.g.

$$\|\underset{\sim}{U}^n\| = \sup_k |U_k^n| . \tag{8.68}$$

The relative convergence test is satisfied when

$$|\delta U_i| / \|\underset{\sim}{U}^n\| < \eta , \quad 1 \leq i \leq M , \tag{8.69}$$

where again η is a fixed positive constant. As an example of application of Newton's method, the linearized system obtained on the basis of (8.50) with the application of (8.66) is

$$\sum_{j=1}^{M} \left[A_{ij} - \sum_{k=1}^{M} (B_{ijk} + B_{ikj}) U_k^n \right] \delta U_j = F_i - \sum_{j=1}^{M} \left[A_{ij} - \sum_{k=1}^{M} B_{ijk} U_k^n \right] U_j^n . \tag{8.70}$$

When Newton's method converges, the rate of convergence is (usually) quadratic; for the definition of quadratic convergence, the reader is referred to §5.3. A key to its success is that the initial guess lies within the *domain of convergence* around the exact solution.

In some problems which will be discussed in Chapter 9, the calculation of the matrix $\partial Z_i/\partial U_j$ in (8.66) may present great analytical and computational difficulties, and it may be necessary to resort to another type of linearization. Typically, some of the unknowns on the left-hand side of (8.62) are replaced by their values after the nth iteration, in such a way that the system becomes linear in the U_j^{n+1}'s. Consider again (8.50); a linearization procedure might be

$$\sum_{j=1}^{M} A_{ij} U_j^{n+1} = F_i + \sum_{j,k=1}^{M} B_{ijk} U_j^n U_k^n , \tag{8.71}$$

or

$$\sum_{j=1}^{M} \left(A_{ij} - \sum_{k=1}^{M} B_{ijk} U_k^n \right) U_j^{n+1} = F_i \quad . \tag{8.72}$$

Several possibilities are in general available (cf. §5.3); however, the rate of convergence is linear if there is convergence at all.

It is clear that, whether the problem is linear or non-linear, the search for the best approximation will result in the solution of large linear systems of algebraic equations. While iterative procedures are commonly used in finite difference problems, most finite element programs proceed with the Gaussian elimination. A problem of major concern is the bandwidth of the stiffness matrix and its storage; when a problem contains a few thousand variables, the storage of the stiffness matrix as such is ruled out, and, moreover, most of its terms vanish identically. Reviewing the numerical techniques for solving large linear systems associated with finite elements lies beyond the scope of the present book; the reader is referred to specialized texts for details (see e.g. Gladwell and Wait 1979, Wait 1979). A very efficient technique is the *frontal elimination* procedure, introduced by Irons (1970), and which is now a standard tool of several finite element codes.

8.7 EXAMPLES

The examples solved in Chapter 5 by means of finite differences are now considered again by means of finite elements. Use is made of the simple linear element where the shape functions are given by (8.5). The discrete solution of the linear examples 1 and 2 is such that the nodal values are exact, apart from the round-off error produced by the Gaussian elimination. It was observed by Strang and Fix (1973) that this is true in general when the shape functions can solve exactly the homogeneous differential equation. Of course, this does not mean that the finite element solution is identical to the analytical solution, since the former is given by a linear interpolation between the nodal values.

The third example, which is non-linear, corresponds to (5.4) with $\alpha = 1$, with (5.6) as the solution. When the domain of integration is divided into 5 elements, and use is made of Newton's method, it is found that the norm defined by (8.67) decreases as follows: 0.26 after 2 iterations, $0.70\ 10^{-2}$ after 3 iterations, and $0.52\ 10^{-5}$ after 4 iterations. Table 8.2 compares the analytical to the finite element solution; the error relative to the maximum value of u is less than 0.04%.

TABLE 8.2
Example 3. Comparison between the analytical and the finite element solution.

x	0	$\pi/10$	$2\pi/10$	$3\pi/10$	$4\pi/10$	$\pi/2$
$1 - \sin x$	1.	0.6910	0.4122	0.1910	0.0489	0.
$\tilde{u}$	1.	0.6905	0.4118	0.1908	0.0489	0.

In order to gain a better insight into the finite element behaviour, we want to consider a fifth example, which is posed as follows,

$$u_{,xx} + \frac{1}{2}(u_{,x}^2 + u^2) = 1 \ , \qquad 0 \leq x \leq \pi \ ,$$
$$u(0) = 0 \ , \qquad u(\pi) = 2 \ ; \tag{8.73}$$

it is verified easily that $u = 1 - \cos x$ is a solution of (8.73). The problem has been solved successively with a division of the domain of integration into 4, 8, 16 and 32 elements; the results are shown graphically on Fig. 8.6a and are summarized as Table 8.3, together with the number of iterations ($\varepsilon = 10^{-4}$).

TABLE 8.3
Example 5. Comparison between the analytical and the finite element solutions.

x	0	$\pi/4$	$\pi/2$	$3\pi/4$	π	Iterations
$1 - \cos x$	0.	0.2929	1.0000	1.7071	2.	-
4 elements	0.	0.0892	0.6873	1.4740	2.	6
8 elements	0.	0.1821	0.8376	1.5897	2.	6
16 elements	0.	0.2350	0.9168	1.6476	2.	7
32 elements	0.	0.2622	0.9562	1.6760	2.	7

If we consider in particular the column $x = \pi/2$, we find that the error of the finite element solution relative to the maximum value of u is 16% for 4 elements, 8% for 8, 4% for 16 and 2% for 32 elements. It is thus obvious that the discretization error in solving (8.73) with simple linear elements varies like the size h of the elements.

Finally, it is worth mentioning a sixth example, which is posed as follows,

$$u_{,xx} + \frac{1}{2}(u_{,x}^2 + u^2) = 1 \ , \qquad 0 \leq x \leq \pi \ ,$$
$$u(0) = 0 \ , \qquad u_{,x}(\pi) = 0 \ , \tag{8.74}$$

which is similar to (8.73) except that the essential boundary condition at $x = \pi$ has been replaced by a natural boundary condition. An analytical solution is again given by $u = 1 - \cos x$.

Starting from $\tilde{u} = 0$ as an initial guess, it is found that Newton's method converges rapidly to a solution shown on Fig. 8.6b which is totally different from the expected solution. It is clear that the discretized form of (8.74) has multiple solutions, and that the initial guess $\tilde{u} = 0$ does not lie in the circle of convergence around the expected solution.

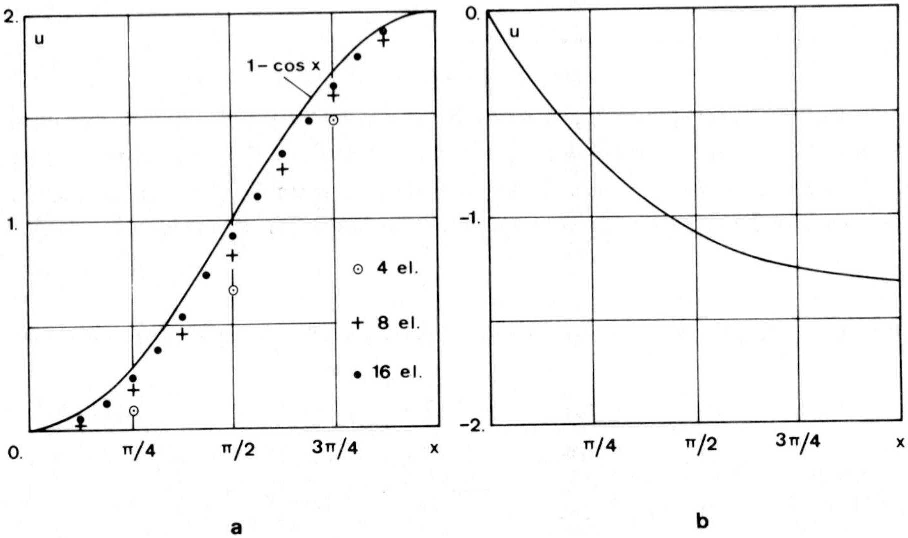

Fig. 8.6 Finite element solution of example 5 and example 6.

8.8 TWO-DIMENSIONAL PROBLEMS. TRIANGULAR AND RECTANGULAR ELEMENTS

In order to facilitate the understanding of the finite element method, we have until now limited our presentation to one-dimensional problems. The main ideas behind the finite element method applied to two-dimensional problems are the same as those which we have explained in the previous sections: i) a weighted residuals formulation of the governing partial differential equation is sought for discretizing the problem; ii) the two-dimensional domain Ω is covered by a mesh of finite elements; iii) a type of approximation satisfying the continuity criterion imposed by the weighted residuals formulation is selected for the unknown function; iv) nodes and nodal values of the unknown function are associated with the elements; v) the application of the weighted residuals formulation results in an algebraic system in terms of the nodal values. The main difference between one-dimensional and two-dimensional problems lies in the nature of the finite elements used for approximating a function. We will first elaborate on the type of finite element approximations used in fluid mechanics.

Let us consider, on Fig. 8.7, a domain Ω covered by a mesh of triangular elements. We assume for the time being that the boundary Γ of Ω consists of straight edges; curved boundaries will be discussed at a later stage. The question to be answered is how to define an approximation of some function u in terms of piecewise polynomials defined over the triangular elements Ω_e.

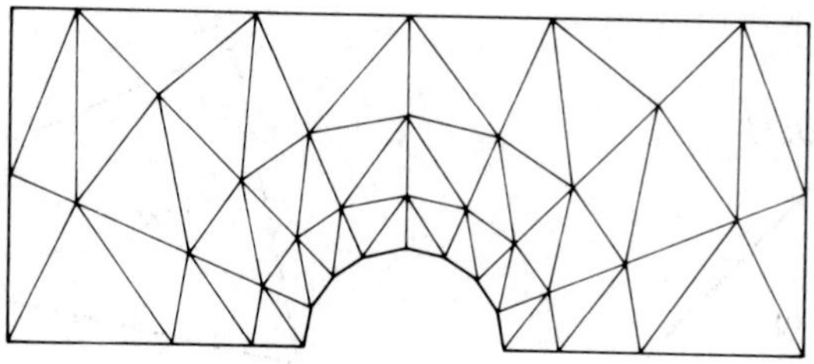

Fig. 8.7 Domain Ω covered by a mesh of triangular elements.

The trivial case is of course an approximation in terms of zero-order piecewise polynomials. The approximate function $\tilde{u}$ is then a constant in each element, and is in general piecewise-continuous over Ω; we will say that such an approximation is of the P^0-C^{-1} type, where P^0 refers to the order of the polynomial and C^{-1} to the discontinuous nature of the approximation.

More interesting is the case of an approximation in terms of piecewise first-order polynomials. A first-order polynomial is characterized by three coefficients, which means that three nodal values will be sufficient to define the approximation over an element, and values of u at the vertices are obvious candidates. The shape functions associated with the nodal values are then first-order polynomials ϕ_i^t, $i = 1,2,3$ which assume the value one at the appropriate node and vanish at the other two. In order to obtain easily the form of the shape functions, let us consider a 1 : 1 affine mapping of the triangular element Ω_e onto a parent triangular element ω (Fig. 8.8) defined in the (ξ,η) plane. Let i,j,k denote the vertices of Ω_e, with respective coordinates (x_i,y_i), (x_j,y_j), (x_k,y_k). The correspondence between ω and Ω_e is given by

$$x = (1 - \xi - \eta)x_i + \xi x_j + \eta x_k \ ,$$
$$y = (1 - \xi - \eta)y_i + \xi y_j + \eta y_k \ . \qquad (8.75)$$

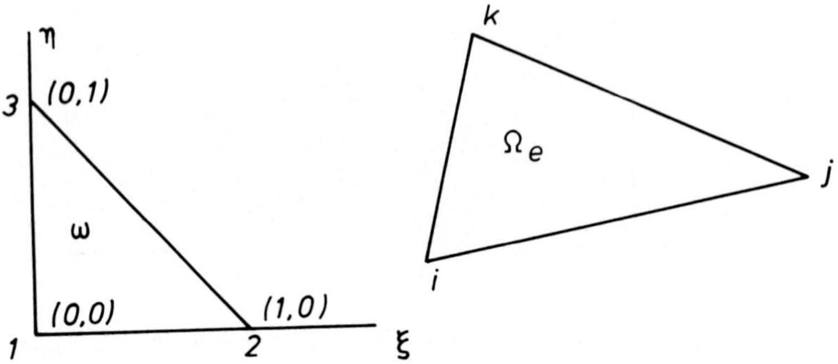

Fig. 8.8 Mapping of a triangular element onto its parent element.

The shape functions on the parent element are first-order polynomials which take the unit value at one node and vanish at the others; one obtains easily

$$\phi_1^t = 1 - \xi - \eta \ , \qquad \phi_2^t = \xi \ , \qquad \phi_3^t = \eta \ . \tag{8.76}$$

The shape functions on Ω_e are then given by $\phi_m^t[\xi(x,y),\ \eta(x,y)]$, $m = 1,2,3$, where $\xi(x,y)$ and $\eta(x,y)$ denote the inverse of (8.75). In future, we will not indicate explicitly whether a shape function is given in terms of the (ξ,η) or the (x,y) coordinates. An interesting feature of this element is the continuity property of the approximation which it generates. Consider on Fig. 8.9a two neighbouring elements with two common nodes, and a view of the function $\tilde{u}$ identified at the nodes; it is clear that the value of $\tilde{u}$ between two nodes depends solely upon the nodal values of $\tilde{u}$ at these nodes, and thus $\tilde{u}$ is continuous across element boundaries. The triangular element that we have just described may thus generate an approximation of the P^1-C^0 type.

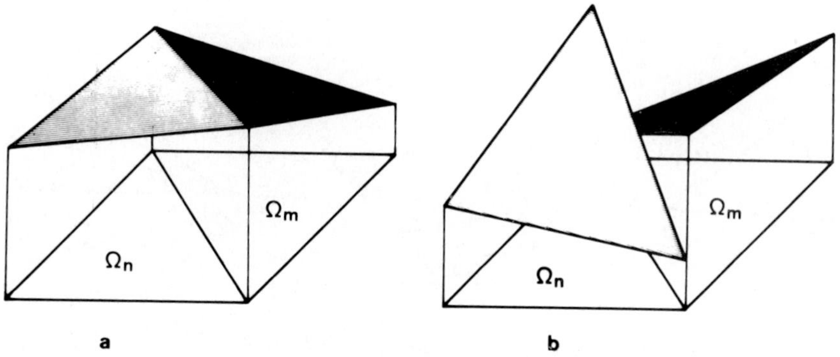

Fig. 8.9 Triangular elements of the P^1-C° and P^1-C^{-1} types.

We may think of other first-order polynomial approximations over an element. We might for example identify the nodal values at the midside nodes, and write the related shape functions as

$$\phi_1' = 1 - 2\eta \ , \qquad \phi_2' = -1 + 2(\xi + \eta) \ , \qquad \phi_3' = 1 - 2\xi \ ; \tag{8.77}$$

each of them takes the unit value at one midside node and vanishes at the others. There is however a fundamental difference between the first-order finite element approximations defined by the shape functions (8.76) and (8.77), respectively. Consider on Fig. 8.9b the common edge between two adjacent triangles; when the approximation is a linear interpolation between midside nodes, the function $\tilde{u}$ may be discontinuous, since along an edge it depends upon nodal values which this time are not located on the edge. We will therefore assign the P^1-C^{-1} label to the approximation generated by that element.

Let us now turn to second-order polynomials, which are identified by means of six coefficients; it is thus desirable to select six nodal values which will allow the calculation of the coefficients. An immediate choice is the value of $\tilde{u}$ at the vertices and at the midside nodes of the triangular element; the shape functions, which form the basis for these second-order polynomials, take the unit value at the related node and vanish at the others. On Fig. 8.10 we show the parent element and the numbering of the nodes, together with two typical shape functions. In the parent element, they are given by

$$\begin{aligned} &\psi_1^t = 1 - 3(\xi+\eta) + 2(\xi+\eta)^2 \ , \quad &&\psi_2^t = \xi(2\xi - 1) \ , \quad &&\psi_3^t = \eta(2\eta - 1) \ , \\ &\psi_4^t = 4\xi(1 - \xi - \eta) \ , &&\psi_5^t = 4\xi\eta \ , &&\psi_6^t = 4\eta(1 - \xi - \eta) \ . \end{aligned} \tag{8.78}$$

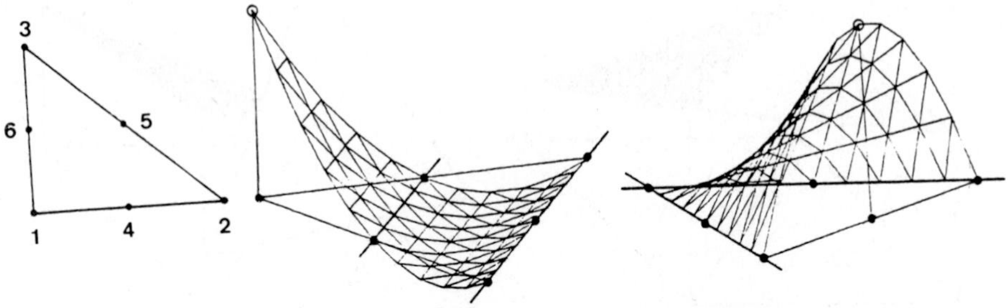

Fig. 8.10 Triangular element of the P^2- C° type.

Again, one finds easily that the value of u along a side of a triangle is fully determined by its nodal values along that side, and we conclude that the element generates for $\tilde{u}$ a continuous approximation, which is of the P^2-C^0 type.

There exist more elaborate elements than those which we have just presented; they have not as yet been used in fluid mechanics, and we will not discuss triangular elements any further. In generating finite element meshes, it is customary to mix triangular and quadrilateral elements; several examples will be given in the next two chapters. The generation of shape functions on arbitrary quadrilaterals will be considered in the next section; first, however, we will study some classical shape functions on rectangular elements. Consider on Fig. 8.11 a rectangular element Ω_m, identified by its vertices i,j,k,ℓ, and the square parent element ω such that $-1\leq\xi,\eta\leq 1$. The coordinates (x,y) and (ξ,η) are related by the equations

$$\xi = (2x - x_i - x_j)/(x_i - x_j) , \qquad \eta = (2y - y_j - y_k)/(y_j - y_k) . \tag{8.79}$$

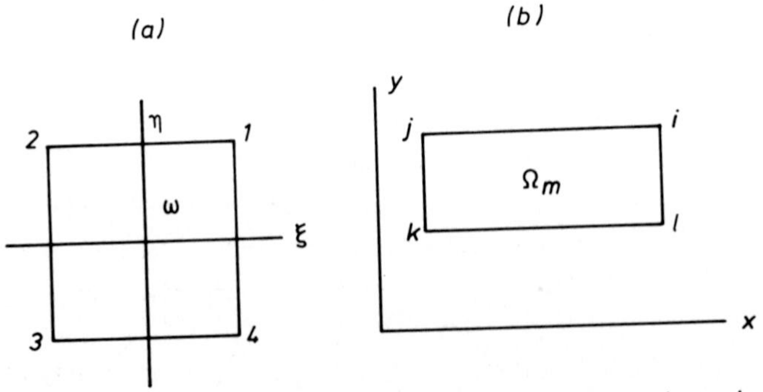

Fig. 8.11 Isomorphism between a rectangular element and its parent element.

An approximation of the P^0-C^{-1} type is trivial and corresponds to constant shape functions on Ω_m. We note immediately that if the value of the function u is given at the vertices i,j,k,ℓ, it is impossible to construct a first-order polynomial as an interpolation, since it contains only three coefficients. Consider, however, on the parent element bilinear shape functions defined as follows,

$$\phi_1^q = (1+\xi)(1+\eta)/4 \;, \qquad \phi_2^q = (1-\xi)(1+\eta)/4 \;,$$
$$\phi_3^q = (1-\xi)(1-\eta)/4 \;, \qquad \phi_4^q = (1+\xi)(1-\eta)/4 \;. \qquad (8.80)$$

The shape function ϕ_i^q takes the value one at node i and vanishes at the others, and varies linearly along the edges; the functions (8.80) constitute therefore a perfect candidate for a continuous approximation. Moreover, it is easy to show that an arbitrary first-order polynomial may be generated as a linear combination of the ϕ_i's. We may therefore classify the approximation generated by these shape functions as one of the P^1-C^0 type.

The generation of two-dimensional shape functions by means of the product of one-dimensional shape functions, as may be found in (8.80), is the characteristic feature of the *Lagrangian family* of rectangular elements. In Fig. 8.12a, we show a nine-node Lagrangian element; its shape functions are obtained by the cross-multiplication of the functions (8.6) in the directions ξ and η, respectively. The shape functions associated with the nodes numbered on Fig. 8.12a are easily shown to be

$$\psi_1^q = \xi(1+\xi)\eta\,(1+\eta)/4 \;, \quad \psi_2^q = -\xi(1-\xi)\eta\,(1+\eta)/4 \;, \quad \psi_3^q = \xi(1-\xi)\eta\,(1-\eta)/4 \;,$$
$$\psi_4^q = -\xi(1+\xi)\eta\,(1-\eta)/4 \;, \quad \psi_5^q = (1-\xi^2)\eta\,(1+\eta)/2 \;, \quad \psi_6^q = -\xi(1-\xi)\,(1-\eta^2)/2 \;,$$
$$\psi_7^q = -(1-\xi^2)\eta\,(1-\eta)/2 \;, \quad \psi_8^q = \xi(1+\xi)(1-\eta^2)/2 \;, \quad \psi_9^q = (1-\xi^2)(1-\eta^2) \;.$$
$$\ldots \qquad (8.81)$$

Any second-order polynomial can be generated by a linear combination of the shape functions (8.81); however, they contain third- and fourth-order terms which may be considered as parasitic since complete third- and fourth-order polynomials cannot be formed by a linear combination of (8.81). On an edge of the rectangle, the approximation is fully characterized by the three nodal values on that edge, and we conclude that the approximation is of the P^2-C^0 type.

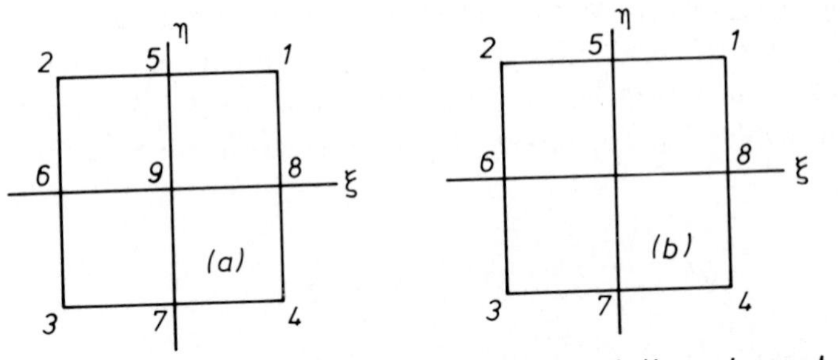

Fig. 8.12 Lagrangian element and serendipity element of the P^2- C° type.

It is also possible to generate an approximation of the P^2-C^0 type by means of an element with no internal node, which belongs to the so-called *serendipity family*. The nodes are numbered on Fig. 8.12b, and the shape functions are given by

$$\begin{aligned}
\psi'_1 &= (1+\xi)\,(1+\eta)\,(\xi+\eta-1)/4 \;, & \psi'_5 &= (1-\xi^2)\,(1+\eta)/2 \;,\\
\psi'_2 &= (1-\xi)\,(1+\eta)\,(-\xi+\eta-1)/4 \;, & \psi'_6 &= (1-\xi)\,(1-\eta^2)/2 \;,\\
\psi'_3 &= (1-\xi)\,(1-\eta)\,(-\xi-\eta-1)/4 \;, & \psi'_7 &= (1-\xi^2)\,(1-\eta)/2 \;,\\
\psi'_4 &= (1+\xi)\,(1-\eta)\,(\xi-\eta-1)/4 \;, & \psi'_8 &= (1+\xi)\,(1-\eta^2)/2 \;.
\end{aligned} \tag{8.82}$$

Again, arbitrary second-order polynomials can be generated by a linear combination of these eight shape functions, which do also contain terms of the $\xi^2\eta$ and $\xi\eta^2$ type.

8.9 ISOPARAMETRIC ELEMENTS

In the previous section, we have shown how a function may be approximated on a mesh made of triangular and rectangular elements. The rectangular shape of an element is a serious limitation in the applications; moreover, it is sometimes desirable to insert elements with curvilinear boundaries. Arbitrary quadrilaterals and curvilinear elements may be generated by means of *isoparametric transformations* of the rectangular and triangular elements which we will now explain.

Let us return to the mapping of a triangular element onto its parent element expressed by (8.75); in view of (8.76), (8.75) may also be written as follows,

$$x = \sum_{i=1}^{3} x_i \phi_i^t(\xi,\eta) \;, \qquad y = \sum_{i=1}^{3} y_i \phi_i^t(\xi,\eta) \;, \tag{8.83}$$

and we find that the coordinate transformation is expressed in terms of the shape functions. In fact, these relations allow us to transform the parent element into a triangle of arbitrary shape. The same idea is applied for transforming a square parent element into an arbitrary quadrilateral, shown for example on Fig. 8.13.

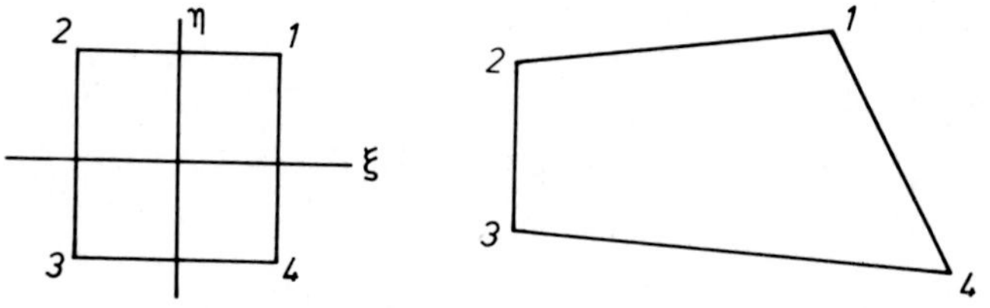

Fig. 8.13 Parent element and arbitrary quadrilateral.

Let (x_i, y_i) denote the coordinates of the i-th corner of the arbitrary quadrilateral; the 1 : 1 polynomial mapping of the parent element onto the arbitrary quadrilateral may be written as follows

$$x = \sum_{i=1}^{4} x_i \phi_i^q(\xi,\eta) , \qquad y = \sum_{i=1}^{4} y_i \phi_i^q(\xi,\eta) . \tag{8.84}$$

The ϕ_i^q's, defined by (8.80), contain a $\xi\eta$ term; the transformation (8.84) is not affine, and the Jacobian

$$J = \det \left[\frac{\partial(x,y)}{\partial(\xi,\eta)} \right] \tag{8.85}$$

is not a constant in general. It will be assumed that the arbitrary quadrilateral is such that J in (8.85) does not vanish; the element is known as the Taig quadrilateral. Since (8.84) may then be inverted (at least formally), it is possible to obtain, for the quadrilateral, shape functions of the form $\phi_i^q[\xi(x,y), \eta(x,y)]$; it is verified easily that ϕ_i^q takes the unit value at the i-th corner and vanishes at the others, and varies linearly between the corners. Moreover, any first-order polynomial can be generated by a linear combination of these shape functions. We have thus obtained the necessary shape functions for an approximation of the type P^1-C^0 on an arbitrary quadrilateral.

The transformation (8.84) may also be used for defining higher-order shape functions on the arbitrary quadrilateral shown on Fig. 8.12. For example, we may use the shape functions ψ_i^q given by (8.81), and obtain, in the x-y plane, the shape functions $\psi_i^q[\xi(x,y), \eta(x,y)]$. Again, it is easy to show that the approximation is C^0-continuous, and that an arbitrary second-order polynomial may be generated as a linear combination of these shape functions. In order to see this, let us consider an arbitrary second-order polynomial given by $(a + bx + cy + dx^2 + exy + fy^2)$; if the assertion is true, we must be able to find coefficients $a_1, \ldots, a_9$ such that

$$a + bx + cy + dx^2 + exy + fy^2 \equiv \sum_{i=1}^{9} a_i \psi_i^q[\xi(x,y), \eta(x,y)], \tag{8.86}$$

or, in view of (8.84),

$$a+b\, x(\xi,\eta) + c\, y(\xi,\eta) + d\, x(\xi,\eta)^2 + e\, x(\xi,\eta)\, y(\xi,\eta) + f\, y(\xi,\eta)^2 \equiv \sum_{i=1}^{9} a_i \psi_i^q(\xi,\eta) . \tag{8.87}$$

It is easy to see, in view of (8.84) and (8.80), that the left-hand side of (8.87) is written in terms of the nine monomials

$$1,\ \xi,\ \eta,\ \xi^2,\ \xi\eta,\ \eta^2,\ \xi^2\eta,\ \xi\eta^2,\ \xi^2\eta^2\ , \tag{8.88}$$

which are also found in the functions $\psi_i^q(\xi,\eta)$; it is therefore possible to identify term by term both sides of (8.87), and to determine the nine coefficients a_i as a function of the coefficients of the polynomial.

The same type of transformation allows the generation of serendipity elements with an arbitrary quadrilateral shape. However, it is not possible to generate an arbitrary second-order polynomial through a linear combination of the shape functions, since the ψ_i'''s given by (8.82) do not contain terms of the type $\xi^2\eta^2$.

The coordinate transformations (8.83) and (8.84) have established a 1 : 1 correspondence between the parent element and an arbitrary straight-edged triangular or quadrilateral element. Consider now the transformations

$$x = \sum_{i=1}^{6} x_i\ \psi_i^t(\xi,\eta)\ , \qquad y = \sum_{i=1}^{6} y_i\ \psi_i^t(\xi,\eta)\ , \tag{8.89}$$

and

$$x = \sum_{i=1}^{9} x_i\ \psi_i^q(\xi,\eta)\ , \qquad y = \sum_{i=1}^{9} y_i\ \psi_i^q(\xi,\eta)\ . \tag{8.90}$$

When the coordinates (x_i,y_i) of the nodes are given (see for example Fig. 8.14), (8.89) and (8.90) transform the straight edges of the parent element into parabolic segments.

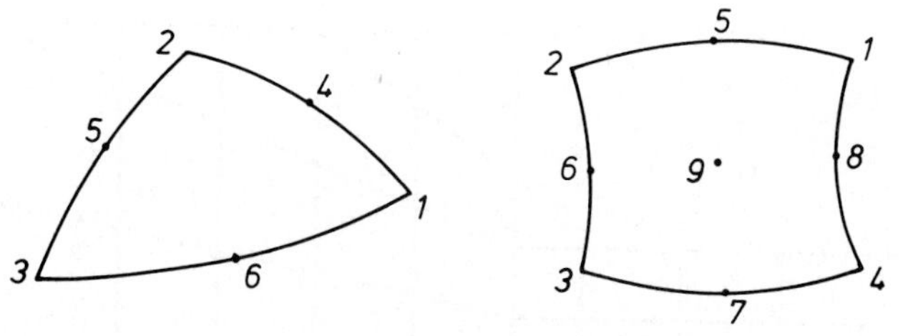

Fig. 8.14 Curvilinear isoparametric elements.

It is easy to verify that two adjacent elements generated by means of (8.89) or (8.90) will enjoy common curvilinear edges (without overlap or holes). The shape functions on the distorted elements are again given by $\psi_i^t[\xi(x,y), \eta(x,y)]$ and $\psi_i^q[\xi(x,y), \eta(x,y)]$, respectively. The approximation based on such shape functions is C^0-continuous; however, it is not possible in general to form an arbitrary second-order polynomial as a linear combination of the shape functions.

A finite element obtained from the parent element by means of the shape functions used for defining the approximation on the element is called an *isoparametric element*. If the order of the shape functions used for the coordinate transformation is lower than the order of those used for the approximation, the element is *subparametric*.

8.10 METHOD OF WEIGHTED RESIDUALS

In the previous sections, we have shown how a function may be interpolated between its nodal values identified at a finite number of nodes on an element. In view of what we have explained for one-dimensional problems, it is now easy to introduce a global representation of the approximation over the whole domain Ω. Consider on Fig. 8.15 a mesh of triangular and quadrilateral elements. A specific node i belongs to several elements; within each of these elements, we have defined a shape function which takes the unit value at that node and vanishes at the others. The union of these shape functions forms a global shape function τ_i associated with node i, and we may write the approximation over Ω as

$$\tilde{u} = \sum_{i=1}^{M} U_i \tau_i , \tag{8.91}$$

where M is the number of nodes and U_i is the nodal value at node i.

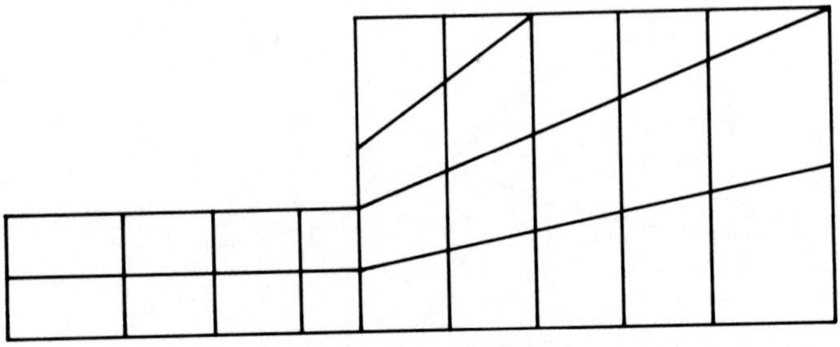

Fig. 8.15 Typical finite element mesh.

Before going further, we need some definitions.

The space $L^2(\Omega)$ of square-integrable functions contains those functions w which have a finite L^2-norm,

$$\|w\|_{0,\Omega} = \left[\int_\Omega w^2 \, d\Omega \right]^{1/2} . \tag{8.92}$$

The approximations of the C^{-1} type which we have introduced earlier clearly belong to $L^2(\Omega)$. The Sobolev-space $H^1(\Omega)$ contains functions w which have a finite H^1-norm, defined by

$$\|w\|_{1,\Omega} = \left[\int_\Omega (w^2 + w_{,x}^2 + w_{,y}^2) \, d\Omega \right]^{1/2} . \tag{8.93}$$

The approximations of the C^0 type over triangles and quadrilaterals belong to $H^1(\Omega)$. Finally, the Sobolev-space $H^2(\Omega)$ contains functions w with a finite H^2-norm, defined by

$$\|w\|_{2,\Omega} = \left[\int_\Omega (w^2 + w_{,x}^2 + w_{,y}^2 + w_{,xx}^2 + w_{,xy}^2 + w_{,yy}^2) \, d\Omega \right]^{1/2} . \tag{8.94}$$

None of the approximations introduced in sections 8.8 and 8.9 belongs to $H^2(\Omega)$; essentially, elements ensuring C^1-continuity would be required if the approximation were to belong to $H^2(\Omega)$.

Let us now consider the particular example of the Poisson equation to make further progress, i.e. how can we determine a function u such that

$$-u_{,xx} - u_{,yy} = f ; \qquad u = 0 \text{ on } \partial\Omega , \tag{8.95}$$

f being for example piecewise continuous.

For the collocation method, we should select an approximating subspace $\tilde{H}_0^2(\Omega)$, the elements of which belong to $H^2(\Omega)$ and vanish on $\partial\Omega$; the approximation $\tilde{u}$ belongs to $\tilde{H}_0^2(\Omega)$. In general, we will find that

$$R(\tilde{u}) = -\tilde{u}_{,xx} - \tilde{u}_{,yy} - f \neq 0 \; . \tag{8.96}$$

In order to determine the nodal values, we select a number of collocation points equal to the number of unknown nodal values, and impose that $R(\tilde{u})$ vanish at these points. The difficulty of the collocation method lies in the construction of the shape functions, which are quite elaborate when the approximation belongs to $H^2(\Omega)$.

In order to reduce the continuity requirements on the global shape functions, let us consider an equivalent interpretation of (8.95). Finding the solution of (8.95) is equivalent to the following problem,

$$\text{find } u \in H_0^2(\Omega) : \langle -u_{,xx} - u_{,yy} - f \, ; v \rangle = 0 \; , \qquad \forall \; v \in L^2(\Omega) \; , \tag{8.97}$$

where $\langle \; \rangle$ denotes the L^2-scalar product,

$$\langle v ; w \rangle = \int_\Omega vw \, d\Omega \; . \tag{8.98}$$

Let us now restrict our attention to functions v in (8.97) which belong to $H_0^1(\Omega)$; we may then perform an integration by parts, and write (8.97) as follows,

$$\langle u_{,x} \, ; v_{,x} \rangle + \langle u_{,y} \, ; v_{,y} \rangle - \langle f \, ; v \rangle = 0 \; . \tag{8.99}$$

In (8.99), u does not need to belong to $H_0^2(\Omega)$; $H_0^1(\Omega)$ is sufficient, and we may introduce a *weak formulation* of (8.95) as follows,

$$\text{find } u \in H_0^1(\Omega) : \langle u_{,x} \, ; v_{,x} \rangle + \langle u_{,y} \, ; v_{,y} \rangle = \langle f \, ; v \rangle \; , \; \forall \; v \in H_0^1(\Omega) \; . \tag{8.100}$$

The Galerkin method for solving (8.95) consists of selecting an approximation given by (8.91) in $\tilde{H}_0^1(\Omega)$; the nodal values are selected in such a way that (8.99) is satisfied for all $\tilde{v}$'s belonging to $\tilde{H}_0^1(\Omega)$, i.e. find

$$\tilde{u} \in \tilde{H}_0^1(\Omega) : \langle \tilde{u}_{,x} \, ; \tilde{v}_{,x} \rangle + \langle \tilde{u}_{,x} \, ; \tilde{v}_{,y} \rangle = \langle f \, ; \tilde{v} \rangle \; , \; \forall \; \tilde{v} \in \tilde{H}_0^1(\Omega) \; . \tag{8.101}$$

By introducing (8.91) into (8.101), we obtain

$$\sum_{j=1}^{M} U_j[<\tau_{j,x} ; \tau_{i,x}> + <\tau_{j,y} ; \tau_{i,y}>] - <f ; \tau_i> = 0 , \qquad (8.102)$$

or, in abridged form,

$$\sum_{j=1}^{M} A_{ij} U_j = F_i , \qquad (8.103)$$

with the definitions

$$A_{ij} = <\tau_{j,x} ; \tau_{i,x}> + <\tau_{j,y} ; \tau_{i,y}> ,$$
$$F_i = <f ; \tau_i> . \qquad (8.104)$$

Thus, an approximate solution to (8.95) may be found by solving a linear system of algebraic equations (8.103). The successive steps that we have explained for the Poisson equation may be carried out similarly with other equations. Essentially they involve: (i) Find a weak form of the differential equation for reducing the order of the highest derivative. (ii) Select an approximating subspace for the approximate solution. (iii) Calculate the coefficients of the resulting linear (or non-linear) algebraic system. (iv) Solve the system. There remains one important step to be elucidated for two-dimensional problems, which is the numerical integration of coefficients like those appearing in (8.104). Indeed, an analytical integration must in general be ruled out, and one must resort to numerical means.

8.11 NUMERICAL INTEGRATION

Consider for example the calculation of A_{ij} given by (8.104), which in explicit form reads

$$A_{ij} = \int_{\Omega} (\tau_{j,x}\, \tau_{i,x} + \tau_{j,y}\, \tau_{i,y})\, d\Omega . \qquad (8.105)$$

The τ_i's are global shape functions which vanish identically everywhere except in the elements to which node i belongs. It is clear therefore that, for a large finite element mesh, most of the A_{ij}'s will vanish identically; they will not vanish in elements to which both nodes i and j belong. Let Ω_m be such an element, and let us concentrate on its contribution,

$$A_{ij}^{m} = \int_{\Omega_m} (\tau_{j,x}\, \tau_{i,x} + \tau_{j,y}\, \tau_{i,y})\, d\Omega . \qquad (8.106)$$

It will be much simpler in general to calculate the integral over the parent element ω in the ξ-η plane. We have,

$$\tau_{j,x} = \tau_{j,\xi}\,\xi_{,x} + \tau_{j,\eta}\,\eta_{,x}\ ,\quad \tau_{j,y} = \tau_{j,\xi}\,\xi_{,y} + \tau_{j,\eta}\,\eta_{,y}\ ; \tag{8.107}$$

the partial derivatives $\xi_{,x}$, $\xi_{,y}$, ... are not known explicitly. However, in view of the transformations (8.83), (8.84) or (8.89), (8.90) we have,

$$\xi_{,x} = y_{,\eta}/J\ ,\quad \xi_{,y} = -x_{,\eta}/J\ ,\quad \eta_{,x} = -y_{,\xi}/J\ ,\quad \eta_{,y} = x_{,\xi}/J\ , \tag{8.108}$$

where J is the Jacobian of the transformation from ω to Ω_m . In view of (8.107) and (8.108), (8.106) becomes

$$A^m_{ij} = \int_\omega [(\tau_{j,\xi}\,y_{,\eta} - \tau_{j,\eta}\,y_{,\xi})(\tau_{i,\xi}\,y_{,\eta} - \tau_{i,\eta}\,y_{,\xi})$$

$$+ (-\tau_{j,\xi}\,x_{,\eta} + \tau_{j,\eta}\,x_{,\xi})(-\tau_{i,\xi}\,x_{,\eta} + \tau_{i,\eta}\,x_{,\xi})]\,\frac{1}{J}\,d\omega\ . \tag{8.109}$$

This last form of A^m_{ij} shows that the integrand is an explicit function of the (ξ,η) coordinates; it also shows that analytical integration will be difficult in general, because of the presence of the Jacobian J in the denominator. The calculation of the coefficients does reduce to the numerical calculation of an expression of the type

$$I = \int_\omega k(\xi,\eta)\,d\omega\ . \tag{8.110}$$

When ω is the square parent element, we may rewrite (8.110) as follows,

$$I = \int_{-1}^{1} d\eta \int_{-1}^{1} k(\xi,\eta)\,d\xi \tag{8.111}$$

and use twice the Gauss-Legendre quadrature rule (8.61), i.e.

$$I = \int_{-1}^{1}\left[\sum_{j=1}^{n} k(\xi,\xi_j)w_j\right] d\xi = \sum_{i,j=1}^{n} k(\xi_i,\xi_j)w_i w_j\ . \tag{8.112}$$

We have seen in §8.5 that such a rule will integrate exactly powers of ξ and η up to (2n-1). The correct choice of the number of integration points is not trivial in general, since the integrand in (8.109) is a rational function rather than a polynomial.

When the parent element is a triangle, the integration may not be performed in the same fashion. However, quadrature rules for an expression like (8.110) are also available in the form

$$I = \sum_{i=1}^{m} w_i^T \, k(\xi_i^T, \eta_i^T) \, , \tag{8.113}$$

where w_i^T , ξ_i^T , η_i^T are respectively the weights and coordinates used for the numerical integration. The selected number of points depends again upon the degree of the complete polynomial which must be integrated exactly by the rule (8.113). In Table 8.4, we show some rules together with the order of the polynomial which is integrated exactly.

TABLE 8.4
Coordinates and weights for the numerical integration on a parent element

Number of points	Order of the polynomials	Coordinate ξ	Coordinate η	Weights
1	linear	1/3	1/3	1/2
3	quadratic	0.5	0	1/6
		0.5	0.5	1/6
		0	0.5	1/6
4	cubic	1/3	1/3	-27/96
		0.2	0.2	25/96
		0.6	0.2	25/96
		0.2	0.6	25/96

For further details, the reader is referred to Strang and Fix (1973) or Zienkiewicz (1977).

8.12 EXAMPLE. CONVERGENCE OF THE FINITE ELEMENT METHOD

Since Chapters 9 and 10 will deal with two-dimensional problems in Fluid Mechanics, we will not elaborate here on typical examples. Let us consider however one simple problem of a Poisson equation with essential boundary conditions, i.e.

$$-u_{,xx} - u_{,yy} = -1 \; ; \quad 0 \leq x \leq 2 \, , \quad 0 \leq y \leq 2 \, , \tag{8.114}$$

$$u(0,y) = u(2,y) = u(x,0) = u(x,2) = 0 \, .$$

The analytical solution of the problem is given by a Fourier series,

$$u(x,y) = - \sum_{m,n \text{ odd}} \frac{64}{\pi^4 (m^2 + n^2) mn} \sin \frac{m\pi x}{2} \sin \frac{n\pi y}{2} \, . \tag{8.115}$$

In view of the symmetry of the problem, we may limit ourselves in the numerical integration to one eighth of the square domain (Fig. 8.16), and impose a natural boundary condition on the lines of symmetry. The problem has been solved on the reduced domain shown on Fig. 8.16 with, respectively, 1, 4 and 16 triangles. The tests were run with three node-triangles, where the shape functions are given by (8.76), and with six node-triangles, where the shape functions are given by (8.78). We show in Table 8.5 the values of u at the centre of the square region obtained on the three meshes, and the relative error with respect to the analytical value, which is 0.2946.

TABLE 8.5

Error as a function of the number of elements in solving (8.114)

	Linear shape functions		Quadratic shape functions	
	value	error (%)	value	error (%)
1 element	-0.3333	13.1	-0.3000	1.83
4 elements	-0.3125	6.1	-0.2950	0.14
16 elements	-0.3013	2.3	-0.2947	0.03
64 elements	-0.2969	0.8		

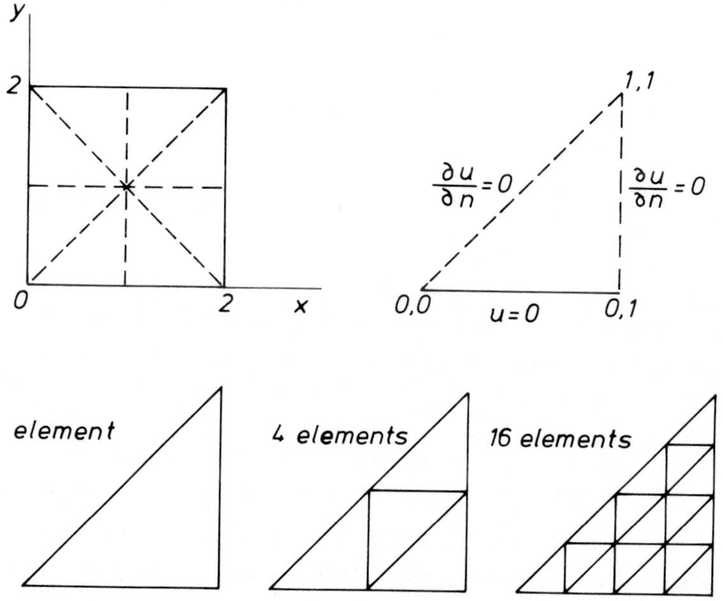

Fig. 8.16 Finite element meshes for solving Poisson's equation on a square domain.

Before closing this introductory chapter to the finite element method, it is appropriate to comment briefly on the problems related to the rate of convergence of the finite element solution towards the exact solution of the differential equation. The problem is difficult and requires an extensive mathematical background. Several books have been devoted to the topic; among them we cite Strang and Fix (1973), Oden and Reddy (1976), and Ciarlet (1978).

In section 8.3, we gave two examples of one-dimensional problems where the functional to be minimized could be written in the form

$$\mathcal{J}(v) = a(v,v) - 2 < f ; v > , \tag{8.116}$$

where a(v,v) is called the *strain energy*; this nomenclature refers to the use of finite elements in elasticity. In (8.19), we have

$$a(v,v) = < v_{,x} ; v_{,x} > , \tag{8.117}$$

while in (8.23) we have

$$a(v,v) = < v_{,xx} ; v_{,xx} > . \tag{8.118}$$

Similarly, in order to obtain the discretized form (8.102) of the Poisson equation, we might have minimized the functional

$$\mathcal{J}(v) = \int_\Omega \frac{1}{2} (v_{,x}^{\;2} + v_{,y}^{\;2})\, d\Omega - \int_\Omega vf\, d\Omega , \tag{8.119}$$

with

$$a(v,v) = < v_{,x} ; v_{,x} > + < v_{,y} ; v_{,y} > . \tag{8.120}$$

Let m be the order of the highest derivative in a(v,v) and let us select v in the space $H^m(\Omega)$. We assume that the strain energy a(v,v) is positive definite in the following sense : there exists a positive constant σ such that

$$a(v,v) \geqslant \sigma \, \|v\|^2_{m,\Omega} \quad ; \tag{8.121}$$

the problem is then said to be elliptic. Let us now divide the domain Ω into finite elements for which a typical dimension is h, and let u_h be the finite element solution for that mesh. We select shape functions which can generate complete polynomials of degree n. We may now wonder how the finite element solution tends to the analytical solution when h becomes small. Let u be the exact solution; for elliptic problems, one may prove the following inequality,

$$a(u - u_h \, , u - u_h) \leqslant C^2 \, h^{2(n+1-m)} |u|^2_{n+1} \quad ; \tag{8.122}$$

here, $|u|_{n+1}$ is the mean-square value of the (n+1)th derivative of u, and C is a constant. In view of (8.121), we may also write

$$\| u - u_h \|_{m,\Omega} \leqslant C/\sigma^{1/2} \, h^{(n+1-m)} |u|_{n+1} \, , \tag{8.123}$$

and we note that the left-hand side contains mth derivatives of $(u - u_h)$. Thus, we observe that the rate of convergence varies with the degree of the polynomial used for the approximation; the factor $|u|_{n+1}$ depends upon the problem itself. The bound (8.123) applies to the mth derivative of $(u - u_h)$; the bound will be lower for the norm $\| u - u_h \|_{p,\Omega}$, $p<m$. Finally, it must be observed that the error is calculated in the mean.

In the example given above, we had $m = 1$ and $n = 1$ for linear shape functions and $n = 2$ for quadratic shape functions. For the norm $\| u - u_h \|_{0,\Omega}$, we might thus expect a factor h^2 for linear elements and h^3 for quadratic elements. Table 8.5 shows that the calculations exhibit the appropriate tendency. On Fig. 8.17 we plot the logarithm of the local error as a function of the logarithm of h for linear elements; it is found that the slope tends to the theoretical value -2 when h decreases. However, it is worth recalling that, in solving the non-linear problem (8.73) in §8.9, we found a rate of convergence proportional to h, while the corresponding linear problem would have led to an error proportional to h^2; thus, the non-linearity of the problem may impair the rate of convergence expected from linear arguments.

It is clear that the knowledge of the rate of convergence is a fundamental problem of the finite element theory; under unfavourable circumstances, it may very well happen that a given algorithm may not converge at all when h decreases. Unfortunately, the mathematical difficulties associated with convergence proofs for the non-Newtonian problems that we wish to treat are formidable, and, apart from a few rules extrapolated from linear problems, we will not be able to rely upon a solid mathematical background for developing the elements we need.

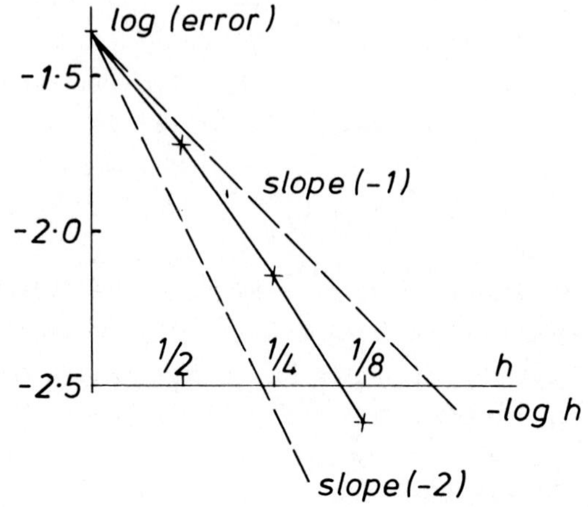

Fig. 8.17 Rate of convergence for linear triangular elements.

Chapter 9

Finite Element Calculation of Generalized Newtonian Flow

9.1 INTRODUCTION

Most of the literature on the application of finite elements to fluid mechanics has been devoted to the solution of the Navier-Stokes equations, given by (2.11), together with the incompressibility constraint (2.3). In the present chapter, we wish to consider the slightly more general problem of the flow of a generalized Newtonian fluid with a constitutive equation given by (3.1); we will soon find that the dependence of the viscosity upon the second invariant I_2 does not pose any new conceptual problem. We will not review here all the available finite element techniques which have been developed for a number of different problems ranging from creeping flow to high Reynolds number flow. We will rather concentrate on those techniques which have proved to be useful in non-Newtonian fluid mechanics, and are in fact widely used in engineering problems.

When the constitutive equations (3.1) are introduced into the momentum equations (2.6), we obtain the system

$$-\frac{\partial p}{\partial x_i} + \frac{\partial}{\partial x_k}\left[\eta(I_2)\left(\frac{\partial v_i}{\partial x_k} + \frac{\partial v_k}{\partial x_i}\right)\right] + \rho F_i = \rho \frac{Dv_i}{Dt} \quad , \tag{9.1}$$

where the velocity components must satisfy the incompressibility constraint (2.3), which we repeat for later convenience,

$$\frac{\partial v_m}{\partial x_m} = 0 \quad . \tag{9.2}$$

When η is a constant, (9.1) reduces to the Navier-Stokes equations (2.11). In this chapter, we will restrict ourselves to steady-state two-dimensional problems. For plane problems, the range of the indices in (9.1) and (9.2) is limited to 2 rather than 3. For axisymmetric problems, we need to use cylindrical coordinates, and the equations differ from (9.1), which are limited to rectangular Cartesian coordinates; the case of cylindrical coordinates will be considered in a specific section.

An inspection of the basic equations (9.1) and (9.2) will already give us some useful hints as to the choice of a finite element method. First, we note that an expansion of (9.1) will contain second-order partial derivatives of the velocity components; if we were to solve the system by means of the collocation method, in which we would simply define an approximation $\tilde{v}_i$ for the velocity components and $\tilde{p}$ for the pressure field, and satisfy the equations (9.1) and (9.2) at a finite number of points, we would need global shape functions of the C^1-type for the velocity components. The relevant finite elements are rather complicated, and the collocation method has seldom been used for solving the Navier-Stokes equations; several examples of applications may be found in Chang et al. (1979). The collocation method will not receive further consideration here.

We will find out in the next few sections that the incompressibility constraint (9.2) is not an easy matter to handle in finite element calculations; indeed, once we select an approximation $\tilde{v}_i$ for the velocity components, it should in some sense satisfy the scalar constraint (9.2). It was shown in Chapter 6 that, with finite differences, one often resorts to a stream function for calculating the velocity components in order to satisfy (9.2) identically. The use of a stream function rather than the velocity components is not widespread in the finite element literature, because of the higher order derivatives which will necessarily appear in the equations of motion (see for example Olson 1975). The order of the highest partial derivatives may also be lowered by the use of a stream function-vorticity formulation together with a so-called mixed finite element method (see, for example, Campion and Crochet, 1978). However, the variable viscosity prevents the transformation of (9.1) into the classical equation for the diffusion of vorticity.

These arguments make for the fact that most of the finite element literature for solving flow problems has adopted the choice of the velocity components as the unknown variables together with the pressure, which may however be sometimes avoided within the frame of the penalty method which will be explained in a later section. In the present chapter, we will exclusively deal with the velocity components and the pressure as the unknowns, although we will show how the stream function may be reconstructed *a posteriori*. Before presenting the Galerkin form of (9.1) and (9.2) within the spirit of Chapter 8, we will find it useful to recall a variational theorem for the *creeping flow* of a generalized Newtonian fluid, because an alternative form of the variational theorem will lead us naturally to the weighted residuals formulation.

9.2 A VARIATIONAL THEOREM FOR CREEPING GENERALIZED NEWTONIAN FLOW

Let us consider on Fig. 9.1 a region Ω of the two-dimensional (or three-dimensional) space, bounded by a curve (or surface) Γ. On the part Γ_t of Γ, surface forces $\bar{t}_i$ are imposed, i.e.

$$\underset{\sim}{x} \in \Gamma_t : P_{ki}\, n_k = \bar{t}_i \quad , \tag{9.3}$$

while on Γ_v (which is such that $\Gamma = \Gamma_v \cup \Gamma_t$) one imposes velocity components $\bar{v}_i$, i.e.

$$\underset{\sim}{x} \in \Gamma_v : v_i = \bar{v}_i \quad . \tag{9.4}$$

Before going further, let us reconsider the constitutive equations (3.1) of the generalized Newtonian fluid, where we recall that I_2 is given by

$$I_2 = (2d_{pq}\, d_{pq})^{1/2} \quad ; \tag{9.5}$$

the form (9.5) of I_2 is such that $I_2 = \gamma$ in a simple shear flow.

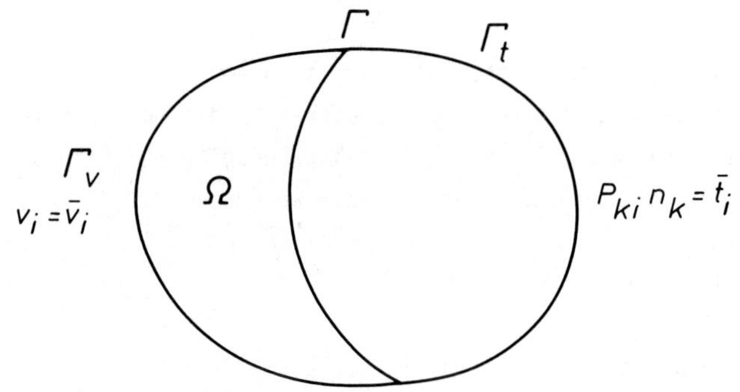

Fig. 9.1 Domain of integration and boundary conditions.

In view of (9.5), one obtains

$$\partial I_2/\partial d_{ik} = 2d_{ik}/I_2 \quad , \tag{9.6}$$

and we rewrite (3.1) as follows :

$$P_{ik} = - p\delta_{ik} + \eta(I_2) I_2\, \partial I_2/\partial d_{ik} \quad . \tag{9.7}$$

Defining a function $\nu(I_2)$ as follows,

$$\nu(I_2) = \int_0^{I_2} u\, \eta(u)\, du \quad , \tag{9.8}$$

we obtain the following form of the constitutive equations :

$$P_{ik} = -p\delta_{ik} + \partial\nu(I_2)/\partial d_{ik} \quad , \tag{9.9}$$

in which one detects an obvious analogy with the theory of incompressible elastic materials if one replaced the infinitesimal strain tensor by the rate of deformation tensor.

Let us now select in Ω a velocity field $\underset{\sim}{v}$ which satisfies the boundary conditions (9.4), and a pressure field p. We will not worry for the time being about the spaces to which $\underset{\sim}{v}$ and p should belong; we will simply assume that $\underset{\sim}{v}$ is twice differentiable in Ω and that p is once-differentiable. We may then construct a *functional* $J(\underset{\sim}{v},p)$ which has the following form

$$J(\underset{\sim}{v},p) = \int_\Omega [\nu(I_2) - \rho F_i v_i - p v_{k,k}]\, d\Omega - \int_{\Gamma_t} \bar{t}_i v_i \; d\Gamma \quad . \tag{9.10}$$

It is then easy to prove the following theorem :
Let ζ be a scalar parameter, $\underset{\sim}{w}$ a vector-valued function which is once differentiable and vanishes on Γ_v and q a scalar function which is square integrable; if, for an arbitrary choice of $\underset{\sim}{w}$ and q we have

$$\left.\frac{d}{d\zeta}\right|_{\zeta=0} J(\underset{\sim}{v} + \zeta\underset{\sim}{w},\ p + \zeta q) = 0 \ , \tag{9.11}$$

then $\underset{\sim}{v}$ and p are the solution of the system (9.1-2).

It is indeed easy, on the basis of (9.9), to see that

$$\begin{aligned}
&\left.\frac{d}{d\zeta}\right|_{\zeta=0} J(\underset{\sim}{v} + \zeta\underset{\sim}{w},\ p + \zeta q) \\
&= \int_\Omega (T_{ki}\, w_{i,k} - \rho F_i w_i - q v_{k,k} - p w_{k,k})\, d\Omega - \int_{\Gamma_t} \bar{t}_i w_i \; d\Gamma \\
&= - \int_\Omega [(-p_{,i} + T_{ki,k} + \rho F_i) w_i + v_{k,k}\, q]\, d\Omega \\
&\quad + \int_{\Gamma_t} [(-p\delta_{ki} + T_{ki}) n_k - \bar{t}_i] w_i \; d\Gamma \quad ,
\end{aligned} \tag{9.12}$$

where we have used the divergence theorem and where T_{ki} stands for the product $2\eta(I_2)d_{ki}$. If (9.11) is required to be valid for an arbitrary choice of the functions $\underset{\sim}{w}$ and q, we find from (9.12) that the momentum equations, the incompressibility equation and the boundary conditions on Γ_t are identically satisfied.

Noting that $\underset{\sim}{w}$ and q in (9.11) may be selected independently, we may rewrite the equality when either $q = 0$ or $\underset{\sim}{w} = \underset{\sim}{0}$, and obtain, from (9.12),

$$< -p\delta_{ki} + 2\eta(I_2)d_{ki} \; ; \; w_{i,k} > - <\rho F_i \; ; \; w_i > = \int_{\Gamma_t} \bar{t}_i w_i \, d\Gamma \quad ,$$

$$<v_{k,k} \; ; \; q > = 0 \quad , \tag{9.13}$$

where < ; > stands for the L^2 scalar product,

$$< f \; ; \; g > = \int_\Omega fg \, d\Omega \quad . \tag{9.14}$$

We will find in the next section that the equations (9.13) are identical to the weak form of the equations of motion (for creeping flow) of a generalized Newtonian fluid, which will be needed for constructing the Galerkin form of these equations; we will thus establish the intimate relationship between the variational approach and the method of weighted residuals.

9.3 GALERKIN FORMULATION OF THE EQUATIONS OF MOTION; PLANE FLOW

We will now limit ourselves to plane flow and write the explicit form of the governing equations, with the purpose of a detailed finite element formulation of the problem. For steady flow we have

$$- p_{,x} + T_{xx,x} + T_{yx,y} + \rho F_x - \rho(uu_{,x} + vu_{,y}) = 0 \quad ,$$

$$- p_{,y} + T_{xy,x} + T_{yy,y} + \rho F_y - \rho(uv_{,x} + vv_{,y}) = 0 \quad , \tag{9.15}$$

while the conservation of mass is written as

$$u_{,x} + v_{,y} = 0 \quad . \tag{9.16}$$

The constitutive equations are

$$T_{xx} = 2\eta(I_2)u_{,x} \quad , \quad T_{yy} = 2\eta(I_2)v_{,y} \quad , \quad T_{xy} = \eta(I_2)(u_{,y} + v_{,x}) \quad , \tag{9.17}$$

where, in view of (9.5), I_2 is now given by

$$I_2 = [2u_{,x}^2 + 2v_{,y}^2 + (u_{,y} + v_{,x})^2]^{1/2} \quad . \tag{9.18}$$

The velocity components u and v are imposed on Γ_v, where we have the *essential boundary conditions*,

$$\underset{\sim}{x} \in \Gamma_v \; : \; u = \bar{u} \; , \quad v = \bar{v} \; ; \tag{9.19}$$

on Γ_t, we impose *natural boundary conditions* which have the form

$$\underset{\sim}{x} \in \Gamma_t : t_x = (-p + T_{xx})n_x + T_{yx}n_y = \bar{t}_x \quad ,$$
$$t_y = T_{yx}n_x + (-p + T_{yy})n_y = \bar{t}_y \quad . \tag{9.20}$$

In the present chapter, we solve the problem in terms of the velocity components and the pressure. Whenever we use the symbols T_{xx}, T_{yy} and T_{xy}, we mean in fact the expressions (9.17) in terms of the velocity components. In that sense, the system (9.15) contains second-order derivatives of the velocity components. We note also that, in problems where the normal component of the velocity is imposed everywhere on the boundary, it is necessary to impose *one* nodal pressure to avoid an indeterminacy, this imposition constitutes a further essential boundary condition to which we will not however refer in our theoretical developments.

Following the line of thought established in Chapter 8, we will first replace the problem by an equivalent formulation. It is entirely equivalent to solve either (9.15-20) or the following problem :
Find functions u and v which belong to $H^2(\Omega)$, defined in section 8.10, and $p \in H^1(\Omega)$ such that, for any arbitrary functions w and $q \in L^2(\Omega)$ we have

$$< -p_{,x} + T_{xx,x} + T_{yx,y} + \rho F_x - \rho(uu_{,x} + vu_{,y}) \; ; w > = 0 \quad ,$$
$$< -p_{,y} + T_{xy,x} + T_{yy,y} + \rho F_y - \rho(uv_{,x} + vv_{,y}) \; ; w > = 0 \quad , \tag{9.21}$$
$$< u_{,x} + v_{,y} \; ; q > = 0 \quad ;$$

it is assumed that u, v and p satisfy the boundary conditions (9.19) and (9.20).

In order to obtain a weak formulation of the problem, as in section 8.10, we will now restrict the space to which w belongs, and enlarge the space where we may find u and v. Consider a function w which belongs to $H_0^1(\Omega)$; here, $H_0^1(\Omega)$ means a subspace of $H^1(\Omega)$, the elements of which vanish on Γ_v. Through the use of the divergence theorem, one easily obtains

$$< -p_{,x} + T_{xx,x} + T_{yx,y} \; ; w > = - < -p + T_{xx} \; ; w_{,x} > - < T_{yx} \; ; w_{,y} > + \int_{\Gamma_t} t_x w \, d\Gamma \; ,$$
$$< -p_{,y} + T_{xy,x} + T_{yy,y} \; ; w > = - < T_{xy} \; ; w_{,x} > - < -p + T_{yy} \; ; w_{,y} > + \int_{\Gamma_t} t_y w \, d\Gamma \; .$$

(9.22)

We are now able to formulate the weak form of the flow problem; our aim in doing this is a relaxation of the rather strong continuity requirements of (9.21). The weak form may be formulated as follows :
Find functions u and v which belong to $H^1(\Omega)$ and $p \in L^2(\Omega)$ such that, for any arbitrary functions $w \in H_0^1(\Omega)$ and $q \in L^2(\Omega)$ we have,

$$
\begin{aligned}
&< -p + T_{xx} \; ; \; w_{,x} > + < T_{yx} \; ; \; w_{,y} > + \rho < uu_{,x} + vu_{,y} \; ; \; w > \\
&\qquad\qquad = \rho < F_x \; ; \; w > + \int_{\Gamma_t} \bar{t}_x w \, d\Gamma \quad , \\
&< T_{xy} \; ; \; w_{,x} > + < -p + T_{yy} \; ; \; w_{,y} > + \rho < uv_{,x} + vv_{,y} \; ; \; w > \\
&\qquad\qquad = \rho < F_y \; ; \; w > + \int_{\Gamma_t} \bar{t}_y w \, d\Gamma \quad , \\
&< u_{,x} + v_{,y} \; ; \; q > = 0 \quad .
\end{aligned}
\tag{9.23}
$$

In going from (9.21) to (9.23) with the use of (9.22), we have also introduced a weak form of the boundary conditions on Γ_t, i.e.

$$
\int_{\Gamma_t} t_x w \, d\Gamma = \int_{\Gamma_t} \bar{t}_x w \, d\Gamma \quad , \qquad \int_{\Gamma_t} t_y w \, d\Gamma = \int_{\Gamma_t} \bar{t}_y w \, d\Gamma \quad . \tag{9.24}
$$

The reader will note that, for creeping flow (i.e. $\rho = 0$), the system (9.23) is actually identical to (9.13) obtained from the variational principle, once we express the extra-stress components in terms of the velocity components through (9.17). The main advantage of the weak formulation (9.23) with respect to (9.21) is that, looking for an approximation, we may now construct u, v and p with a much wider class of shape functions. With little loss of generality, let us assume temporarily that the velocity boundary conditions (9.19) are homogeneous, i.e. $\bar{u} = \bar{v} = 0$. Let us define in $H_0^1(\Omega)$ a finite-dimensional approximating subspace $\tilde{H}_0^1(\Omega)$, with a basis consisting of M global shape functions τ_i, $1 \leq i \leq M$, and also in $L^2(\Omega)$ a finite-dimensional approximating subspace $\tilde{L}^2(\Omega)$, with a basis of N global shape functions τ_i', $1 \leq i \leq N$. In order to obtain the finite element formulation of the flow problem, we will now look for an approximation $\tilde{u}$, $\tilde{v}$ of u, v in $\tilde{H}_0^1(\Omega)$, and for an approximation $\tilde{p}$ of p in $\tilde{L}^2(\Omega)$, given respectively by

$$
\tilde{u} = \sum_{i=1}^{M} U_i \tau_i \; , \qquad \tilde{v} = \sum_{i=1}^{M} V_i \tau_i \; , \qquad \tilde{p} = \sum_{i=1}^{N} P_i \tau_i' \quad . \tag{9.25}
$$

In order to determine the nodal values U_i, V_i and P_i, we impose that (9.23) be satisfied when $\tilde{u}$, $\tilde{v}$, $\tilde{p}$ are substituted for u, v, p, and for all $w \in \tilde{H}_0^1(\Omega)$ and all $q \in \tilde{L}^2(\Omega)$. Since $\tilde{H}_0^1(\Omega)$ and $\tilde{L}^2(\Omega)$ have a finite dimension, it is equivalent to say that (9.23) must be satisfied whenever $w = \tau_i$ and $q = \tau_i'$. The Galerkin form of the equations of motion may then be written as follows :

$$
\begin{aligned}
&< -\tilde{p} + 2\eta(\tilde{I}_2)\tilde{u}_{,x} \; ; \; \tau_{i,x} > + < \eta(\tilde{I}_2)(\tilde{u}_{,y} + \tilde{v}_{,x}) \; ; \; \tau_{i,y} > \\
&\quad + \rho < \tilde{u}\,\tilde{u}_{,x} + \tilde{v}\,\tilde{u}_{,y} \; ; \; \tau_i > = \rho < F_x \; ; \; \tau_i > + \int_{\Gamma_t} \bar{t}_x \, \tau_i \, d\Gamma \; , \\
&< \eta(\tilde{I}_2)(\tilde{u}_{,y} + \tilde{v}_{,x}) \; ; \; \tau_{i,x} > + < -\tilde{p} + 2\eta(\tilde{I}_2)\tilde{v}_{,y} \; ; \; \tau_{i,y} > \\
&\quad + \rho < \tilde{u}\,\tilde{v}_{,x} + \tilde{v}\,\tilde{v}_{,y} \; ; \; \tau_i > = \rho < F_y \; ; \; \tau_i > + \int_{\Gamma_t} \bar{t}_y \, \tau_i \, d\Gamma \; , \\
&< \tilde{u}_{,x} + \tilde{v}_{,y} \; ; \; \tau_n' > = 0,
\end{aligned}
\tag{9.26}
$$

where $1 \leqslant i \leqslant M$ and $1 \leqslant n \leqslant N$, with the understanding that whenever an essential boundary condition is imposed at a given node, say k, the corresponding Galerkin equations are replaced by the boundary conditions, i.e. $U_k = 0$, $V_k = 0$ when the boundary conditions are homogeneous. When the boundary conditions are not homogeneous, we would first construct two functions u*, v* which satisfy the essential boundary conditions, and select an approximation given as follows :

$$
\tilde{u} = u^* + \sum_{i=1}^{M} U_i \tau_i \; , \quad \tilde{v} = v^* + \sum_{i=1}^{M} V_i \tau_i \; , \quad \tilde{p} = \sum_{i=1}^{N} P_i \tau_i' \; . \tag{9.27}
$$

Instead of an approximating subspace, we should now talk about an approximating linear manifold; apart from that, we obtain again the discretized form (9.26). Until now, we have also assumed that, whenever we have an essential boundary condition, both velocity components are simultaneously imposed. In general, however, we will also encounter cases where we have an essential boundary condition in one direction and a natural boundary condition in the normal direction; this is the case in particular when part of the boundary consists of a plane of symmetry. Our developments may be generalized easily to such boundary conditions.

From the Galerkin form (9.26) of the equations of motion and incompressibility, we may now sketch the finite element technique. Let us replace $\tilde{u}$, $\tilde{v}$ and $\tilde{p}$ in (9.26) by their representation (9.25), and define the following matrix components,

$$A_{ij} = < 2\eta(\tilde{I}_2)\ \tau_{j,x}\ ;\ \tau_{i,x} > + < \eta(\tilde{I}_2)\ \tau_{j,y}\ ;\ \tau_{i,y} > \quad ,$$

$$B_{ij} = < \eta(\tilde{I}_2)\ \tau_{j,x}\ ;\ \tau_{i,x} > + < 2\eta(\tilde{I}_2)\ \tau_{j,y}\ ;\ \tau_{i,y} > \quad ,$$

$$C_{ij} = < \eta(\tilde{I}_2)\ \tau_{j,x}\ ;\ \tau_{i,y} > \quad , \tag{9.28}$$

$$D_{in} = < \tau_n'\ ;\ \tau_{i,x} > \quad , \quad E_{in} = < \tau_n'\ ;\ \tau_{i,y} > \quad ,$$

$$IX_{ijk} = < \tau_j\tau_{k,x}\ ;\ \tau_i > \quad , \quad IY_{ijk} = < \tau_j\tau_{k,y}\ ;\ \tau_i > \quad ,$$

and the vector components,

$$X_i = \rho < F_x\ ;\ \tau_i > + \int_{\Gamma_t} \bar{t}_x\ \tau_i\ d\Gamma \quad ,$$

$$Y_i = \rho < F_y\ ;\ \tau_i > + \int_{\Gamma_t} \bar{t}_y\ \tau_i\ d\Gamma \quad , \tag{9.29}$$

where $1 \leq i,j,k \leq M$ and $1 \leq n \leq N$; in (9.28), $\tilde{I}_2$ is obtained by substituting $\tilde{u}$, $\tilde{v}$ for u, v in the right-hand side of (9.18).

By introducing (9.25) in (9.26), and in view of the definitions (9.28) and (9.29), one easily obtains the following algebraic system :

$$A_{ij}U_j + C_{ij}V_j - D_{in}P_n + \rho(IX_{ijk}U_j + IY_{ijk}V_j)U_k = X_i \quad ,$$

$$C_{ji}U_j + B_{ij}V_j - E_{in}P_n + \rho(IX_{ijk}U_j + IY_{ijk}V_j)V_k = Y_i \quad , \tag{9.30}$$

$$- D_{jn}U_j - E_{jn}V_j = 0 \quad ,$$

where we have used the summation convention on repeated indices and where $1 \leq i,j,k \leq M$, $1 \leq n \leq N$. The algebraic system (9.30) is non-linear; the non-linearities arise from the convective terms which are quadratic in the velocity components, and from the non-Newtonian character of the fluid which appears through the viscosity function $\eta(\tilde{I}_2)$ in the calculation of the matrices A_{ij}, B_{ij} and C_{ij}.

Our derivation until now has been fairly general; there remains the essential and delicate task of selecting the proper approximating subspaces for $\tilde{u}$, $\tilde{v}$ and $\tilde{p}$, or, in other words, the types of elements which will secure a successful calculation. Before considering that problem, however, we want to study briefly the important case of axisymmetric flows, which will lead in fact to a set of algebraic equations formally equivalent to (9.30).

9.4 GALERKIN FORMULATION OF THE EQUATIONS OF MOTION; AXISYMMETRIC FLOW

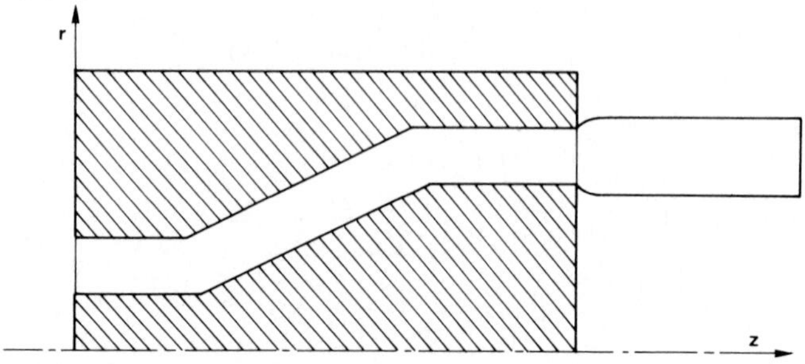

Fig. 9.2 Typical domain of an axisymmetric flow.

We show on Fig. 9.2 a typical axisymmetric flow domain, for the simulation of the flow through an annular die; we wish to consider axisymmetric flows without torsion, where we denote by u and v the velocity components in the r and z directions, respectively. Axisymmetric flows are important in technical applications; moreover, they provide unique examples of truly three-dimensional flows which may be calculated by means of two-dimensional techniques.

For axisymmetric flow without torsion, the momentum and incompressibility equations are

$$\begin{aligned} & -p_{,r} + T_{rr,r} + T_{rz,z} + (T_{rr} - T_{\theta\theta})/r + \rho F_r - \rho(uu_{,r} + vu_{,z}) = 0 \ , \\ & -p_{,z} + T_{rz,r} + T_{rz}/r + T_{zz,z} + \rho F_z - \rho(uv_{,r} + vv_{,z}) = 0 \ , \\ & u_{,r} + u/r + v_{,z} = 0 \ . \end{aligned} \tag{9.31}$$

The extra-stress components are given by the following constitutive equations :

$$\begin{aligned} & T_{rr} = 2\eta(I_2)u_{,r} \ , \qquad T_{zz} = 2\eta(I_2)v_{,z} \ , \\ & T_{\theta\theta} = 2\eta(I_2)u/r \ , \qquad T_{rz} = \eta(I_2)(u_{,z} + v_{,r}) \ , \end{aligned} \tag{9.32}$$

where the invariant I_2 is now given by the equation

$$I_2 = [2u_{,r}^2 + 2(u/r)^2 + 2v_{,z}^2 + (u_{,z} + v_{,r})^2]^{1/2} \ . \tag{9.33}$$

The system is completed by essential and natural boundary conditions which are essentially similar to (9.19) and (9.20).

It has now been shown several times that the first step towards the finite element solution of a problem is the obtaining of a weak formulation which will allow the reduction of the order of the highest partial derivatives. For plane flow, we found a close relationship between the weak formulation and the variational theorem presented in section 9.2. Before going further, it is useful to have a further look at the expression (9.10) of the functional $J(\underset{\sim}{v},p)$; we find indeed that the integrals are calculated over the domain occupied by the flow. For plane flow, it is sufficient to calculate the integrals on a two-dimensional domain, since it amounts to considering a volume per unit length in the direction normal to the flow. For axisymmetric flow, the situation is different; indeed, the integration must be performed over the three-dimensional axisymmetric domain surrounding the axis of symmetry, and we have

$$d\Omega = 2\pi\ r\ d\bar{\Omega}\quad , \tag{9.34}$$

where $d\bar{\Omega}$ is the differential element of the area dr dz in a meridian plane. The equivalent form of (9.31) is now

$$\langle -p_{,r} + T_{rr,r} + T_{rz,z} + (T_{rr} - T_{\theta\theta})/r + \rho F_r - \rho(uu_{,r} + vu_{,z})\ ;\ 2\pi rw\rangle = 0,$$

$$\langle -p_{,z} + T_{rz,r} + T_{rz}/r + T_{zz,z} + \rho F_z - \rho(uv_{,r} + vv_{,z})\ ;\ 2\pi rw\rangle = 0\ , \tag{9.35}$$

$$\langle u_{,r} + u/r + v_{,z}\ ;\ 2\pi rq\rangle = 0\ ,$$

where w and q have the same meaning as in (9.21).

We will not go into detail on the spaces to which u, v, p, w and q should belong; when the divergence theorem is applied on the left-hand side of (9.35), one obtains

$$\langle -p + T_{rr}\ ;\ rw_{,r}\rangle + \langle T_{rz}\ ;\ rw_{,z}\rangle + \langle -p + T_{\theta\theta}\ ;\ w\rangle$$
$$+\ \rho\ \langle uu_{,r} + vu_{,z}\ ;\ rw\rangle = \rho\ \langle F_r, rw\rangle + \int_{\Gamma_t} \bar{t}_r\ rw\ d\Gamma\quad ,$$

$$\langle T_{rz}\ ;\ rw_{,r}\rangle + \langle -p + T_{zz}\ ;\ rw_{,z}\rangle$$
$$+\ \rho\ \langle uv_{,r} + vv_{,z}\ ;\ rw\rangle = \rho\ \langle F_z, rw\rangle + \int_{\Gamma_t} \bar{t}_z\ rw\ d\Gamma\quad ,$$

$$\langle u_{,r} + u/r + v_{,z}\ ;\ rq\rangle = 0\quad , \tag{9.36}$$

where we have omitted the factor 2π on both sides of the equations. The rest of the development follows closely that of the previous section. The approximation (9.25) is introduced into the constitutive equations (9.32), and the weak form (9.36) is then expressed in terms of the nodal velocity components U_i, V_i and the nodal pressures P_i. The final algebraic system is identical to (9.30), but the matrices and vectors are now different; it is easy to obtain the following expressions :

$$
\begin{aligned}
A_{ij} &= \langle 2\eta(\tilde{I}_2)\tau_{j,r} \,;\, r\tau_{i,r} \rangle + \langle \eta(\tilde{I}_2)\tau_{j,z} \,,\, r\tau_{i,z} \rangle + \langle 2\eta(\tilde{I}_2)\tau_j/r \,;\, \tau_i \rangle \,, \\
B_{ij} &= \langle \eta(\tilde{I}_2)\tau_{j,r} \,;\, r\tau_{i,r} \rangle + \langle 2\eta(\tilde{I}_2)\tau_{j,z} \,;\, r\tau_{i,z} \rangle \,, \\
C_{ij} &= \langle \eta(\tilde{I}_2)\tau_{j,r} \,;\, r\tau_{i,z} \rangle \,, \qquad\qquad (9.37) \\
D_{in} &= \langle \tau'_n \,;\, r\tau_{i,r} + \tau_i \rangle \,, \qquad E_{in} = \langle \tau'_n \,;\, r\tau_{i,z} \rangle \,, \\
IX_{ijk} &= \langle \tau_j\tau_{k,r} \,;\, r\tau_i \rangle \,, \qquad IY_{ijk} = \langle \tau_j\tau_{k,z} \,;\, r\tau_i \rangle \,,
\end{aligned}
$$

while for the nodal forces we obtain

$$
\begin{aligned}
X_i &= \rho \langle F_r \,,\, r\tau_i \rangle + \int_{\Gamma_t} \bar{t}_r \, r\tau_i \, d\Gamma \,, \\
Y_i &= \rho \langle F_z \,,\, r\tau_i \rangle + \int_{\Gamma_t} \bar{t}_z \, r\tau_i \, d\Gamma \,. \qquad\qquad (9.38)
\end{aligned}
$$

It is now easy to see that, with the finite element technique, the methods for calculating plane and axisymmetric flows are essentially equivalent. The only modification appears in the calculation of the integrals in (9.37) and (9.38); in a finite element program, it is an easy matter to introduce a parameter which, say, vanishes for plane flow and takes the unit value for axisymmetric flow. It is then possible to reformulate (9.28-29) and (9.37-38) into a single set which covers both plane and axisymmetric solutions.

9.5 FINITE ELEMENTS FOR SOLVING THE NAVIER-STOKES EQUATIONS

Now that we have explained the finite element technique applied to flow problems, there remains the crucial question of selecting an appropriate finite element; up to this point we have neither specified the shape of an element, nor made explicit the global shape functions τ_i and τ'_i found in the definition of the matrices (9.28) and (9.37). Although we have until now considered the case of the generalized Newtonian fluid, all the available theoretical developments deal with the Navier-Stokes equations, where the shear viscosity η has a constant value. The results presented in this section are based on the treatment of the Navier-Stokes equations; we will find in a later section that the relevant finite elements may also be used for solving the flow of generalized Newtonian fluids.

We already know that the global shape functions τ_i for approximating the velocity components must belong to $H^1(\Omega)$, while the shape functions τ_i' for approximating the pressure are required to belong to $L^2(\Omega)$ only. If we transfer these requirements to the elements which we studied in Chapter 8, we find that the shape functions τ_i must be of the C^0-type, while the τ_i''s may possibly be of class C^{-1}.

There is another serious requirement which relates the order of the highest polynomial which can be constructed with the τ_i's as a basis to the similar order based on the τ_i''s. In the early work by Taylor and Hood (1973-74) on finite elements for solving the Navier-Stokes equations, it was noticed that the use of piecewise polynomials of the same order for approximating the velocity components and the pressure leads to spurious oscillatory modes in the pressure field. This anomaly has been further investigated by Sani et al (1981a,b) who have confirmed that the order of the τ_i''s should be at least one unit lower than the order of the τ_i's.

These conditions do in fact restrict our possible choices of elements considerably if we wish to simultaneously limit to two the highest order of the polynomials. Let us start with shape functions τ_i's of the type P^1-C^0. It is then clear that the approximation for the pressure should be of the P^0-C^{-1} type. On Fig. 9.3a, we show the simplest element that one might think of: on a triangle, the velocity field is expressed in terms of the functions ϕ_i^t in (8.76), while the pressure is a constant over the element; it is also a simple example of an element which does not work. In order to see this, let us make a calculation which is useful in the discussion of new elements. On Fig. 9.3b, we show a large mesh of triangular finite elements; whenever we add two triangles to the mesh, we obtain simultaneously two more element pressures, and two nodal velocity components. Globally, for a large mesh, the number of element pressures is equal to the number of velocity components. Since the number of discretized incompressibility constraints, expressed in terms of the velocity components, equals the number of nodal pressures, we find that the velocity field is fully determined by the condition of incompressibility; the system is *locked*, and the element is useless.

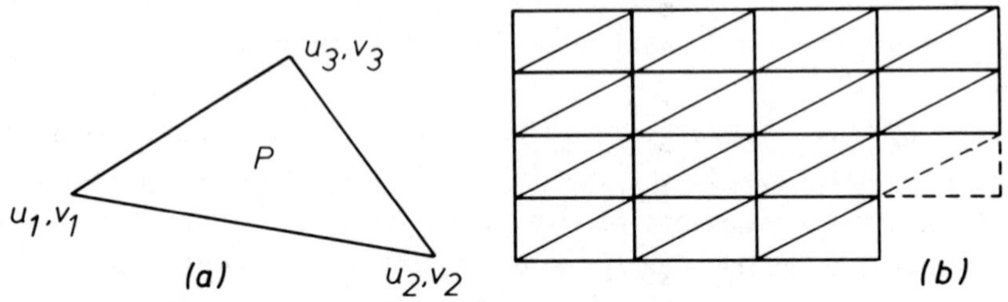

Fig. 9.3 A simple element which does not work.

It has been argued, on a heuristic basis, that the ratio between the number of discrete momentum equations and the number of equations expressing mass conservation should approach 2 in order to emulate the continuum representation in which a vector momentum equation and a scalar equation for mass conservation must be satisfied everywhere. This ratio is one for the useless element of Fig. 9.3. The element shown on Fig. 9.4 is quite satisfactory in this respect. Here, we consider an isoparametric quadrilateral on which the velocity components are given by bilinear polynomials in the parent element, while the pressure is a constant. The velocity/pressure ratio is now precisely two in a large mesh; moreover, the constant value of the pressure over an element guarantees global conservation of mass over the element, since, over element ω, say, the discretized form of the conservation of mass is

$$< \tilde{u}_{,x} + \tilde{v}_{,y} \; ; \; q > \; = q < \tilde{u}_{,x} + \tilde{v}_{,y} \; ; \; 1 > \; = q \int_{\omega} (\tilde{u} n_x + \tilde{v} n_y) d\partial\omega = 0 \; . \qquad (9.39)$$

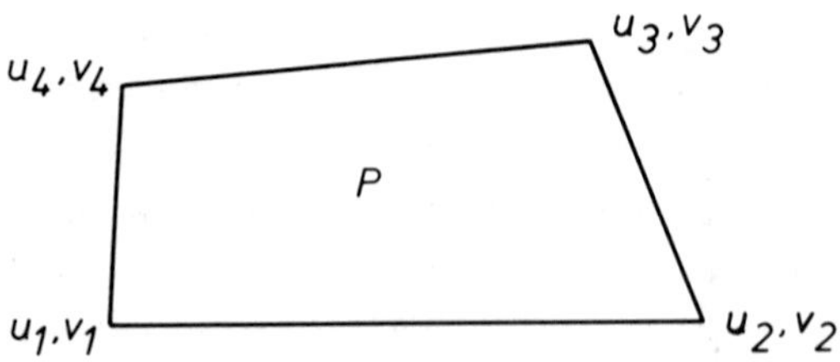

Fig. 9.4 Quadrilateral finite element with spurious pressure modes.

This property of mass conservation per element is important; in particular, it allows for an easy and accurate calculation of the stream function along the element boundaries. Unfortunately, the quadrilateral element of Fig. 9.4 does suffer from a defect; under some boundary conditions it may be shown that oscillatory spurious pressure modes occur as a result of the calculation. Sani et al (1981) discuss the problem at length and show how the oscillations may be filtered.

By far the most popular elements applied to viscous flow problems are shown on Fig. 9.5. The three of them share a pressure field of the type P^1-C^0 and a velocity field of the type P^2-C^0; the nodal pressures are defined at the corners of the elements. The shape functions $\tau_i^!$ are then replaced in triangles by the functions ϕ_i^t (8.76) and in quadrilaterals by the functions ϕ_i^q (8.80). For the triangular element, the velocity components are given by complete second-order polynomials; the nodal values are selected at the vertices and at the midside nodes, and the τ_i 's are replaced by the functions ψ_i^t (8.78). This triangular element was introduced by Nickell et al (1974) and used extensively by Tanner, Caswell and their co-workers (see later references). These authors use the triangular element as part of a composite quadrilateral shown on Fig. 9.6. Given an arbitrary quadrilateral, a central node is selected with its coordinates being the arithmetic average of the coordinates of the vertices; the central node is joined to the vertices in order to form four triangles. With this arrangement, it is possible to eliminate the velocity components and the pressures at the internal nodes before assembling the global matrix. However, the generalized use of the *frontal elimination* technique introduced by Irons (1970) has devalued the advantages that one may find in macro-elements such as the composite quadrilateral; one might as well use a mesh made of triangles rather than composite quadrilaterals.

In the element b shown on Fig. 9.5, the pressure is interpolated by means of bilinear polynomials ϕ_i^q (8.80) in the parent element, while the velocity components, identified at 9 nodes, are interpolated by means of bi-quadratic Lagrangian polynomials ψ_i^q (8.81) in the parent element. This *Lagrangian element* has become quite popular over the last few years; its performance with respect to other elements for solving the Navier-Stokes equations was rated excellent in a review paper by Huyakorn et al (1978). The element c of Fig. 9.5, called the *serendipity element* is quite similar to the Lagrangian element. The interpolation for the pressure is the same; however, since there is no central node, the velocity components are approximated by means of the 8-node serendipity shape functions (8.82). In problems where convective terms are important, it appears that the Lagrangian element offers a superior performance as compared to the serendipity element.

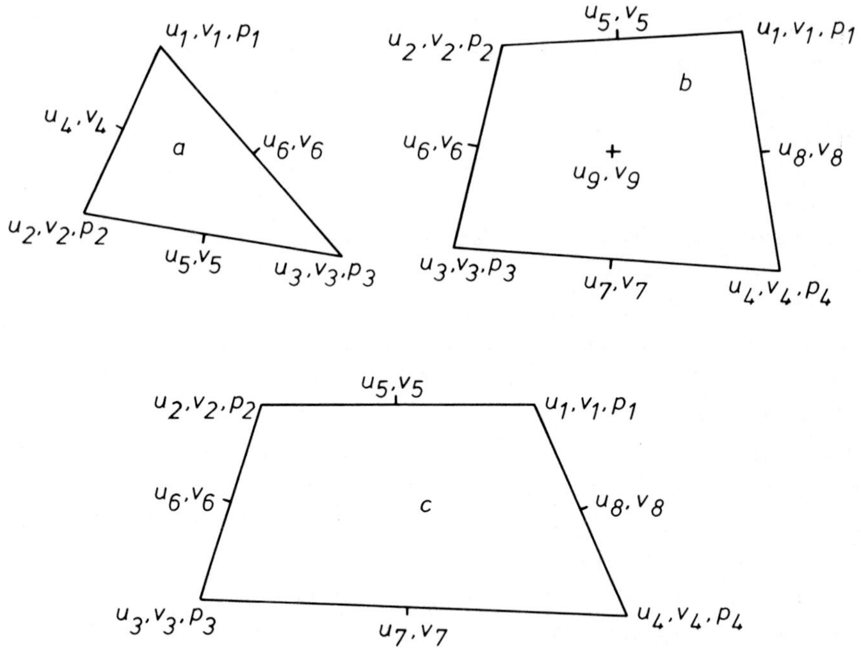

Fig. 9.5 Successful triangular and quadrilateral elements a. Triangular element, b. Lagrangian element, c. Serendipity element.

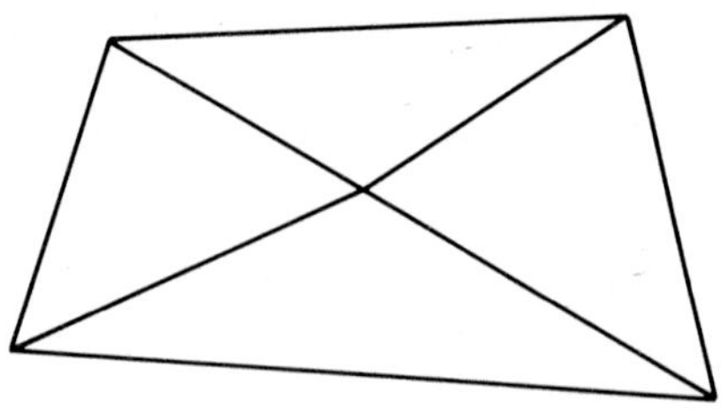

Fig. 9.6 Composite quadrilateral.

A weak point of the elements shown on Fig. 9.5 is the rather high ratio between the number of nodal velocity components and the number of nodal pressures. For the elements a and b, the ratio is 8/1 while for the serendipity element we obtain a 6/1 ratio. In some problems where the velocity field undergoes drastic changes within short distances, or in the neighbourhood of corners and singularities in general, the conservation of mass may be only weakly satisfied with such a high ratio. An interesting element mentioned by Zienkiewickz (1980) and devoid of spurious pressure modes (Sani et al, 1981) is a 9-node element with bi-quadratic velocity components in the parent element, while the pressure is discontinuous and approximated by means of a complete first-order polynomial within the element; thus, the element is of the P^2-C^0 type for the velocity components, and of the P^1-C^{-1} type for the pressure. In a large mesh, any additional element brings 8 velocity components and three element pressures, with an 8/3 ratio which is close to the optimal value of 2. The optimal value of 2 would be obtained with a quadrilateral element of the serendipity type for the velocity components, and of the P^1-C^{-1} type for the pressure; however, the phenomenon of spurious pressure modes is again present for that element.

9.6 PENALTY FORMULATION FOR SOLVING THE NAVIER-STOKES EQUATIONS

The use of the penalty formulation has found widespread use over recent years. While Bernstein et al (1981) and Nakazawa et al (1982) have been the only advocates of the penalty formulation in numerical non-Newtonian fluid mechanics, the subject is important and deserves further study, because of the sensible economy of computer time which may result from its use. For brevity, we will not consider the inertia terms in the present section, with the understanding that whatever we show for creeping flow remains valid in the general case.

Let us return to the Stokes form of the momentum equations (9.15) for plane flow, i.e.

$$
\begin{aligned}
- p_{,x} + T_{xx,x} + T_{yx,y} + \rho F_x &= 0 \quad , \\
- p_{,y} + T_{xy,x} + T_{yy,y} + \rho F_y &= 0 \quad .
\end{aligned}
\tag{9.40}
$$

In the penalty formulation, rather than considering the conservation of mass (9.16), one assumes that the pressure p is given by a constitutive equation of the form

$$
p = - \Lambda(u_{,x} + v_{,y}) \quad , \tag{9.41}
$$

where Λ is a large parameter. Thus, the Cauchy stress tensor is now given by

$$P_{xx} = \Lambda(u_{,x} + v_{,y}) + 2\eta u_{,x} \;,$$

$$P_{yy} = \Lambda(u_{,x} + v_{,y}) + 2\eta v_{,y} \;, \qquad (9.42)$$

$$P_{xy} = \eta(u_{,y} + v_{,x}) \;.$$

It must be realized that these equations *are not* the constitutive equations of a compressible Newtonian fluid, where p is a function of the specific volume. Rather, it is interesting to recall the analogy between the equations of creeping flow and those of linear elasticity; when the velocity components are replaced by displacements, (9.42) become indeed the constitutive equations of a linear compressible elastic material.

There are now two ways of implementing the penalty formulation in a finite element method. In a first approach, the weak formulation of the problem is still given by the first two equations (9.23), while the third one is replaced by

$$< p + \Lambda(u_{,x} + v_{,y}) \;;\; q > = 0 \;. \qquad (9.43)$$

The discretization (9.25) is adopted for the velocity components and the pressure and, for creeping flow, one easily obtains

$$A_{ij} U_j + C_{ij} V_j - D_{in} P_n = X_i \;,$$

$$C_{ji} U_j + B_{ij} V_j - E_{in} P_n = Y_i \;, \qquad (9.44)$$

$$-D_{jm} U_j - E_{jm} V_j - F_{nm} P_n / \Lambda = 0 \;.$$

where the matrices A_{ij}, B_{ij}, C_{ij}, D_{in}, E_{in} are given by (9.28) while the matrix F_{nm} is given by

$$F_{nm} = < \tau'_m \;;\; \tau'_n > \;. \qquad (9.45)$$

When the pressure variables P_n are defined *inside* an element, they are easy to eliminate from the system (9.44). Indeed we may, at the element level, calculate

$$P_n = -\Lambda F^{-1}_{mn} (D_{jm} U_j + E_{jm} V_j) \;, \qquad (9.46)$$

where F^{-1}_{mn} are the components of the inverse of F_{mn}; when (9.46) is substituted for P_n in the first two equations (9.44), one obtains

$$(A_{ij} + \Lambda D_{in} F^{-1}_{mn} D_{jm})U_j + (C_{ij} + \Lambda D_{in} F^{-1}_{mn} E_{jm})V_j = X_i \quad ,$$

$$(C_{ji} + \Lambda E_{in} F^{-1}_{mn} D_{jm})U_j + (B_{ij} + \Lambda E_{in} F^{-1}_{mn} E_{jm})V_j = Y_i \quad . \tag{9.47}$$

Thus, the pressure variables have been removed from the system; the state of incompressibility will be approached provided the parameter Λ is large enough. For the penalty formulation (9.44) to be valid, the selection of the shape functions for the pressure must follow the same rules as those given in the previous section; they must be of at least one order less than the shape functions for the velocity components (Sani et al. 1981).

The second approach to the penalty formulation (Hughes et al. 1979) consists of substituting the expression (9.41) for p into (9.23) *before* the discretization process; it is then easy to show that the Galerkin form of the field equations is given by

$$(A_{ij} + \Lambda A'_{ij})U_j + (C_{ij} + \Lambda C'_{ij})V_j = X_i \quad ,$$

$$(C_{ji} + \Lambda C'_{ji})U_j + (B_{ij} + \Lambda B'_{ij})V_j = Y_i \quad , \tag{9.48}$$

where

$$A'_{ij} = < \tau_{i,x} \; ; \; \tau_{j,x} > , \qquad B'_{ij} = < \tau_{i,y} \; ; \; \tau_{j,y} > ,$$

$$C'_{ij} = < \tau_{i,x} \; ; \; \tau_{j,y} > . \tag{9.49}$$

Let us rewrite the system (9.48) in the following form :

$$(\underline{K} + \Lambda \underline{K}')\underset{\sim}{U} = \underset{\sim}{X} \quad , \tag{9.50}$$

where $\underset{\sim}{U}$, $\underset{\sim}{X}$ are the vectors of nodal velocities and nodal forces, respectively. When the integrals given by (9.49) are calculated exactly, the matrix $\underline{K}'$ in (9.50) will generally be non-singular. It is then clear that, when Λ increases, the velocity components will eventually vanish identically. The matrix $\underline{K}'$ must in fact be singular; in order to satisfy that condition, the integrals in (9.49) are calculated numerically with a lower number of quadrature points than those in (9.28) (Malkus and Hughes 1978). The procedure is called *reduced numerical integration*. Examples of the use of the penalty formulation may be found in

Hughes et al. (1979) and Heinrich and Marshall (1981).

9.7 CALCULATION OF THE STREAM FUNCTION

We explained earlier the reasons why we concentrate on the finite-element formulation in terms of the velocity components and the pressure. However, in most practical applications, the stream function is a quantity of primary interest, since the knowledge of its value at the nodes allows us to draw streamlines. When the flow problem is solved in terms of the velocity components, it is necessary to calculate the stream function *a posteriori*; in the present section, we show briefly how the calculation may be performed.

It was recalled in section 2.2 that, in plane flow, the velocity components u, v are related to the stream function ψ by means of the relations

$$u = \partial\psi/\partial y \quad , \qquad v = -\partial\psi/\partial x \quad . \tag{9.51}$$

An easy way of calculating ψ would be to integrate (9.51) along the boundary of the elements. Consider on Fig. 9.7 a typical quadrilateral; the velocity field is known along the boundary of the element, together with the value of ψ at one point, coming from its calculation in a previous element, or an essential boundary value.

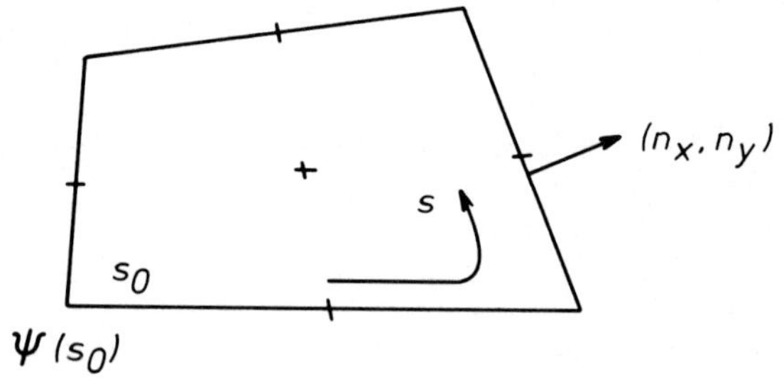

Fig. 9.7 Calculation of the stream function on a quadrilateral element.

Let s be a line coordinate along the boundary of the element; the value of ψ at s is then given by

$$\begin{aligned}\psi(s) &= \psi(s_0) + \int_{s_0}^{s} (d\psi/ds)ds \\ &= \psi(s_0) + \int_{s_0}^{s} (u\,n_x + v\,n_y)\,ds \quad ,\end{aligned} \tag{9.52}$$

where use has been made of (9.51), and where n_x, n_y denote the components of the unit vector normal to the boundary. When the choice of shape function for the pressure is such that incompressibility is satisfied at the element level, the integral in (9.52) vanishes once the integration is performed along the entire boundary of the element; we have shown in section 9.5 that some elements satisfy the incompressibility constraint at the element level, and (9.52) may then be used for evaluating ψ at the nodes.

However, the incompressibility constraint is generally satisfied only in the mean; this is true in particular for all the elements endowed with a continuous approximation for the pressure field. Then, the integral in (9.52) does not vanish on a closed contour, and ψ may not be evaluated by means of (9.52). We know from (9.51) that ψ satisfies the following Poisson equation,

$$\partial^2\psi/\partial x^2 + \partial^2\psi/\partial y^2 = \partial u/\partial y - \partial v/\partial x \quad . \tag{9.53}$$

On the part of the boundary where the velocity field is imposed, the value of ψ may also be determined at the outset, and be imposed as a Dirichlet boundary condition. In many problems (like entry flow), the tangent velocity component vanishes on the other parts of the boundary, and we may then impose a Neumann boundary condition, i.e. the vanishing of the normal derivative of ψ along that part of the boundary.

The calculation of the stream function then proceeds along the philosophy of the finite element method. Since the equations of motion and incompressibility are solved in the mean with the use of the Galerkin formulation, we might as well use the same procedure for calculating the stream function. We showed in section 8.10 how to solve Poisson's equation; the same procedure can be used for solving (9.53), where the right-hand side is now given by $(\partial\tilde{u}/\partial y - \partial\tilde{v}/\partial x)$, with $\tilde{u}$ and $\tilde{v}$ resulting from the finite element calculation of the flow. In practice, one will use for ψ a representation

$$\tilde{\psi} = \Sigma \ \psi_i \ \tau_i \quad , \tag{9.54}$$

where the τ_i's are the shape functions selected for the velocity field.

An entirely similar procedure may be used for the case of axisymmetric flow; in view of $(9.31)_3$, there exists a stream function ψ such that

$$ru = \partial\psi/\partial z \quad , \qquad rv = -\partial\psi/\partial r \quad , \tag{9.55}$$

and the Poisson equation for calculating ψ becomes

$$\partial^2\psi/\partial r^2 + \partial^2\psi/\partial z^2 = \partial(ru)/\partial z - \partial(rv)/\partial r \quad . \tag{9.56}$$

In a finite element code making use of the velocity components and the pressure, it is appropriate to include at the end of the program a module for calculating the stream function on the basis of the nodal velocity components; the calculation is cheap and provides an essential input to the flow visualisation process.

9.8 SOLVING THE GENERALIZED NEWTONIAN FLOW

We have discussed in section 9.5 the selection of shape functions for solving the Navier-Stokes equations, and we will now assume that a similar choice may be made for solving generalized Newtonian flow. Once the shape functions τ_i and τ_i' have been selected, it is possible to obtain the final form (9.30) of the algebraic system which is non-linear in the nodal velocity components. The non-linearity of the system (9.30) has a double origin, i) the inertia terms, which contain ρ as a factor, are quadratic in the velocity components, ii) the matrices A_{ij}, B_{ij} and C_{ij}, defined in (9.28), depend upon the shear viscosity which is itself a function of the nodal velocity components through the invariant I_2.

These two types of non-linearities are usually treated by different techniques, and we will consider them separately. Let us first consider the case of the Navier-Stokes equations, where all the coefficients in (9.30) are independent of the nodal velocity components. An efficient way of attaining a given value of the Reynolds number is to start from Stokes flow and to increase the Reynolds number by successive steps; an iterative procedure of Newton's type is used at each step until a satisfactory convergence test has been satisfied. Newton's method, which was explained in section 8.6, is easy to apply to a system of the form (9.30), because the non-linear terms are quadratic in the nodal velocity components. Indeed, let U_j^n, V_j^n, P_j^n denote the values of the unknowns after the n-th iteration. A correction δU_j, δV_j, δP_j is sought, such that the vectors $U_j^n + \delta U_j$, $V_j^n + \delta V_j$, $P_j^n + \delta P_j$ satisfy the system (9.30) up to the first order in the increments. It is easy to show that the algebraic system for the increments is given by

$$\begin{aligned}
&[A_{ij} + \rho(IX_{ijk} + IX_{ikj})U_k^n + \rho IY_{ikj}V_k^n]\delta U_j + (C_{ij} + IY_{ijk}U_k^n)\delta V_j - D_{im}\delta P_m \\
&\qquad = X_i - A_{ij}U_j^n - C_{ij}V_j^n + D_{im}P_m^n - \rho(IX_{ijk}U_j^n + IY_{ijk}V_j^n)U_k^n \quad , \\
&(C_{ji} + \rho IX_{ijk}V_k^n)\delta U_j + [B_{ij} + \rho(IY_{ijk} + IY_{ikj})V_k^n + \rho IX_{ikj}U_k^n]\delta V_j - E_{im}\delta P_m \\
&\qquad = Y_i - C_{ji}U_j^n - B_{ij}V_j^n + E_{im}P_m^n - \rho(IX_{ijk}U_j^n + IY_{ijk}V_j^n)V_k^n \quad , \\
&- D_{jm}\delta U_j - E_{jm}\delta V_j = D_{jm}U_j^n + E_{jm}V_j^n \quad .
\end{aligned} \tag{9.57}$$

It is quite clear that the values of the right-hand sides of the system (9.57) give a precise indication of how well the equations are satisfied after the n-th iteration, since these right-hand sides reproduce exactly the left-hand sides of the equations which should be satisfied by the solution. The standard Galerkin technique is usually satisfactory for reaching moderate values of the Reynolds number, which are typical in applications dealing with polymeric solutions and polymer melts. It will usually be found that, above some value of the Reynolds number, spatial oscillations (or wiggles) appear. The way of handling the difficult problem of high Reynolds number flow is still a question of debate. Special techniques, called *upwinding techniques*, have been proposed by Heinrich and Zienkiewicz (1979), and by Hughes and Brooks (1979), which allow us to obtain smooth solutions even on coarse meshes and at high values of the Reynolds number. On the other hand, it is argued by Gresho and Lee (1979) that the wiggles are simply a sign that the finite element mesh is too coarse and should be graded in some regions of the flow; it may be shown that upwinding introduces a sizable amount of artificial viscosity, which dampens the spatial oscillations while degrading the accuracy of the results. A full discussion of the finite element methods for convection-dominated flows may be found in Hughes (1979).

Let us now come to the second source of non-linearity, i.e. the dependence of the coefficients A_{ij}, B_{ij} and C_{ij} appearing in (9.30) upon the velocity components. We will first limit ourselves to the creeping flow of a generalized Newtonian fluid. For plane as well as axisymmetric flow, we may then write the system (9.30) as follows :

$$
\begin{aligned}
&A_{ij}(U_k,V_k)U_j + C_{ij}(U_k,V_k)V_j - D_{im}P_m = X_i \quad , \\
&C_{ji}(U_k,V_k)U_j + B_{ij}(U_k,V_k)V_j - E_{im}P_m = Y_i \quad , \qquad (9.58) \\
&-D_{jm}U_j - E_{jm}V_j = 0 \quad .
\end{aligned}
$$

Again, we need an iterative algorithm for solving the system (9.58). Although it is possible in general to use Newton's method, our experience is that it is often difficult to obtain a converging sequence (U_i^n, V_i^n, P_m^n). On the other hand, one may devise a simple iterative procedure, first suggested by Tanner et al. (1975), which has been found successful in all problems. Quite simply, the viscosities which appear in the calculation of the coefficients in (9.58) are calculated in terms of the old nodal velocity components; the system (9.58) is then written as :

$$A_{ij}(U_k^n,V_k^n)U_j^{n+1} + C_{ij}(U_k^n,V_k^n)V_j^{n+1} - D_{im}P_m^{n+1} = X_i \quad ,$$

$$C_{ji}(U_k^n,V_k^n)U_j^{n+1} + B_{ij}(U_k^n,V_k^n)V_j^{n+1} - E_{im}P_m^{n+1} = Y_i \quad , \tag{9.59}$$

$$-D_{jm}U_j^{n+1} - E_{jm}V_j^{n+1} = 0 \quad .$$

The iterations are stopped when, for a given value of ε, we have

$$\frac{|U_i^{n+1} - U_i^n|}{V_{max}^n} < \varepsilon \quad , \qquad \frac{|V_i^{n+1} - V_i^n|}{V_{max}^n} < \varepsilon \quad , \qquad \frac{|P_m^{n+1} - P_m^n|}{P_{max}^n} < \varepsilon \quad , \tag{9.60}$$

where

$$V_{max}^n = \sup(|U_i^n|,|V_i^n|,\ 1 \leqslant i \leqslant M) \ , \qquad P_{max}^n = \sup(|P_m^n|,\ 1 \leqslant m \leqslant N) \ ; \tag{9.61}$$

ε is typically of the order of 10^{-3} or 10^{-4}, and convergence is then guaranteed up to three or four significant digits for the velocity and pressure fields.

In order to exhibit the type of convergence that one may expect from the algorithm, let us consider a simple problem of entry flow for a power-law fluid. The geometry and boundary conditions of the problem are shown on Fig. 9.8. The domain is a rectangular plane channel with a plane of symmetry; in the entry section, we impose a vanishing vertical velocity component, while the horizontal velocity component is flat everywhere except near the lower boundary where the fluid sticks to the wall. On the right-hand side, we impose a vanishing vertical velocity component and a vanishing normal contact force. The finite element mesh is graded in the vertical direction with small spacing close to the wall where high velocity gradients are met, and in the horizontal direction with small spacing close to the entry section, where the flow development occurs.

The solution presented here has been obtained with the rectangular finite elements shown on Fig. 9.5b, with bi-quadratic shape functions for the velocity components, and bilinear shape functions for the pressure. The fluid is of the power-law type; in (3.1), the shear viscosity is then given by

$$\eta(I_2) = K\ I_2^{p-1} \quad , \tag{9.62}$$

where p is the power-law index, and K is the consistency factor. Fig. 9.9 shows the development of the velocity profile along the channel for $p = 1$ (Newtonian case), $p = 0.6$ and $p = 0.2$. For the Newtonian case, the initially flat velocity profile becomes parabolic in the exit section; for the power-law fluid, the

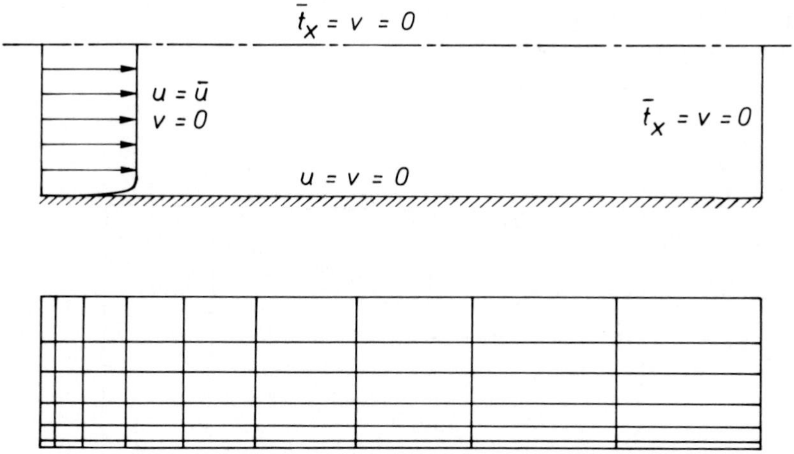

Fig. 9.8 Geometry, boundary conditions and finite element mesh for the entry flow problem.

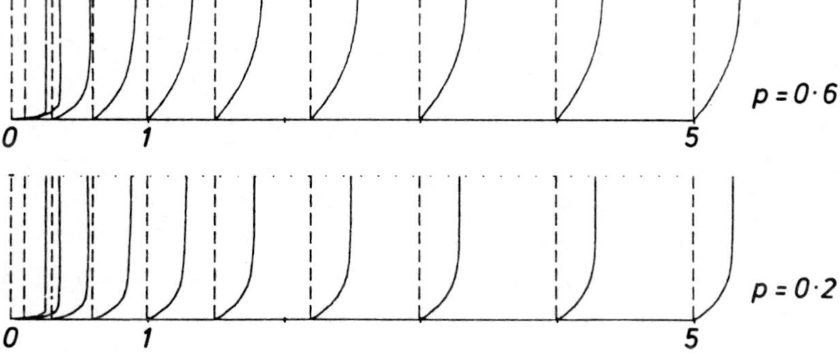

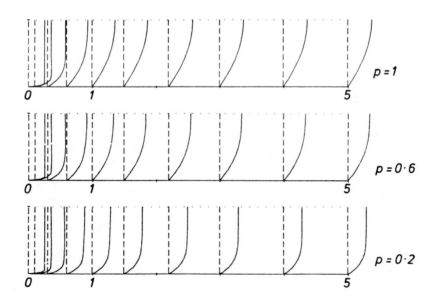

Fig. 9.9 Development of the velocity profile for the flow of a power-law fluid in a plane channel.

fully developed profile gets flatter as p decreases. Fig. 9.10 shows the ratio, to the half-width H of the channel, of the lengths L in the plane of symmetry needed to reach a horizontal velocity which is equal, respectively, to 95 and 98% of the fully developed value.

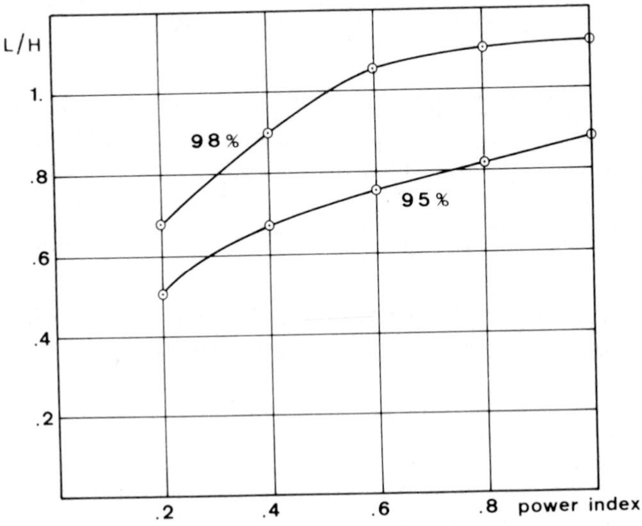

Fig. 9.10 Entry length as a function of the power-index.

Our main object here in studying the entry flow is to show how the number of iterations needed for reaching convergence is affected by the nature of the fluid. On Fig. 9.11 we show the variation of the relative error defined in (9.60) as a function of the number of iterations for respective values of the power-law index equal to 0.8, 0.6 and 0.4. Fig. 9.11 shows that the rate of convergence decreases when the power-law index decreases. The number of iterations needed for reaching a relative error of 10^{-4} increases appreciably when p becomes small; however, a lower degree of accuracy, say 10^{-2}, may be obtained within a few iterations. The rate of convergence obtained for the present problem is representative of most flows of generalized Newtonian fluids.

When Newton's method is being used for solving (9.58), it is found that the success of the iterative procedure depends upon the viscosity law and the power index. In calculating the entry flow of a power-law fluid, it is found that Newton's algorithm converges when the power index is higher than 0.5; the

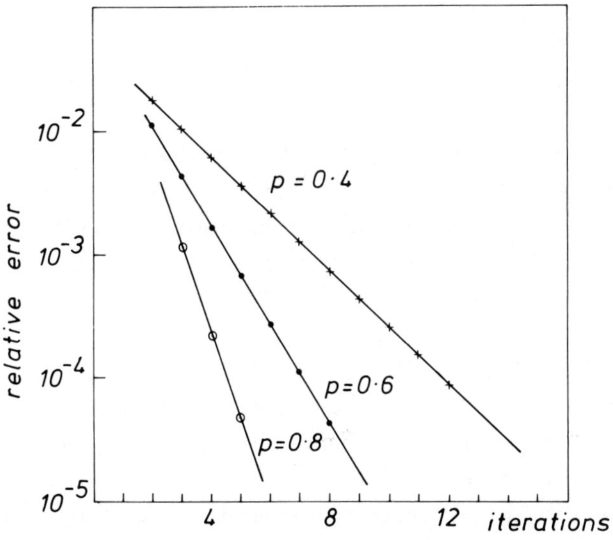

Fig. 9.11 Relative error as a function of the number of iterations for various values of the power-index.

convergence may sometimes be lost for lower values. With a Bird-Carreau law and the associated finite value of the viscosity for a vanishing shear-rate, Newton's method may converge for values of the power index lower than 0.5.

It must be pointed out that the method given by (9.59) for calculating a generalized Newtonian flow is valid whatever be the law used for relating the shear viscosity to the shear-rate. As a typical example, consider the matrix components C_{ij} given in (9.28). We have shown in section 8.11 that the integrals on the right-hand side of (9.28) will usually be calculated by means of a numerical integration scheme over the parent element; the calculation of C_{ij} reduces to a sum of the form,

$$C_{ij} = \sum_{k=1}^{K} w_k \, \eta[I_2(\xi_k,\eta_k)] \, g_{ij}(\xi_k,\eta_k) \quad , \qquad (9.63)$$

where ξ_k,η_k are the coordinates of the K integration points in the parent element, w_k are the associated weights and g_{ij} is representative of the factors other than η in the integrand. It is clear that, in calculating the right-hand side of (9.63), the degree of complexity is the same whether the viscosity is given by a power law or by, say, a Bird-Carreau law of the type

$$\eta(I_2) = K\,\bar{\lambda}^{1-p}(1 + \bar{\lambda}\, I_2)^{p-1} \tag{9.64}$$

where $\bar{\lambda}$ is a time constant.

All one needs to do is to declare the function $\eta(I_2)$ within the finite element program.

Finally, we mention briefly how to deal with the flow of generalized Newtonian fluids when inertia effects may not be neglected; an efficient algorithm involves using an iterative technique similar to (9.57), where the coefficients which depend upon the viscosity are calculated in terms of the nodal velocity components U_i^n, V_i^n .

9.9 ENTRY FLOW IN A TUBULAR CONTRACTION

An important problem in rheology is the prediction of entry losses when a viscous fluid flows through a contraction at the junction of two tubes of different diameters, and of the exit losses when the fluid leaves the tube and enters a jet. The problem is important in rheometry as well as in technological applications, where contractions and free surfaces are present in most forming devices. A numerical approach to the problem has been presented by Boger et al. (1978), while an extended review of problems associated with entry flow has been provided by Boger (1982), where a list of related references may be found. In the present section, we wish to show how the numerical simulation allows us to calculate the flow of a generalized Newtonian fluid in an axisymmetric contraction, while the prediction of the shape of the jet and the exit losses will be considered in the next section. In Chapter 10, we shall return to the same problems taking viscoelastic effects into account.

Fig. 9.12 shows the geometry of the problem and the boundary conditions, together with a finite element mesh. We consider an axisymmetric 4/1 contraction, where 4 is the ratio between the radii of the upstream and downstream tubes, respectively. The boundary ABCD is a fixed wall on which the viscous fluid is assumed to stick, while EF is an axis of symmetry. While considering creeping flow of an inelastic fluid, we will adopt long enough upstream and downstream tubes so that we may impose a fully developed velocity profile in the entry and in the exit sections. The velocity profile depends of course upon the type of fluid being considered; for a power-law fluid, the profiles may be found in Bird et al. (1977, section 5.2). The mesh shown on Fig. 9.12 is typical of that used in several problems of fluid mechanics solved by means of finite elements. We know that high velocity and high pressure gradients will occur in the neighbourhood of the re-entrant corner; if we wish to reach a reasonable accuracy, it is therefore essential to surround the corner area and the contraction with small elements. On the other hand, the fully developed

Poiseuille flow in the entry and exit sections may be correctly simulated with a relatively small number of elements. We see here the advantage of isoparametric elements of an arbitrary shape which allow the implementation of graded meshes. On Fig. 9.12, we go from three elements across the entry section to eight just before the contraction. In the present case, the use of rectangular elements only would necessarily lead to a much larger number of elements and nodal values. In the present calculation, we use nine-node quadrilaterals with biquadratic velocity components and bilinear pressures, together with six-node triangles and the associated shape functions.

The entry and the exit sections of the mesh shown on Fig. 9.12 are divided into three elements of equal size. Such a lay-out is suitable for the flow of a Newtonian fluid since the fully developed parabolic profile may be reproduced exactly within the element. When the power index p decreases, the fully developed profile tends to a plug-flow, and the arrangement shown on Fig. 9.8 is preferable. Still, it is found that, with three equal elements across the channel, the error of the velocity field with respect to the analytical solution at p = 0.5 is less than 0.2%, and the error on the pressure gradient is less than 0.02%. At p = 0.25, these bounds become 0.4% for the velocity field, and 0.08% for the pressure gradient.

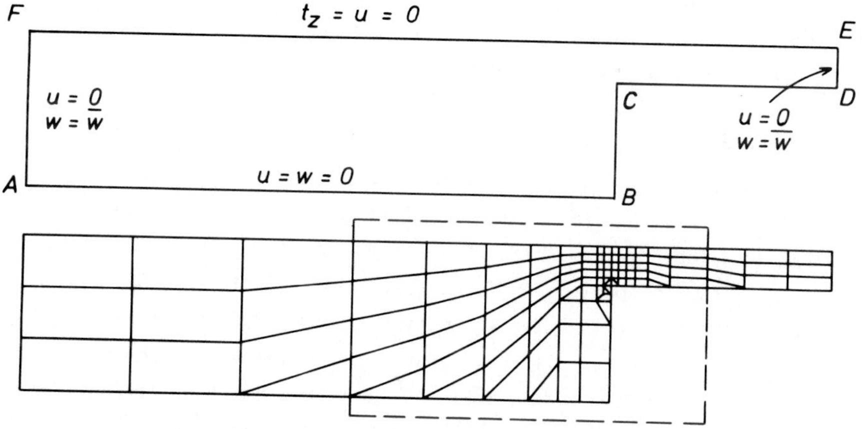

Fig. 9.12 Geometry, boundary conditions and finite element mesh for the 4/1 contraction flow.

Let us first examine creeping Newtonian flow. On Fig. 9.13, we show the streamlines, where we find in particular the corner vortex which is known to occur experimentally (see, e.g., Nguyen and Boger, 1979); the stream function has been normalized to the (0,1) interval, and the streamlines are drawn on the subdomain marked by the dashed frame on Fig. 9.12.

An important global quantity is the excess pressure drop in the axisymmetric duct which is due to the tubular contraction and which may be calculated as follows. Let p_F and p_E denote respectively the pressure at point F and point E shown on Fig. 9.12, and let ∇p_u, ∇p_d denote respectively the pressure gradients in the fully developed upstream and downstream tubes. The excess pressure drop in the contraction is defined by

$$\Delta p_{en} = (p_F - p_E - \overline{AB} \times \nabla p_u - \overline{CD} \times \nabla p_d) / 2\tau_w \quad , \qquad (9.65)$$

where τ_w denotes the wall shear-stress in the fully developed downstream flow. With the mesh shown on Fig. 9.12, one obtains, for Stokes flow, a value of Δp_{en} equal to 0.548; Viryayuthakorn and Caswell (1980) obtain 0.566 with a different mesh, while 0.555 is obtained with the mesh shown on Fig. 9.18. These numbers show that, even with refined meshes, one finds a dispersion of the order of a few percent. The value of the excess pressure drop is even more affected by the continuity conditions which have been implicitly assumed at the re-entrant corner. It was mentioned in section 6.2.4 that, on the basis of the work done by Moffatt (1964), one expects a stress singularity at the re-entrant corner; in particular, the pressure is infinite at that point. However, the finite element approximation characterized by (9.25) together with continuous representation for the pressure is unable to approach the singular behaviour of the solution near the corner. It is therefore interesting to consider a special element which allows for a discontinuous pressure at the corner, if not an infinite value.

The special element is shown on Fig. 9.14. Consider a quadrilateral which, through an isoparametric transformation, is mapped into a triangle by the superposition of two nodes. If we impose that the velocity components be the same at the two superposed nodes, while the pressures may be different, we obtain the desired behaviour. When such an element is used, the excess pressure drop becomes 0.582, while the streamlines cannot be distinguished from those shown on Fig. 9.13. Thus, we find values of Δp_{en} lying between 0.55 and 0.58, which are in very good agreement with those mentioned by Boger (1982) in his review; it is also fairly obvious that the inclusion of a third digit in the present evaluations of Δp_{en} is useless, since Δp_{en} depends slightly upon the mesh refinement and the type of element used for the calculation.

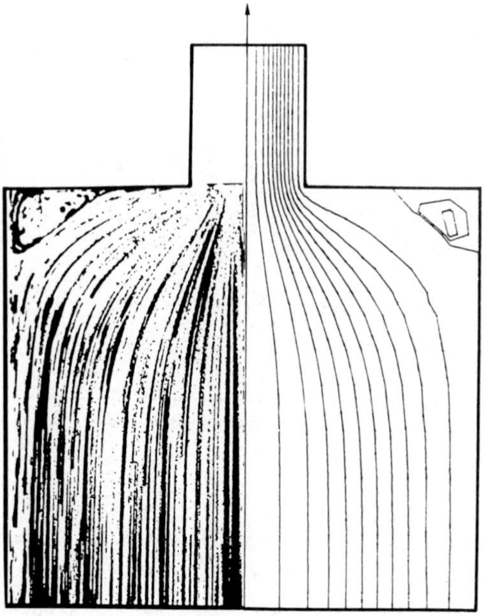

Fig. 9.13 Streamlines of the Newtonian flow in a 4/1 contraction. On the left: experimental data obtained by Boger. On the right: numerical solution. The stream function takes the unit value on the axis of symmetry and vanishes on the wall. The streamlines shown on the Figure correspond to $\psi = 0{\cdot}05, 0{\cdot}15, \ldots, 0{\cdot}95$ in the main stream; $\psi = -0{\cdot}0005$ and $\psi = -0{\cdot}001$ in the corner vortex.

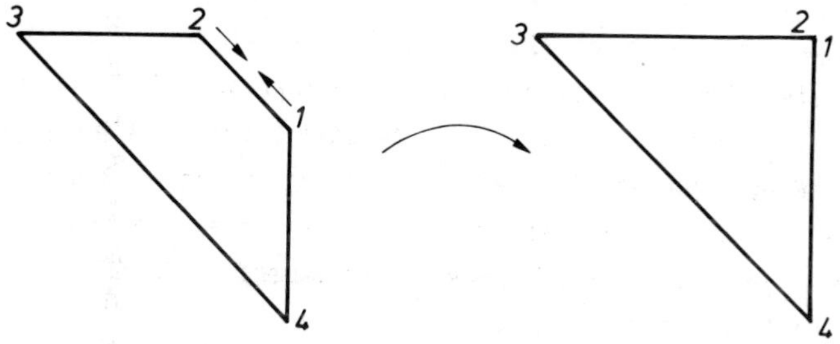

Fig. 9.14 Special element for obtaining a locally discontinuous pressure.

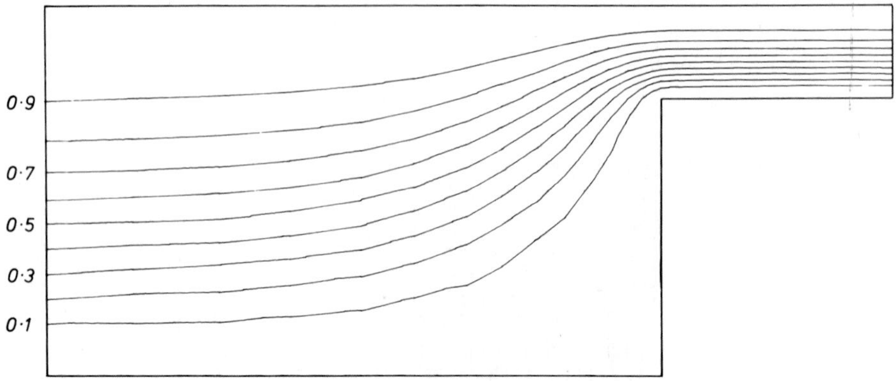

Fig.9.15 Streamlines for the flow of a power-law fluid, p = 0·25.

We will now consider the flow of a power-law fluid through the 4/1 axisymmetric contraction; we make use of the special corner element which we have just described. The calculations have been conducted for a power-law index equal, respectively, to 0.75, 0.50 and 0.25, for which one finds a rate of convergence similar to that shown on Fig. 9.11. We show on Fig. 9.15 the streamlines for $p = 0.25$. They do not differ appreciably from those shown on Fig. 9.13, except that the recirculating vortex has disappeared; this is already true for $p = 0.75$, and has been confirmed by several other numerical tests (see, e.g., Bezy 1982). Fig. 9.16 shows the development of the axial velocity along the axis of symmetry, and the velocity profiles across the tube for $p = 0.25$. It is found that the final velocity is already reached at the entrance to the tube, and that the developed velocity profile is attained within a fraction of the downstream radius. On Fig. 9.17, we show the entry length as a function of the power index; the entry length is the downstream tube length needed for the axial velocity profile to attain either 98 or 99% of its fully developed value.

Fig. 9.17 also shows the value of Δp_{en} as a function of the power index p for a 4/1 and an 8/1 contraction ratio. On the same diagram, we also report the values obtained by Boger et al. (1978) for the entry from a large reservoir to a circular tube. It is fairly clear that the excess pressure drop at the entry increases when the power index decreases, and depends slightly upon the geometry when the contraction ratio increases.

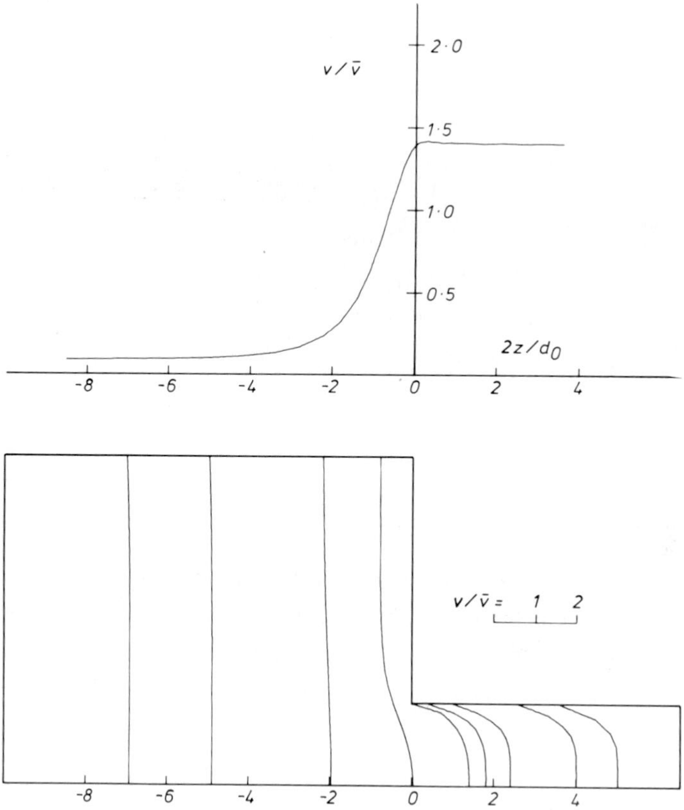

Fig. 9.16 Development of the axial velocity along the axis of symmetry, and of the velocity profiles across the tube, p = 0·25.

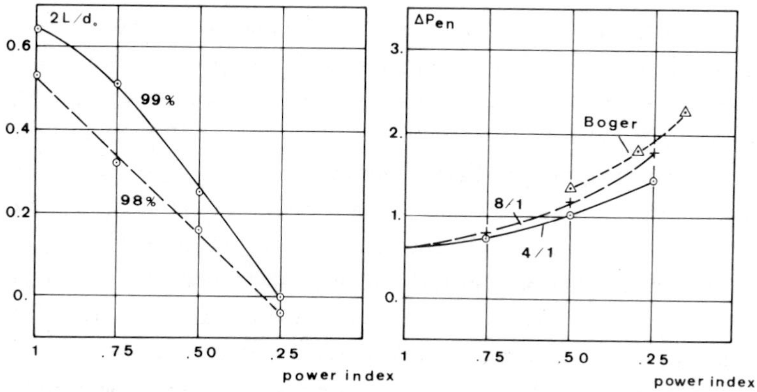

Fig. 9.17 Entry length and excess pressure drop for the entry flow of a power-law fluid.

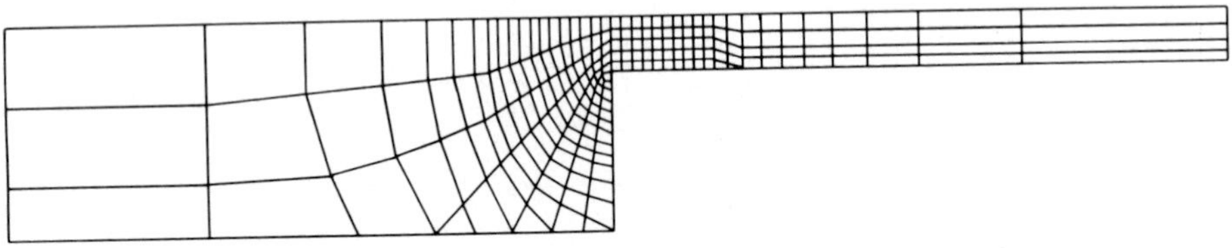

Fig. 9.18 Refined finite element mesh used for the 4/1 contraction.

In order to show the efficiency of the finite element method for solving problems where inertia effects do become important, we have also considered the problem of the entry flow in a 4/1 contraction of a Newtonian fluid for a range of Reynolds numbers between 0 and 100, with the definition

$$R = \frac{\rho \bar{w} d_0}{\eta} , \tag{9.66}$$

where $\bar{w}$ is the mean downstream velocity and d_0 is the diameter of the downstream tube. The calculations have been conducted on the refined mesh[†] shown on Fig. 9.18, which was designed for convergence experiments, with the boundary conditions shown on Fig. 9.12. For values of the Reynolds number higher than 100, it is necessary to modify the downstream boundary conditions and impose a vanishing normal force rather than the fully developed velocity profile; this is evident on Fig. 9.19, where the development of the axial velocity profile is shown for $R = 0$ and $R = 100$. Quite obviously, at $R = 100$, a length of 20 radii for the downstream tube is barely sufficient to reach a fully developed flow in the end section. Fig. 9.20b shows the excess pressure drop as a function of the Reynolds number together with the results obtained by Kestin et al. (1973) for a 10/1 contraction, and on the basis of an equation suggested by Boger (1982); the correlation is excellent. Fig. 20a shows the entry length as a function of R, together with the results obtained previously by Vrentas and Duda (1973) with finite differences. These are represented in Fig. 20a by triangles.

[†]We wish to thank Dr. R. Keunings for performing these calculations.

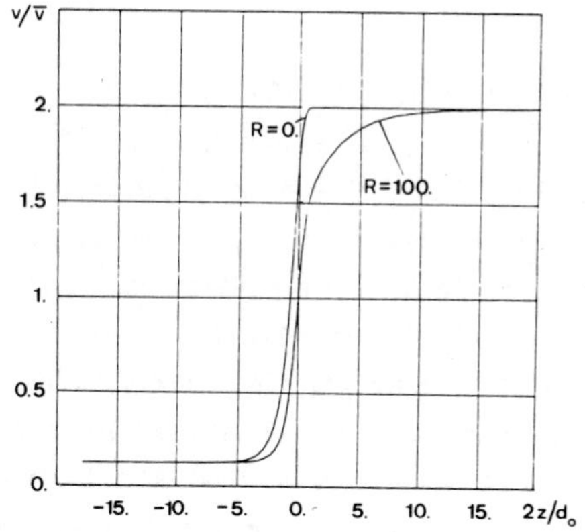

Fig. 9.19 Development of the axial velocity along the axis of symmetry for R=0 and R=100, Newtonian fluid.

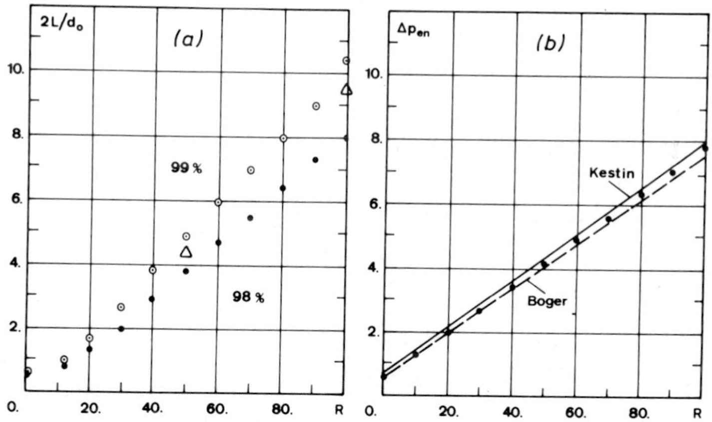

Fig. 9.20 Entry length and excess pressure drop for the entry flow of a Newtonian fluid as a function of the Reynolds number.

It must be pointed out that these results for moderate values of the Reynolds number have been obtained without any recourse to upwinding techniques, and that, up to the highest value, convergence is reached with three to four iterations of Newton's algorithm.

9.10 DIE-SWELL OF A GENERALIZED NEWTONIAN FLUID

The problem of the extrusion of a viscous fluid from a slit or a circular die presents formidable difficulties from an analytical point of view (see, e.g., Trogdon and Joseph 1980, Sturges 1979, 1981); however, progress in numerical techniques has allowed us to obtain a considerable amount of data on the flow behaviour in problems related to extrusion. A review of the field is given by Tanner (1983). It is clear that the flexibility of the finite element technique for handling complicated geometries has been an important asset in solving extrusion problems; some recent work by Ryan and Dutta (1981) provides an indication however that finite differences may also be considered for solving extrusion problems.

Fig. 9.21 shows the geometry and boundary conditions associated with the die-swell problem; in the present section, we limit ourselves to circular dies while many references cited in the text treat slit as well as circular dies. The line AB is the entry section where we assume that the flow is fully developed, and that we may impose the associated velocity profile; this is acceptable provided the tube length AE is large enough. The fluid sticks to the wall AE, where the velocity components vanish identically. On DC we assume that the free jet conditions are attained, and that the surface force components vanish identically. The main difficulty is related to the part ED of the boundary where the shape of the jet is not known, while three boundary conditions must be taken into account. More precisely, if n_r, n_z are the components of the unit normal to the free surface, we must have at every point of ED,

$$u\, n_r + v\, n_z = 0 \quad , \tag{9.67}$$

$$t_r n_r + t_z n_z = \sigma\left(\frac{1}{\rho_1} + \frac{1}{\rho_2}\right) \quad , \tag{9.68}$$

$$t_r n_z - t_z n_r = 0 \quad ; \tag{9.69}$$

in (9.68), ρ_1 and ρ_2 are the principal radii of curvature, while σ is the surface tension coefficient. Thus, the shape of the domain is unknown as well as the velocity and pressure fields.

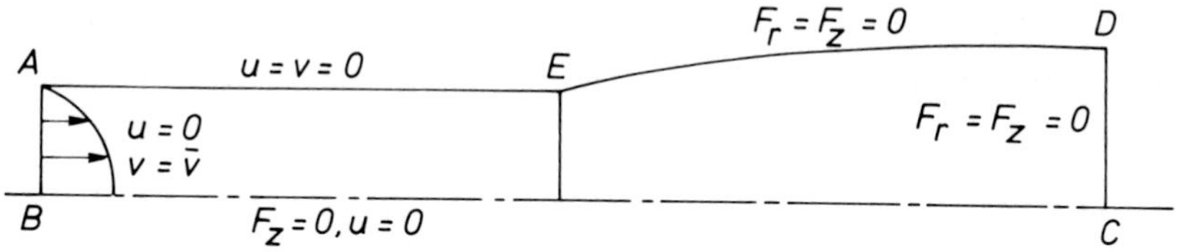

Fig. 9.21 Geometry and boundary conditions for the die-swell problem.

The finite element solution of the problem was considered for the first time by Tanner (1973) and by Nickell et al. (1974) for the case of a Newtonian fluid in the absence of surface tension. Their algorithm, which is now widely used in the literature, involves an iterative procedure which may be described as follows :

i. an initial guess is made of the shape of the free surface.

ii. the finite element problem is solved on the assumed domain of integration, while the force boundary conditions (9.68) and (9.69) are imposed on the free surface.

iii. the kinematic condition (9.67) is generally not satisfied on the free surface; thus, a new shape is sought by drawing a streamline originating at the edge. This is easily obtained by stating that the slope of the free surface should be parallel to the velocity field.

iv. a regridding of the mesh occurs in view of the new boundary, and the procedure is restarted. When σ in (9.68) is small, the location of the free surface is obtained within 3 to 4 iterations. When the surface tension becomes a dominant parameter of the flow behaviour it has been shown by Silliman and Scriven (1980) that an iterative procedure based on the satisfying of the kinematic condition (9.67) has a decreasing efficiency; they show indeed that it is better to impose the essential boundary condition (9.67) and the natural boundary condition (9.69) for calculating the flow on the trial domain, while the normal stress condition (9.68) is used as a criterion for relocating the free surface. It has also been shown by Ruschak (1980) that it is possible to solve for the location of the free surface together with the velocity and pressure fields by considering that the nodal coordinates of the free surface are explicit unknowns of the problem.

Two different approaches have been used for correcting an assumed free surface on the basis of the kinematic condition (9.67). Consider on Fig. 9.22a an assumed free surface, where the velocity is known as the output of a calculation based on the natural boundary conditions; the velocity vector will not in general be tangent to the free surface, which should be corrected. In Nickell et al. (1974) the ratio u/v is calculated at the nodes of the boundary (the value at the edge may be obtained through a limit process); if $r = F^n(z)$ is the equation of the free surface before the n-th iteration, the new surface is obtained by a numerical integration of the equation

$$dF^{n+1}(z)/dz = (u/v)^n \quad , \tag{9.70}$$

with the use of Simpson's rule. Caswell and Viryayuthakorn (1983) use a different procedure which is summarized on Fig. 9.22b. When the assumed free surface is not a streamline, it is possible to calculate the flow rate through each segment between two successive nodes. Each node is assigned a new position along the normal to the old surface by the assumption of local uniform flow in conjunction with a cumulative balance of mass in order to absorb the excess flow rate. Caswell and Viryayuthakorn find that their new scheme converges more rapidly and smoothly than the former.

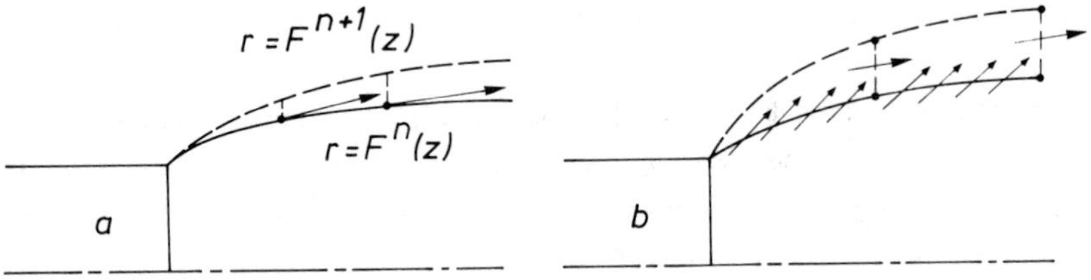

Fig.9.22 Two schemes for updating the free surface; a: the new free surface is obtained by integration of the slope on the old free surface; b: the displacement of the old free surface is such that a vanishing mass flux is obtained on the new surface.

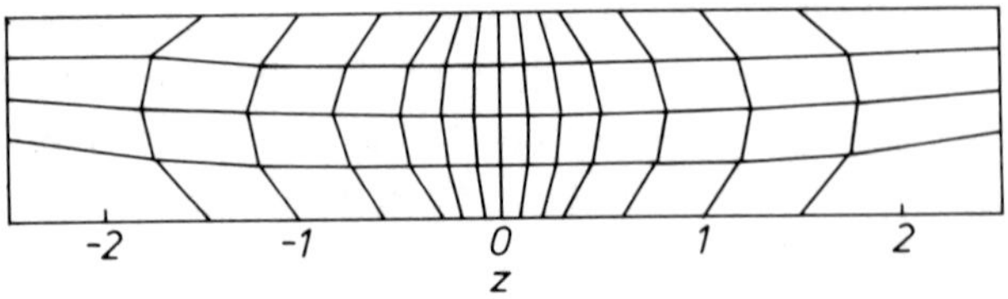

Fig.9.23 Mesh used for calculating the die-swell of a generalized Newtonian fluid (Boger et al 1978).

Several figures related to the streamlines and the pressure field for the Newtonian case will be shown in section 10.12 when we discuss the die-swell of viscoelastic fluids. The die-swell of a power-law fluid has been studied by Tanner et al. (1975) and Boger et al. (1978) with a finite element mesh shown on Fig. 9.23, where the quadrilaterals are composite elements made of four triangles with a quadratic velocity field and a linear pressure field. The excess pressure loss is calculated as follows : let ∇p be the fully developed pressure gradient in the tube; referring to Fig. 9.21, we define the excess pressure drop by the following relation,

$$\Delta p_{ex} = (p_A - p_D - \overline{AE} \times \nabla p) \; / \; 2\tau_w \quad , \tag{9.71}$$

where τ_w denotes the wall shear-stress in the fully developed flow. Fig. 9.24 shows the excess pressure drop as a function of the power index found by Boger et al. (1978) and the dependence of the swelling ratio upon the same index. The swelling ratio is defined by

$$S_w = d_J/d_0 \quad , \tag{9.72}$$

where d_J and d_0 are the jet diameter and the tube diameter, respectively. One finds that the swelling ratio decreases with p; this might be expected since the fully developed upstream flow field tends to a plug flow.

The effects of surface tension and the Reynolds number upon die-swell have been studied by Reddy and Tanner (1977); Fig. 9.25 shows the swelling ratio as a function of the Reynolds number in the absence of surface tension.

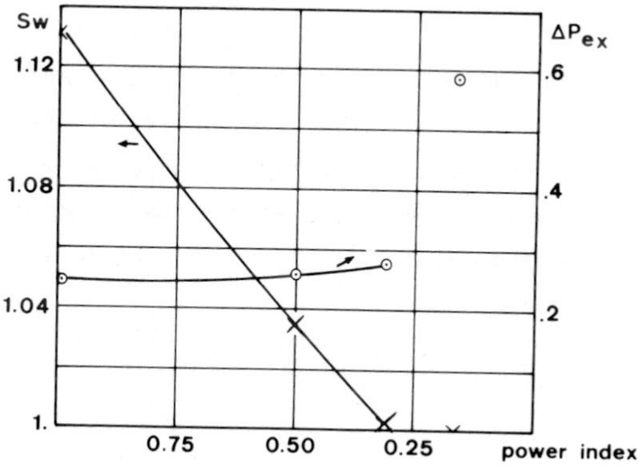

Fig. 9.24 Exit pressure loss and swelling ratio for a power-law fluid (Boger et al 1978)

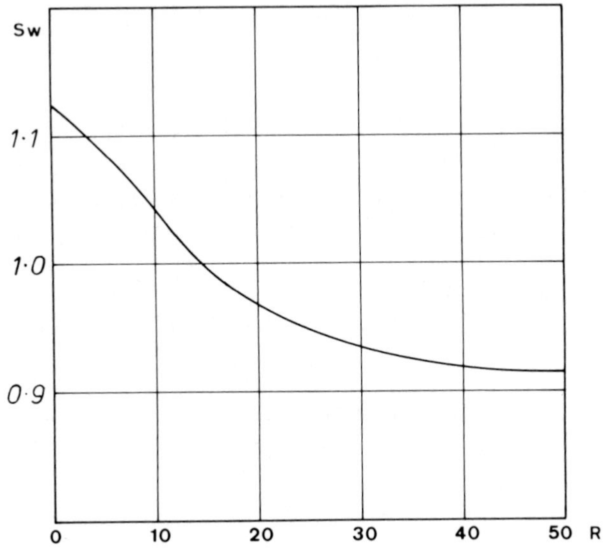

Fig.9.25 Swelling ratio as a function of the Reynolds number (Reddy and Tanner 1977).

9.11 THE FLOW OF A POWER-LAW FLUID AROUND A SPHERE

As a last example of the use of finite elements for solving the flow of a generalized Newtonian fluid, we will now consider briefly the case of a sphere falling in a cylindrical tube. The creeping motion of a sphere through a power-law fluid has been considered by several authors who have used approximate solutions based on a variational principle; a list of references and a synthesis of the results on the drag evaluation may be found in Chhabra et al. (1980). The finite element method allows for a rather cheap and easy way of obtaining a detailed solution for the velocity field as well as for global quantities such as the drag on the moving sphere.

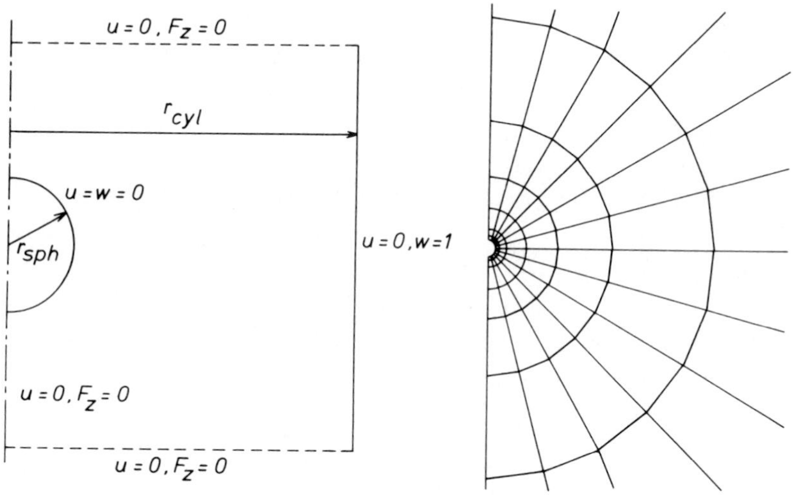

Fig. 9.26 Geometry, boundary conditions, and finite element for the flow of a viscous fluid around a sphere.

The geometry of the problem and the boundary conditions are shown on Fig. 9.26; we assume that the sphere of radius r_s is fixed in space, and that a cylinder of radius r_c moves along the vertical axis with unit velocity. We have selected the value 50 for the ratio r_c/r_s. The end planes are located at a distance of $50r_s$ from the centre of the sphere; in order to approach the case of a very long cylinder, we assume that, on the planar boundaries, the radial velocity vanishes together with the axial surface force. The finite element mesh used for the calculations is also shown on Fig. 9.26; each meridian is divided into 12 equal parts, while the size of the elements increases together with their distance from the centre of the sphere. On the surface of the sphere, we use isoparametric elements; the spherical surface is made up of a series of parabolic segments which have a common tangent at the joint nodes. It is easy to calculate the drag on the sphere; indeed, once the approximate velocity field is known, the system (9.30) allows us to calculate the nodal forces on the surface. In view of (9.36), the sum of the nodal forces in the axial direction multiplied by 2π provides the value of the drag.

Let D be the drag, V the velocity of the sphere, d its diameter and ρ the mass density of the fluid. The drag coefficient C_D is defined by

$$C_D = 8D/(\rho V^2 \pi d^2) \quad , \tag{9.73}$$

while the Reynolds number for the flow of a power-law fluid is given by

$$R = d^p V^{2-p} \rho/K \quad . \tag{9.74}$$

The Stokes solution for the slow flow of a Newtonian fluid provides the relation

$$C_D = 24/R \quad . \tag{9.75}$$

However, the value of C_D given by (9.75) is not correct for a power-law fluid, and is expressed as follows,

$$C_D = (24/R)X \ , \tag{9.76}$$

where X is a correction factor with the value 1 for the Newtonian case.

The creeping flow calculation has been performed for values of the index p equal, respectively, to 1, 0.8, 0.6, 0.4, 0.2 and 0.1. Let us first examine some results related to the velocity field. On Fig. 9.27, we show the axial velocity along the axis of symmetry as a function of the distance to the centre of the sphere, and of the axial velocity in the plane of symmetry against the same variable, for $p = 1$, 0.6, 0.4, 0.2. The problem has also been run on a refined mesh, where the sphere is surrounded by 10 layers of elements rather than 8; the refined results are also shown on Fig. 9.27. We see clearly that the size of the flow domain perturbed by the sphere becomes very small when p decreases; indeed, for a highly shear-thinning fluid, the viscosity becomes quite large at some distance from the sphere, and the fluid moves like a rigid body with the cylindrical wall. Table 9.1 gives the distance of the centre of the sphere where the axial velocity reaches 99% of its long distance value, on the axis of symmetry and in the equatorial plane.

It is remarkable to observe that, when $p < 0.6$, the disturbed region becomes so small that an important overshoot appears in the velocity profile on the plane of symmetry. The overshoot is evident for $p = 0.2$, and its size has been confirmed by a further numerical test with a refined mesh in the radial direction. The results of Fig. 9.27 suggest that the drag correction due to the presence of the cylindrical wall is much less important for a shear-thinning fluid than for a Newtonian fluid (Brenner, 1962).

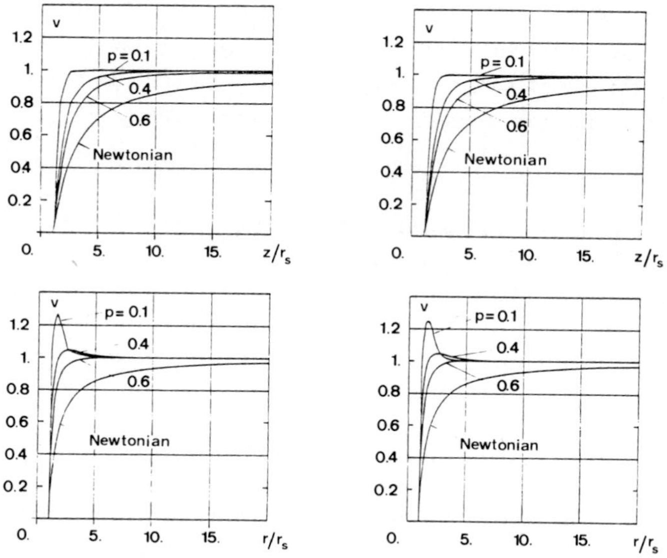

Fig. 9.27 Axial velocity along the axis of symmetry and in the plane of symmetry; on the left, 8 layers of elements; on the right, 10 layers.

TABLE 9.1

Distance of the centre on the axis and on the plane of symmetry where the axial velocity reaches 99% of its long range value

p	1	0.8	0.6	0.4	0.2	0.1
axis of symmetry, z/r_s	30.8	22.7	14.1	7.7	4.2	2.5
plane of symmetry, r/r_s	30.9	14.4	3.6	1.7	1.2	1.1

Table 9.2 gives the correction factor X in (9.76) as a function of the power index p. For the Newtonian case, the exact solution for the flow of a sphere in an infinite space would provide $X = 1$; if the cylindrical wall is taken into account, Brenner's correction gives a value of $X = 1.04$ when $r_c/r_s = 50$. Thus, the error of the numerical results is of the order of 2% for the drag calculation, which is reasonable for a relatively coarse finite element mesh. The

results of Table 9.2 are plotted on Fig. 9.28, which was published by Chhabra et al. (1980), and summarizes available analytical and experimental results. An attractive feature of the finite element calculation is that the viscous behaviour of any experimental fluid may be introduced without difficulty in the program; the use of several viscosity laws might help in explaining the discrepancies in the evaluation of X shown on Fig. 9.28.

TABLE 9.2

Drag correction factor as a function of the power index, for the sphere falling along the axis of a circular cylinder

p	1	0.8	0.6	0.4	0.2	0.1
X	1.02	1.27	1.44	1.51	1.46	1.39

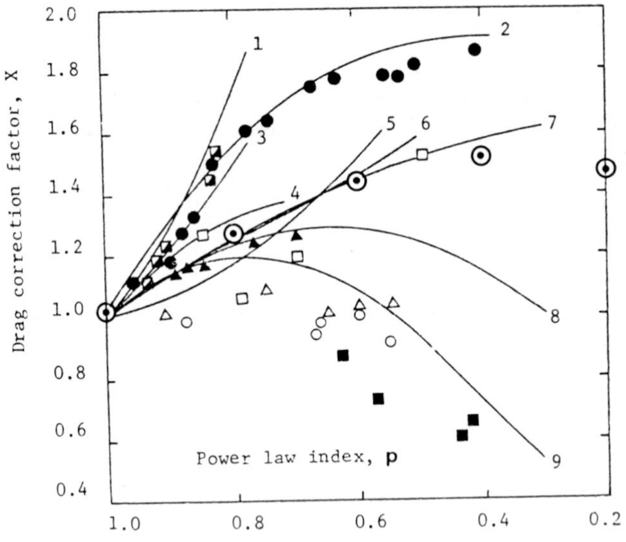

Fig. 9.28 Drag correction factor, X, as a function of the power law index p. The finite element results ⊙ are compared with theoretical and experimental results compiled by Chhabra et al (1980).

Several problems of technological importance involving the flow of Newtonian and generalized Newtonian fluids may be found in the recent literature. The flow through a wire-coating die has been studied by Tanner (1976) and Caswell and Tanner (1978). Thermal effects have been considered by Phuoc and Tanner (1980), Ben-Sabar and Caswell (1979) and Nakazawa et al. (1982). The extrusion from conical dies has been studied by Crochet and Keunings (1981). The die-swell problem for a Newtonian fluid has also been solved by Chang et al. (1979) with the use of the collocation method. The Hele-Shaw approximation has been used by Hieber and Shen (1980) for calculating the mould-filling flow of a power-law fluid. Finally, we wish to mention the work of Zienkiewicz et al. (1974, 1978) which uses the generalized Newtonian fluid for solving metal forming processes.

Chapter 10

Finite Element Calculation of Viscoelastic Flow

10.1 INTRODUCTION

We saw in the previous chapter that the numerical technique for solving the flow of a generalized Newtonian fluid differs little from the one used for solving the Navier-Stokes equations. The reason is that when the shear viscosity is a function of the rate-of-deformation tensor, the equations of motion can still be written explicitly in terms of the velocity components and the pressure. Under these conditions, an approximate solution of the flow problem can be sought in terms of nodal values of the velocity components and the pressure, and the equations of motion are solved by means of a velocity-pressure, or u-v-p, finite element algorithm. We have seen in Chapter 2 that, for slow flow, the general constitutive equations of a simple fluid degenerate into those of an nth-order fluid (2.55-57); there again, the extra-stresses are explicit functions of the velocity components and their space and time derivatives. A new difficulty arises, however, because of the appearance of higher order spatial derivatives; still one may use a u-v-p technique for solving, for example, the flow of second-order fluids (Reddy and Tanner 1977).

Here, we are concerned with the flow of viscoelastic fluids beyond the range of application of the hierarchy equations; our aim in this chapter is to review finite element techniques which will allow us to solve the flow of fluids with constitutive equations like those given in section 2.7, where the extra-stress components are given either in terms of a differential equation or as a time integral where the integrand contains the strain history. We have commented in section 3.7 on the constitutive equations of a Maxwell fluid, given by (3.5); despite its somewhat unrealistic behaviour for simulating the flow of polymeric solutions and polymer melts, the Maxwell fluid is endowed with certain features which make it attractive for the development of a numerical algorithm : i) the fact that the shear viscosity is a constant allows us to separate shear-thinning from elastic effects, ii) the model contains only two material parameters which are clearly related to the viscometric functions, shown in (3.9), iii) the fluid may be studied as a differential model (3.5), or as an integral model (3.6). These features justify why we will first develop algorithms pertaining to the flow of a Maxwell fluid. We will then show that it is easy to generalize them to the flow of more complicated models.

Before starting a detailed analysis, it is useful to comment briefly on the choice between the differential model (3.5) and the integral model (3.6). While the extra-stress components in (3.5) are not explicit expressions of the spatial derivatives of the velocity components, the situation is not better in (3.6) where the full deformation history appears in the integrand. The main difficulty of the differential model lies in the convective terms $\overset{\nabla}{T}_{ij}$ which are difficult to calculate with finite elements. As usual, however, there is no short cut in solving a difficult problem: while (3.6) does not contain stress derivatives, convective terms will arise again in the calculation of the strain tensor along the streamlines. The development of a good algorithm for solving either (3.5) or (3.6) presents at some stage similar difficulties.

Recognizing the fact that, as long as we work with differential models, there is no explicit form for the extra-stress tensor as a function of the velocity field and its derivatives, we will find it instructive to return first to the Navier-Stokes equations, and obtain for them a solution technique as if the constitutive equations of a Newtonian fluid were not explicit. We will then be able to suggest mixed finite element techniques which are applicable for solving the flow of viscoelastic fluids. Next, we will explain the available methods for solving the flow of fluids given by integral models.

Whatever be the algorithm, the inertia terms, when they are needed, may be treated in a way similar to that already explained in Chapter 9. For the sake of simplicity, we will now neglect the inertia terms and concentrate on the creeping flow of viscoelastic fluids for explaining the numerical techniques.

10.2 ANOTHER VARIATIONAL THEOREM FOR CREEPING NEWTONIAN FLOW

In section 9.2, we dealt briefly with a variational theorem for the creeping flow of a generalized Newtonian fluid. It became clear that the variational theorem was not essential for introducing the finite element algorithm; it was useful, however, for building a bridge between classical analytical techniques and the Galerkin method. The same procedure will be used here for introducing the mixed finite element method which will be discussed in the next few sections.

We consider the same general problem as in section 9.2, with the domain and boundary conditions shown on Fig. 9.1. For simplicity, we consider here a Newtonian fluid for which the rate-of-deformation tensor d_{ij} is related to the extra-stress tensor T_{ij} through the equation

$$d_{ij} = \frac{\partial \tau(T_{pq})}{\partial T_{ij}} \quad , \tag{10.1}$$

where $\tau(T_{pq})$ is a stress function defined by

$$\tau(T_{pq}) = T_{ij}\, T_{ij}/4\eta \quad . \tag{10.2}$$

Let us now select in the flow domain Ω a velocity field v_i which satisfies the boundary conditions (9.4), a pressure field p, and an extra-stress field T_{ij} ; we will simply assume that v_i is twice differentiable in Ω, and that p and T_{ij} are once-differentiable. We may construct a functional $\bar{J}(v_k,p,T_{pq})$ as follows :

$$\bar{J}(v_k,p,T_{pq}) = \int_\Omega [-\tau(T_{pq}) + v_{i,j}T_{ij} - \rho F_i v_i - p v_{k,k}]d\Omega - \int_{\Gamma_t} \bar{t}_i v_i \, d\Gamma \, . \qquad (10.3)$$

A variational theorem, which is similar to Reissner's variational theorem in linear elasticity (Reissner 1950) may then be stated as follows. Let ζ be a scalar parameter, w_i a vector-valued function which is once differentiable and vanishes on Γ_v, q a square-integrable scalar function, and t_{pq} a symmetric tensor with square integrable components; if, for an arbitrary choice of w_i, q, and t_{pq}, we have

$$\left.\frac{d}{d\zeta}\right|_{\zeta=0} \bar{J}(v_k + \zeta w_k, p + \zeta q, T_{pq} + \zeta t_{pq}) = 0 \quad , \qquad (10.4)$$

then v_i, p and T_{pq} solve the system (9.1-2) (with $\rho = 0$) and the associated boundary conditions.

With the help of (10.1) and (10.2) it is indeed easy to see, by means of an integration by parts, that

$$\left.\frac{d}{d\zeta}\right|_{\zeta=0} \bar{J}(v_k + \zeta w_k,\ p + \zeta q,\ T_{pq} + \zeta t_{pq})$$
$$= \int_\Omega \left(- \frac{\partial\tau}{\partial T_{ij}} t_{ij} + v_{i,j}\ t_{ij} + w_{i,j}\ T_{ij} - \rho F_i w_i - q v_{k,k} - p w_{k,k}\right) d\Omega - \int_{\Gamma_t} \bar{t}_i w_i \, d\Gamma$$
$$= \int_\Omega \left[\left(- \frac{\partial\tau}{\partial T_{ij}} + d_{ij}\right) t_{ij} - (-p_{,i} + T_{ij,j} + \rho F_i) w_i - v_{k,k}\, q \right] d\Omega$$
$$+ \int_{\Gamma_t} [(-p\delta_{ij} + T_{ij})n_j - \bar{t}_i] w_i \, d\Gamma \, . \qquad (10.5)$$

If (10.4) is valid for an arbitrary choice of w_i, q and t_{ij}, we recover the momentum and the constitutive equations, together with the boundary conditions on Γ_t.

Since w_i, q and t_{ij} are selected independently, we may rewrite (10.4) separately for every non-vanishing component and obtain from (10.5)

$$< -\frac{\partial \tau}{\partial T_{ij}} + d_{ij} \; ; \; t_{ij} > = 0 \quad ,$$

$$< T_{ij} - p\delta_{ij} \; ; \; w_{i,j} > - < \rho F_i \; ; \; w_i > = \int_{\Gamma_t} \bar{t}_i \, w_i \, d\Gamma \quad , \tag{10.6}$$

$$< v_{k,k} \; ; \; q > = 0 \quad .$$

We recover in (10.6) the weak form (9.23) of the inertialess field equations which are now expressed in terms of the extra-stress components; a new relation, $(10.6)_1$, is now needed for relating the extra-stress tensor to the rate-of-deformation tensor.

10.3 A MIXED METHOD FOR SOLVING THE STOKES EQUATIONS

The variational theorem of the previous section has shown the feasibility of an attempt to solve the Stokes equations (in the absence of inertia effects) while using the velocity components, the pressure and the extra-stress components as the unknowns. Although such a procedure is useless for solving Stokes equations in general (since the u-v-p method is quite satisfactory), we will pursue the problem further while having in mind the later treatment of the flow of a Maxwell fluid, where the extra-stress components are not explicit functions of the velocity components and their derivatives. Moreover, before applying an algorithm to the flow of a Maxwell fluid, a minimal requirement is to verify that it is satisfactory for a Newtonian fluid.

Let us construct a finite element method where the velocity components, the pressure and the extra-stress components are the unknowns; we cover the domain Ω with a mesh of finite elements and select an approximation as follows

$$\tilde{v}_i = \sum_{k=1}^{M} v_i^k \, \psi_k \quad , \qquad \tilde{p} = \sum_{k=1}^{N} p^k \, \pi_k \quad , \qquad \tilde{T}_{ij} = \sum_{k=1}^{L} T_{ij}^k \, \phi_k \quad , \tag{10.7}$$

where v_i^k, p^k and T_{ij}^k are nodal values.

For the time being, we leave unspecified the nature of the shape functions ψ_k, π_k and ϕ_k, which will soon be the object of our discussion. One may wonder why advantage should not be taken of the fact that the trace of the extra-stress tensor vanishes for a Newtonian fluid; the reason is that, for most viscoelastic constitutive equations, the extra-stress tensor is not traceless, a good example being the case of the Maxwell fluid.

Once our approximation of the type (10.7) has been selected, we may use the variational formulation (10.6) of the problem for determining the unknowns v_i^k, p^k and T_{ij}^k. Provided the approximations for the pressure and the extra-stresses belong to $L^2(\Omega)$, and the approximation for the velocity field belongs to $H^1(\Omega)$, we may use for the arbitrary functions t_{ij}, w_i and q in (10.6) a representation similar to the one used for T_{ij}, v_i and p in (10.7). Then, the variational statement (10.6) becomes

$$
\begin{aligned}
&< -\frac{1}{\eta}\tilde{T}_{mn} + (\tilde{v}_{m,n} + \tilde{v}_{n,m}) \; ; \; \phi_i > = 0 \quad , && 1 \leqslant i \leqslant L \, , \\
&< \tilde{T}_{mn} - \tilde{p}\delta_{mn} \; ; \; \psi_{i,m} > - <\rho F_n \; ; \; \psi_i > = \int_{\Gamma_t} \bar{t}_n \psi_i \, d\Gamma \quad , && 1 \leqslant i \leqslant M \, , \\
&< \tilde{v}_{m,m} \; ; \; \pi_i > = 0 \quad , && 1 \leqslant i \leqslant N \, .
\end{aligned}
\tag{10.8}
$$

For later applications, it is important to realize that (10.8) may be obtained alternatively through a Galerkin-type formulation of the field and constitutive equations. We may already recognize that the second and third equations (10.8) are identical to (9.23) in the case of plane creeping flow. The first of (10.8) is obtained by calculating the scalar product of the constitutive equations with the shape functions of the extra-stresses.

Let us now substitute the discretized form (10.7) for $\tilde{T}_{mn}$, $\tilde{p}$ and $\tilde{v}_m$ in (10.8). One obtains

$$
\begin{aligned}
&< \phi_i \; ; \; -\frac{1}{\eta}\phi_k > T^k_{mn} + < \phi_i \; ; \; \psi_{k,n} > v^k_m + < \phi_i \; ; \psi_{k,m} > v^k_n = 0 \; , \\
&< \psi_{i,m} \; ; \; \phi_k > T^k_{mn} - < \psi_{i,n} \; ; \; \pi_k > p^k = F^i_n \quad , \\
&< \pi_i \; ; \; \psi_{k,m} > v^k_m = 0 \quad ,
\end{aligned}
\tag{10.9}
$$

where

$$
F^i_n = \int_{\Gamma_t} \bar{t}_n \psi_i \, d\Gamma + < \rho F_n \; ; \; \psi_i > \quad . \tag{10.10}
$$

For the sake of brevity, let us introduce a compact notation for the system (10.9) and designate by $\hat{\underline{T}}$, $\hat{\underset{\sim}{v}}$ and $\hat{p}$ vectors containing the nodal values of, respectively, the extra-stress tensor, the velocity vector and the pressure. The system (10.9) may then be written in compact form as follows :

$$
\begin{aligned}
&K_{TT}\hat{\underline{T}} + K_{Tv}\hat{\underset{\sim}{v}} && = \hat{\underline{0}} \; , \\
&K^T_{Tv}\hat{\underline{T}} \qquad\qquad + K_{vp}\hat{p} && = \hat{\underset{\sim}{F}} \; , \\
&\qquad K^T_{vp}\hat{\underset{\sim}{v}} && = \hat{0} \; .
\end{aligned}
\tag{10.11}
$$

Thus, the nodal values of the extra-stresses, the velocity components and the pressure are obtained at once by solving the linear system (10.11), which is endowed with a symmetric stiffness matrix (we are only considering the linear Stokes problem for the time being).

The discretized form (10.9) of the field and constitutive equations was first derived by Kawahara and Takeuchi (1977) for the case of a Maxwell fluid; however, an inaccurate sentence in their paper did not correspond to their actual findings and led to confusion. Specifically, Kawahara and Takeuchi claimed to have solved the system (10.11) using triangular elements with complete quadratic polynomials for the shape functions ψ_i and complete linear polynomials for the shape functions ϕ_i and π_i. This would appear to be a natural choice in view of our previous experience with u-v-p techniques and the form of the constitutive equations. The confusion came from the fact that, with such shape functions, the global stiffness matrix in (10.11) is singular (after imposing, of course, the boundary conditions). A closer inspection of the figures provided by Kawahara and Takeuchi in their paper leads one to believe, however, that they used complete quadratic polynomials for the shape functions ϕ_k as well as ψ_k. We will now explain why some selections of shape functions lead to singular matrices.

Let us consider the system (10.11). The first equation establishes a constitutive relation between the components of the extra-stress tensor and the velocity components; since the velocity field determines the stress components uniquely, we will require that the square matrix K_{TT} be non-singular. This is usually the case when the matrix components are given by terms of the form $\langle\phi_i;\phi_k\rangle$, provided the integrals are calculated exactly. When numerical integration rules are being used (which is the usual procedure), the matrix K_{TT} may become singular when an insufficient number of integration points are being used; for example, with triangular elements and full quadratic polynomials for the shape functions ϕ_i, it is found that when the number of points of integration is four or less, the matrix K_{TT} is singular. The reason why the matrix becomes singular when the number of integration points is insufficient is discussed by Zienkiewicz (1977, section 8.11).

Once the matrix K_{TT} is non-singular we may, at least formally, solve the first equation (10.11) for $\hat{\underset{\sim}{T}}$ and substitute in the second equation; we obtain

$$\begin{aligned} K_{vv}\hat{\underset{\sim}{v}} + K_{vp}\hat{p} &= \hat{\underset{\sim}{F}} \quad , \\ K_{vp}^T\hat{\underset{\sim}{v}} &= \hat{0} \quad , \end{aligned} \tag{10.12}$$

with

$$K_{vv} = - K_{Tv}^T \, K_{TT}^{-1} \, K_{Tv} \quad . \tag{10.13}$$

It must be pointed out that the matrix K_{vv} in (10.12) differs from the stiffness matrix which would be found with the u-v-p technique, where the extra-stress components are eliminated at the continuum level, *before* the discretization. In order to solve the system (10.12), we must verify that the rank of the matrix $[K_{vv}, K_{vp}]$ equals the number of unknown velocities (after rigid body modes and the indeterminacy of the pressure have been removed through boundary conditions); otherwise, the global stiffness matrix will be singular.

This condition on the rank of the matrix $[K_{vv}, K_{vp}]$ imposes severe restrictions on the choice of shape functions. In order to see this, let us consider, on Fig. 10.1, two large meshes containing triangular or quadrilateral elements for calculating plane Stokes flow.

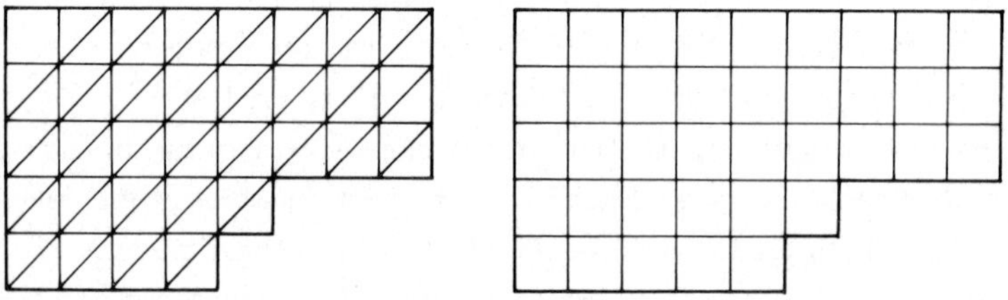

Fig. 10.1 Large meshes containing triangular and quadrilateral elements.

Let M be the number of quadrilaterals and 2M be the number of triangles. From Chapter 9, we know that a good choice of shape functions ψ_i for the velocities would be complete quadratic polynomials with triangles, and biquadratic polynomials with quadratilaterals; similarly, we would adopt, for the pressure shape functions π_i, linear polynomials with triangles and bilinear polynomials with quadrilaterals. Thus, for any additional quadrilateral or for two additional triangles, we would obtain 8 new velocity components and 1 new pressure; the numbers of nodal velocity components and nodal pressures are roughly equal to 8M and M, respectively. Let us now assume that the shape functions ϕ_i for the extra-stress components are linear polynomials for triangles, or bilinear polynomials for quadrilaterals. The total number of nodal stresses would then be 3M, and the rank of the matrix K_{TT} would also be 3M. From (10.13), it is now clear that the rank of the matrix $[K_{vv}, K_{vp}]$ could not possibly exceed 3M (from K_{vv}) + M (from K_{vp}), which is lower than the number of nodal velocities.

In an early paper, Crochet and Bezy (1979) solved the problem by keeping linear polynomials for the stress components, and assuming that the pressure is discontinuous and given by a complete linear polynomial within each element (i.e. P^1-C^{-1}); such an approximation had been used earlier by Thompson and Haque (1973) with the u-v-p technique. The triangular mesh then contains 6M nodal pressures, and it is possible to obtain a large enough rank for the matrix $[K_{vv}, K_{vp}]$ in view of the enlarged matrix K_{vp}. However, the ratio of nodal velocities to the number of incompressibility constraints is now 8M/6M, which is much too low; the element is too stiff and has been forsaken.

The triangular element initially intended by Kawahara and Takeuchi had the right rank properties. Indeed, if the extra-stresses are given by second-degree polynomials, we may define nodal stresses at the vertices of the triangles and at mid-side nodes; the large mesh contains 12M extra-stresses, and the rank condition is satisfied with the use of linear pressures. This triangular element has been used by Crochet and co-workers (1980 to 1982), together with a Lagrangian element on which the extra-stresses are approximated by biquadratic polynomials. The triangular and quadrilateral elements, which can be transformed into curvilinear elements through an isoparametric transformation, are shown on Figure 10.2. The triangular element contains a total of 33 nodal variables for plane flow, while the quadrilateral element contains 49 variables.

For axisymmetric flow, the extra-stress $T_{\theta\theta}$ must also be considered as one of the unknown functions; the triangular element then contains 39 variables, while the quadrilateral element contains 58 variables.

It is obvious that such elements are quite expensive to use, in view of the large number of nodal variables with which they are associated. They should never be used for solving Newtonian flow as such, for which the u-v-p technique is much more economical; we will see in the next section that they are designed for solving the flow of viscoelastic fluids. A much simpler element, shown on Fig. 10.3, has been used by Coleman (1980); it consists of a quadrilateral divided into two triangles. On each triangle, the velocity and stress components are approximated by linear polynomials. In order to avoid locking of the triangular element, a single constant pressure is associated with the 2 triangles forming the quadrilateral. This quadrilateral element contains 21 variables; one of its attractive features is the optimal ratio, 2/1, of the number of nodal velocity components to nodal pressures, and the fact that conservation of mass is satisfied at the element level. One should note, however, that, in general, the use of a simpler element imposes a finer mesh for the same degree of accuracy of the calculations. This composite element is in fact quite similar to a quadrilateral element with bilinear velocities and stresses, and a uniform pressure; in its u-v-p version, the latter element is afflicted with spurious pressure modes when the velocity components are imposed along the entire boundary. In view of the similarity between the elements, one might suspect that the same danger holds with the composite element.

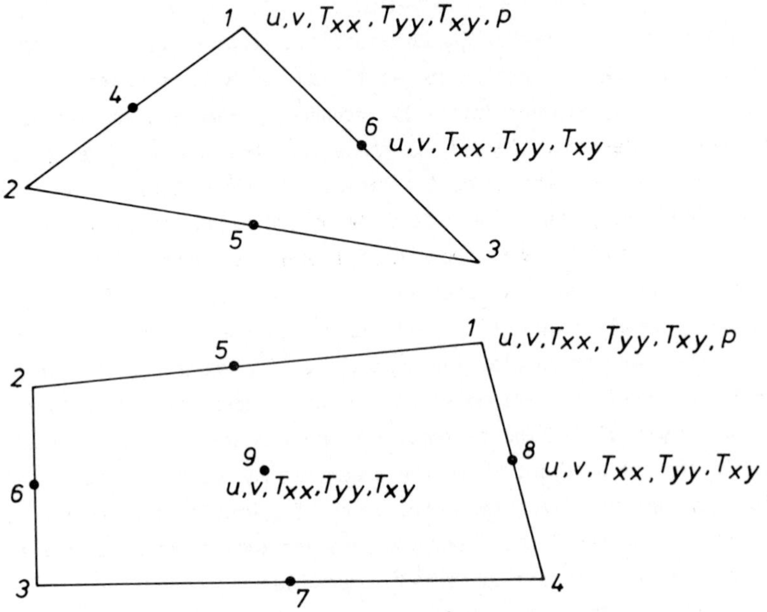

Fig.10.2 Triangular and quadrilateral elements used in the mixed method for plane flow.

Finally, we wish to mention the element used by Richards and Townsend (1981), shown on Fig. 10.3, which is in fact the 8-node serendipity version of the quadrilateral element shown on Figure 10.2. With the u-v-p technique, the 9-node Lagrangian element is presently preferred to the 8-node element when convective terms are important; one should keep in mind that the constitutive relation of a Maxwell fluid contains a convective term which plays a dominant role when elastic effects are important.

Once the stiffness matrix has been built on the left hand side of (10.11), one must solve the linear system in the nodal stresses, velocity components and pressure. When the system is solved by Gaussian elimination, which is part of the frontal elimination technique developed by Irons (1970) and has been used extensively for solving finite element problems, it is found that pivoting is usually not mandatory provided the variables are eliminated in the right order. At the element level, the rule is to eliminate first the nodal stresses, next the nodal velocity components and finally the pressures; also, whenever it is feasible, variables associated with vertices should be eliminated before those associated with midside nodes. Acting in a contrary way generally leads to vanishing diagonal terms in the elimination process.

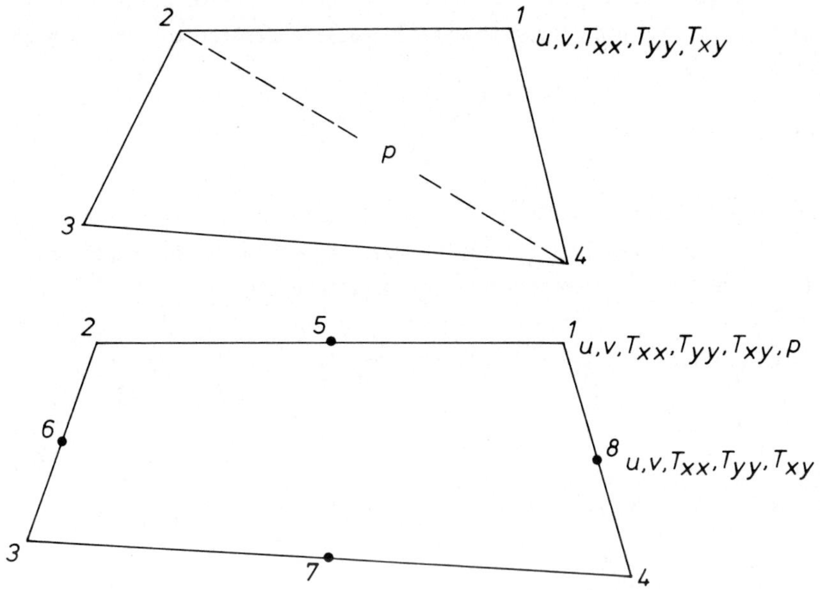

Fig. 10.3 Composite quadrilateral element and serendipity element used in the mixed method for plane flow.

10.4 A MIXED METHOD FOR SOLVING THE FLOW OF A MAXWELL FLUID (MIX1)

The philosophy behind the mixed method was made clear in section 10.3 by means of the equations (10.7) and (10.8). The extra-stress components are discretized together with the velocity components and the pressure; the algebraic system is formed by calculating the scalar products of the constitutive equations, the momentum equations and the continuity equation by the shape functions ϕ_i, ψ_i and π_i, respectively. Precisely the same idea is used for calculating the flow of a Maxwell fluid with, however, the observation that the extra-stress tensor is now given by the constitutive equations (3.5). The presence of stress gradients in (3.5) requires that the approximation for the extra stresses be also selected in $H^1(\Omega)$. Instead of the system (10.8), we write (in dimensional form)

$$< -\frac{1}{\eta_0}\tilde{T}_{mn} - \frac{\lambda}{\eta_0}\overset{\nabla}{\tilde{T}}_{mn} + (\tilde{v}_{m,n} + \tilde{v}_{n,m})\ ;\ \phi_i > = 0\ , \qquad 1 \leqslant i \leqslant L\ ,$$

$$< \tilde{T}_{mn} - \tilde{p}\delta_{mn}\ ;\ \psi_{i,m} > - < \rho F_n\ ;\ \psi_i > = \int_{\Gamma_t} \bar{t}_n \psi_i\, d\Gamma\ , \qquad 1 \leqslant i \leqslant M\ , \qquad (10.14)$$

$$< \tilde{v}_{m,m}\ ;\ \pi_i > = 0\ , \qquad 1 \leqslant i \leqslant N\ .$$

It is worth showing, at least once, the form of the discretized system for plane flow in component form; towards an easier notation we introduce the symbols

$$R = T_{xx}\ , \qquad S = T_{yy}\ , \qquad T = T_{xy}\ , \qquad (10.15)$$

while we recall that u and v denote the velocity components in the x and y directions, respectively. For plane flow, the constitutive equations, the momentum equations and conservation of mass take the form :

$$R + \lambda(uR,_x + vR,_y - 2u,_xR - 2u,_yT) = 2\eta_0 u,_x\ ,$$

$$S + \lambda(uS,_x + vS,_y - 2v,_xT - 2v,_yS) = 2\eta_0 v,_y\ ,$$

$$T + \lambda(uT,_x + vT,_y - u,_xT - u,_yS - v,_xR - v,_yT) = \eta_0(u,_y + v,_x)\ , \qquad (10.16)$$

$$- p,_x + R,_x + T,_y + \rho F_x = 0\ ,$$

$$- p,_y + T,_x + S,_y + \rho F_y = 0\ ,$$

$$- u,_x - v,_y = 0\ .$$

We need a finite element approximation for the functions R, S, T, u, v and p, and write

$$\tilde{R} = \sum_{j=1}^{L} R_j\ \phi_j\ , \qquad \tilde{S} = \sum_{j=1}^{L} S_j\ \phi_j\ , \qquad \tilde{T} = \sum_{j=1}^{L} T_j\ \phi_j\ ,$$

$$\tilde{u} = \sum_{j=1}^{M} U_j\ \psi_j\ , \qquad \tilde{v} = \sum_{j=1}^{M} V_j\ \psi_j\ , \qquad \tilde{p} = \sum_{j=1}^{N} P_j\ \pi_j\ . \qquad (10.17)$$

The discretized set of field and constitutive equations is obtained by replacing R, S, T, u, v, p by their finite element counterparts $\tilde{R}$, $\tilde{S}$, $\tilde{T}$, $\tilde{u}$, $\tilde{v}$, $\tilde{p}$ in equations (10.16), and calculating their scalar product with ϕ_i for the constitutive equations, ψ_i for the momentum equations, and π_i for the equation of mass conservation. If we introduce the matrices

$$m_{ij} = < \phi_i;\phi_j > \;, \qquad a^x_{ij} = < \phi_i;\psi_{j,x} > \;, \qquad a^y_{ij} = < \phi_i;\psi_{j,y} > ,$$

$$b^x_{ij} = < \pi_i;\psi_{j,x} > \;, \qquad b^y_{ij} = < \pi_i;\psi_{j,y} > \;, \qquad c^x_{ijk} = < \phi_i;\psi_j\, \phi_{k,x} > \;, \tag{10.18}$$

$$c^y_{ijk} = < \phi_i;\psi_j\, \phi_{k,y} > , \qquad d^x_{ijk} = < \phi_i,\psi_{j,x}\, \phi_k > , \qquad d^y_{ijk} = < \phi_i;\psi_{j,y}\, \phi_k > \;,$$

we obtain the following system of equations,

$$m_{ij}\, R_j + \lambda[(c^x_{ijk}\, U_j - 2d^x_{ijk}\, U_j + c^y_{ijk}\, V_j)R_k - 2d^y_{ijk}\, U_j\, T_k] = 2\eta_0 a^x_{ij}\, U_j \;,$$

$$m_{ij}\, S_j + \lambda[(c^x_{ijk}\, U_j + c^y_{ijk}\, V_j - 2d^y_{ijk}\, V_j)S_k - 2d^x_{ijk}\, V_j\, T_k] = 2\eta_0 a^y_{ij}\, V_j \;,$$

$$m_{ij}\, T_j + \lambda\{[(c^x_{ijk} - d^x_{ijk})U_j + (c^y_{ijk} - d^y_{ijk})V_j]T_k - d^y_{ijk}\, U_j\, S_k - d^x_{ijk}\, V_j\, R_k\}$$

$$= \eta_0(a^y_{ij}\, U_j + a^x_{ij}\, V_j) \;, \tag{10.19}$$

$$a^x_{ji}\, R_j + a^y_{ji}\, T_j - b^x_{ji}\, P_j = F^x_i \;,$$

$$a^x_{ji}\, T_j + a^y_{ji}\, S_j - b^y_{ji}\, P_j = F^y_i \;,$$

$$- b^x_{ij}\, U_j - b^y_{ij}\, V_j = 0 \;.$$

The nodal force components F^x_i and F^y_i include the contribution of the body forces and the imposed contact forces on the surface of the fluid. Despite the length of the discretized equations (10.19), they show clearly that implementing a finite element procedure for solving the flow of a Maxwell fluid first consists of calculating for each element a set of matrices given by (10.18). The calculation is accomplished by means of numerical integration procedures described in Chapter 8. The left hand sides of (10.19) are second-order polynomials in the unknown nodal variables; the polynomial form of the equations facilitates considerably the implementation of a Newton iterative procedure. In order to show this, let us consider the first equation (10.19), and assume that R^n_i, T^n_i, U^n_i, V^n_i are known after the nth iteration; the incremental form of equation (10.19) becomes

$$[m_{ij} + \lambda(c^x_{ikj}\ U^n_k - 2d^x_{ikj}\ U^n_k + c^y_{ikj}\ V^n_k)]\delta R_j - 2\lambda d^y_{ikj}\ U^n_k\ \delta T_j$$

$$- [2\eta_0 a^x_{ij} - \lambda(c^x_{ijk}\ R^n_k - 2d^x_{ijk}\ R^n_k - 2d^y_{ijk}\ T^n_k)]\delta U_j + \lambda c^y_{ijk}\ R^n_k\ \delta V_j \qquad (10.20)$$

$$= 2\eta_0 a^x_{ij}\ U^n_j - m_{ij}\ R^n_j - \lambda[(c^x_{ijk}\ U^n_j - 2d^x_{ijk}\ U^n_j + c^y_{ijk}\ V^n_j)R^n_k - 2d^y_{ijk}\ U^n_j\ T^n_k]\ .$$

This equation shows that the cost of solving the flow of a Maxwell fluid should not be underestimated. For a given mesh, the mixed method uses more than twice the number of variables as the u-v-p method; moreover, the construction of the stiffness matrix at each iterative step is in general quite long. A modified Newton procedure in which the stiffness matrix is kept the same for a few iterations while the right hand side is updated does not help much, since the rate of convergence is decreased while the calculation of the right hand side of (10.20) remains a time-consuming task.

A good convergence test is obtained by calculating the sum of the absolute values of the right hand sides of the incremental equations similar to (10.20). The system is said to have a solution when the sum is less than some fixed ε, say 10^{-4}. At the same time it is very useful to verify the relative convergence test which was defined in (8.69) for a one-dimensional problem.

An important difference between Newtonian flow and the flow of a Maxwell fluid is the imposition of the boundary conditions, a subject which was briefly considered in section 2.9. It is known that, for solving the plane Newtonian flow problem, one needs two conditions all along the boundary, either two essential conditions in the x and y directions, or two natural conditions, or one of each. The problem is much more complex for a Maxwell fluid in view of the non-linear character of the constitutive equations, and the lack of a uniqueness proof for the solution of the field and constitutive equations. When one considers the flow (Figure 10.4a) of a Maxwell fluid in a region surrounded by a surface through which the fluid does not flow, it seems plausible to impose boundary conditions identical to those used with Newtonian fluids, i.e. velocity components or associated nodal force components. The situation is different, however, when the fluid flows through the boundary; as an example Figure 10.4b shows the entry flow in a contraction.

Two arguments can be used in this case in favour of imposing the extra-stress components along a boundary crossing the streamlines. Assume indeed that the velocity field is known; the first three equations (10.16) form a system of first order linear partial differential equations which requires for uniqueness the knowledge of R, S, T along a line crossing the characteristics which are the streamlines in the present case. The second argument, which illuminates the first, is given by integral form (3.6) of the stress constitutive relations. Consider a fluid particle which enters the flow domain at time $(t-s^*)$; the extra-stress at time t with $a=0$ will be given by

$$T_{ij} = \int_{s^*}^{\infty} \eta_0/\lambda^2 \exp(-s/\lambda)H_{ij}(t-s)ds + \int_0^{s^*} \eta_0/\lambda^2 \exp(-s/\lambda)H_{ij}(t-s)ds \ . \qquad (10.21)$$

The first term on the right hand side of (10.21) will not be found on the basis of the flow within the domain of integration; it must be imposed as an initial condition for the fluid particle, taking into account the previous deformation history. We conclude that all the extra-stress components must be specified as essential boundary conditions along the *inlet* to a flow region. Failing to do so may lead to the propagation of errors throughout the flow domain when the relaxation time λ becomes large, i.e. when stress relaxation occurs over a long portion of the path-lines; this has been confirmed by numerical experiments. In practice, entry regions should be designed in such a way that the local flow may be considered as being a fully-developed flow of some known sort. For example, a long straight duct allows us to impose the extra-stress components calculated for fully-developed Poiseuille flow. A further discussion of the boundary conditions may be found in Shimazaki and Thompson (1981).

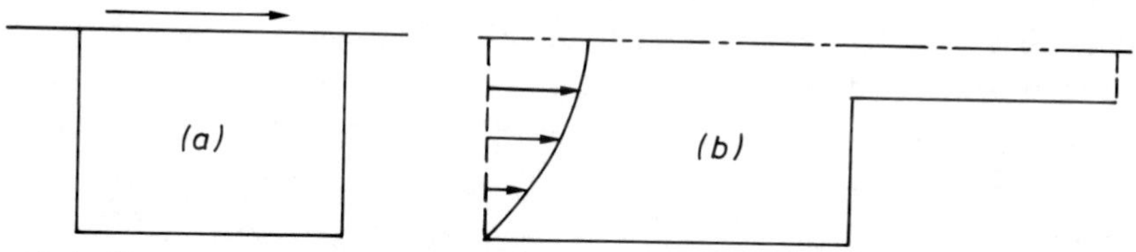

Fig.10.4 Flow region surrounded by a closed boundary (flow in a cavity - a), and by an open boundary (entry flow - b).

Mixed techniques of the type we have just described have been implemented by various authors. Crochet et al. (1980-1982) use triangular six-node elements with second-order polynomials for the velocity and extra-stress components and first-order polynomials for the pressure, together with 9-node Lagrangian elements with the same order for the approximation; they use a Newton iterative technique. Coleman (1980, 1981) makes use of the composite quadrilateral made of two triangles which has been described above; Coleman replaces the Newton scheme by a successive substitution technique. The non-linear algebraic system is rearranged in the form

$$\underset{\sim}{S}(\underset{\sim}{X})\underset{\sim}{X} = \underset{\sim}{b} \quad , \tag{10.22}$$

where $\underset{\sim}{X}$ is the vector of the unknown variables and $\underset{\sim}{b}$ is known, and solved through an iterative process given by

$$\underset{\sim}{S}(\underset{\sim}{X}^{(i)})\underset{\sim}{X}^{(i+1)} = \underset{\sim}{b} \quad , \tag{10.23}$$

where $\underset{\sim}{S}(\underset{\sim}{X})$ is chosen to be independent of the nodal extra-stresses and pressures. The rate of convergence is accelerated by an extrapolation process. Using that technique, Coleman obtains convergence of the iterative technique for higher values of λ than he would reach with the Newton scheme. Richards and Townsend (1981, 1982) use a serendipity eight-node element, with quadratic shape functions for the velocity and extra-stress components, and linear shape functions for the pressure; they also use an iterative procedure of the type (10.23).

The method of calculation which we have just described will from now on be called MIX1, for the sake of brevity.

10.5 A SECOND MIXED METHOD FOR SOLVING THE FLOW OF A MAXWELL FLUID (MIX2, MIX3)

The method MIX1 which we have just described, and which is summarized by the set of equations (10.14), is associated with two important features which deserve further consideration :

i. we found in section 10.3 that the shape functions for the extra-stress components should in general be identical to those used for the velocity components;

ii. the system (10.14) shows that the discrete form of the momentum equations does not contain the velocity components; as a consequence, it is impossible to implement an iterative procedure in which the system would be decoupled into the constitutive equations on the one hand, and the momentum and incompressibility equations on the other. It is clear that these two features lead to important storage requirements when dense finite element meshes are used for solving a problem.

A different mixed method can be constructed by rewriting the constitutive equation (3.5) in a different form, i.e.

$$T_{pq} = 2\eta_0 \, d_{pq} - \lambda \overset{\nabla}{T}_{pq} \quad . \tag{10.24}$$

Going back to the system (10.14), we may now keep the first and third equations, and substitute for T_{pq} in the second equation the right hand side of (10.24). We obtain the system :

$$< -\frac{1}{\eta_0}\tilde{T}_{mn} - \frac{\lambda}{\eta_0}\overset{\nabla}{T}_{mn} + (\tilde{v}_{m,n} + \tilde{v}_{n,m}) \; ; \; \phi_i > = 0 \, , \qquad 1 \leqslant i \leqslant L \, ,$$

$$< \eta_0(\tilde{v}_{m,n} + \tilde{v}_{n,m}) - \tilde{p}\delta_{mn} \; ; \; \psi_{i,m} > - \lambda < \overset{\nabla}{T}_{mn} \; ; \; \psi_{i,m} >$$

$$- < \rho F_n \; ; \; \psi_i > = \int_{\Gamma_t} \bar{t}_n \, \psi_i \, d\Gamma \, , \qquad 1 \leqslant i \leqslant M \, , \tag{10.25}$$

$$< v_{m,m} \; ; \; \pi_i > = 0 \, , \qquad 1 \leqslant i \leqslant N \, .$$

When λ vanishes, the second and third equations reduce to the u-v-p technique developed in section 9.3 for solving Stokes equations; we may consider that, when λ is small, this new mixed method is a perturbation of the original u-v-p method. The discretized form of the constitutive equations and of the incompressibility equation remains unchanged with respect to the set (10.19); however, the discretized form of the momentum equations is now much more complicated. When λ vanishes, we recover the system (9.30) in the absence of inertia terms; when elasticity is taken into account, it is necessary to calculate the terms $< \overset{\nabla}{T}_{mn} \; ; \; \psi_{i,m} >$ in the momentum equations. With respect to the system (10.19), it is necessary to define new matrices, i.e.

$$e^{xx}_{ijk} = < \psi_{i,x} \; ; \; \psi_j \, \phi_{k,x} > \, , \quad e^{yy}_{ijk} = < \psi_{i,y} \; ; \; \psi_j \, \phi_{k,y} > \, ,$$

$$e^{xy}_{ijk} = < \psi_{i,x} \; ; \; \psi_j \, \phi_{k,y} > \, , \quad e^{yx}_{ijk} = < \psi_{i,y} \; ; \; \psi_j \, \phi_{k,x} > \, , \tag{10.26}$$

$$f^{xx}_{ijk} = < \psi_{i,x} \; ; \; \psi_{j,x} \, \phi_k > \, , \quad f^{yy}_{ijk} = < \psi_{i,y} \; ; \; \psi_{j,y} \, \phi_k > \, , \quad f^{xy}_{ijk} = <\psi_{i,x} \; ; \; \psi_{j,y} \, \phi_k> \, ,$$

and make use of the matrices A_{ij}, B_{ij} and C_{ij} defined by (9.28), where $\tau_i = \psi_i$. The discretized form of the field equations becomes

$$A_{ij}U_j + C_{ij}V_j - b^x_{ji}P_j - \lambda\{[(e^{xx}_{ijk} - 2f^{xx}_{ijk})R_k - f^{yy}_{ijk}S_k$$

$$+ (e^{yx}_{ijk} - f^{xy}_{jik} - 2f^{xy}_{ijk})T_k]U_j + [(e^{xy}_{ijk} - f^{xy}_{jik})R_k + (e^{yy}_{ijk} - f^{yy}_{ijk})T_k]V_j\} = F^x_i \, ,$$

$$C_{ji}U_j + B_{ij}V_j - b^y_{ji}P_j - \lambda\{[(e^{yx}_{ijk} - f^{xy}_{ijk})S_k + (e^{xx}_{ijk} - f^{xx}_{ijk})T_k]U_j \tag{10.27}$$

$$+ [-f^{xx}_{ijk}R_k + (e^{xy}_{ijk} - f^{xy}_{ijk} - 2f^{xy}_{jik})T_k + (e^{yy}_{ijk} - 2f^{yy}_{ijk})S_k]V_j\} = F^y_i \, ,$$

$$- b^x_{ij}U_y - b^y_{ij}V_j = 0 \, .$$

As a first comment, we observe that the algebraic system is now quite complicated, and that its intricacy is even more pronounced when Newton's method is used for solving the non-linear system; the calculation of the stiffness matrix takes a large part of the computer time needed for solving a specific problem.

Secondly, we note that the system (10.25) may be decoupled into two sets of equations. Indeed, the system (10.25) may formally be rewritten as follows :

$$
\begin{aligned}
&F_i(U_j , V_j , R_j , S_j , T_j) = 0 , \\
&G_i(U_j , V_j , R_j , S_j , T_j , P_j) = 0 , \qquad (10.28) \\
&H_i(U_j , V_j) = 0 ;
\end{aligned}
$$

these equations represent respectively the constitutive equations, the momentum equations and the incompressibility equation. It is now possible to write an iterative algorithm which would work as follows; starting from a known stress field R_j^n , S_j^n , T_j^n , one calculates a new velocity and pressure field by solving

$$
\begin{aligned}
&G_i(U_j^{n+1} , V_j^{n+1} , R_j^n , S_j^n , T_j^n , P_j^{n+1}) = 0 , \\
&H_i(U_j^{n+1} , V_j^{n+1}) = 0 ;
\end{aligned} \qquad (10.29)
$$

next, the stress field is updated by solving the system

$$
F_i(U_j^{n+1} , V_j^{n+1} , R_j^{n+1} , S_j^{n+1} , T_j^{n+1}) = 0 . \qquad (10.30)
$$

The rate of convergence can be expected to be slower than with Newton's method; however, splitting a large system into two smaller systems may be cost efficient.

A third important remark is that one may now use lower-order polynomials for the extra-stress components than for the velocity components. Indeed, the discretized momentum equations (10.27) contain terms on the diagonal, and the rank condition given in section 10.3 is now irrelevant. We have thus two possibilities, giving rise to two algorithms which we will call MIX2 and MIX3, and which are summarized in table 10.1. Also included is a MIX4 which will be discussed in detail later.

TABLE 10.1

Summary of mixed methods for solving the flow of a Maxwell fluid

Method	System	Velocity	Stresses	Pressure
MIX1	(10.14)	P^2-C^0	P^2-C^0	P^1-C^0
MIX2	(10.25)	P^2-C^0	P^2-C^0	P^1-C^0
MIX3	(10.25)	P^2-C^0	P^1-C^0	P^1-C^0
MIX4	(10.47)	P^2-C^0	P^1-C^0	P^1-C^0

The elements of method MIX3 are shown on Fig. 10.5; it is clear that the number of nodal variables for solving a problem with MIX3 is lower than the related quantity with MIX1 and MIX2.

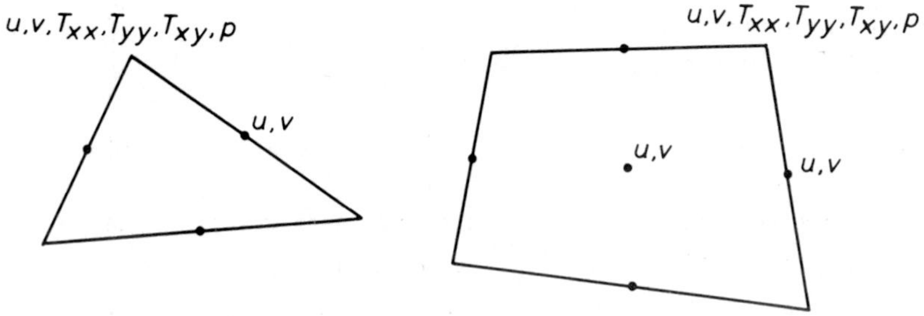

Fig. 10.5 Triangular and quadrilateral elements used in the mixed method MIX3 for plane flow.

The basic procedure of the alternative mixed method was first introduced by Chang et al. (1979), with a different algorithm. In their method, the extra-stress components are calculated at the integration points and are not represented by an approximation of the type (10.7); it follows that the discretized equations contain in fact second-order spatial derivatives of the velocity field, which are approximated by global shape functions of the C^0-type. This may be why some problems seem to occur when elastic effects dominate the flow. The method MIX2 has been used by Crochet and Keunings (1982a) and Crochet (1982a) who found that MIX2 behaves better than MIX1 when the elasticity of the flow is low, in the sense that the results are smoother. When elastic effects tend to dominate the flow, it was found repeatedly that the convergence of the iterative technique when λ (i.e. W) increases is lost earlier with MIX2 than with MIX1 (see also Keunings et al. 1983); a further example will be shown in section 10.7. The method MIX3 has been used by Mendelson et al. (1982), Jackson and Finlayson (1982) and Finlayson and Tuna (1982). Examples of application will be discussed in later sections.

10.6 AXISYMMETRIC FLOW

We have until now limited the explicit developments to the study of the plane flow of a Maxwell fluid, and it seems appropriate to make a few comments on the calculation of axisymmetric flows. In cylindrical coordinates, let us introduce the notation

$$R = T_{rr} \quad , \qquad S = T_{zz} \quad , \qquad T = T_{rz} \quad , \qquad Q = T_{\theta\theta} \quad . \tag{10.31}$$

The momentum and incompressibility equations are now given by

$$\begin{aligned} - p,_r + R,_r + T,_z + (R - Q)/r + \rho F_r &= 0 \quad , \\ - p,_z + T,_r + T/r + S,_z \qquad + \rho F_z &= 0 \quad , \\ - u,_r - u/r - v,_z \qquad\qquad &= 0 \quad . \end{aligned} \tag{10.32}$$

The problem here is that the stress-component $Q = T_{\theta\theta}$ appears explicitly in the momentum equations. On the other hand, the extra-stress tensor is generally not traceless; this may be easily seen by calculating the trace of (3.5). Thus, the stress component Q in (10.32) cannot be calculated in terms of R and S, and it is necessary to add an explicit constitutive equation for obtaining Q (Finlayson and Tuna (1982) use the trace of the extra-stress tensor rather than Q as an additional variable). The constitutive equations for axisymmetric flows are then given in dimensional form by

$$\begin{aligned} R + \lambda(uR,_r + vR,_z - 2u,_r R - 2u,_z T) &= 2\eta_0\, u,_r \quad , \\ S + \lambda(uS,_r + vS,_z - 2v,_r T - 2v,_z S) &= 2\eta_0\, v,_z \quad , \\ T + \lambda(uT,_r + vT,_z - u,_r T - u,_z S - v,_r R - v,_z T) &= \eta_0(u,_z + v,_r) \quad , \\ Q + \lambda(uQ,_r + vQ,_z - 2u/r\ Q) &= 2\eta_0\, u/r \quad . \end{aligned} \tag{10.33}$$

In order to obtain the discretized form of the system of equations, we replace R, S, T, Q, u, v, p by their finite element counterparts $\tilde{R}$, $\tilde{S}$, $\tilde{T}$, $\tilde{Q}$, $\tilde{u}$, $\tilde{v}$, $\tilde{p}$ and calculate the scalar product *over the flow volume* of the constitutive equations with ϕ_i, of the momentum equations with ψ_i and of the incompressibility equation with π_i. The final equations are similar to either (10.19) for MIX1 or (10.27) for MIX2 and MIX3, except that the matrices in (10.18) and (10.26) are slightly modified and calculated over an axisymmetric volume. The same computer code should be used for plane as well as axisymmetric flow, by including in the calculations a parameter which indicates the nature of the flow. The comments which are made in this chapter on the mixed methods are valid for both plane and axisymmetric flows.

10.7 PROBLEMS WITH THE MIXED METHODS

In later sections we will review several results which have been obtained by means of the mixed methods for flow through an abrupt contraction and for extrusion problems. In the meantime it seems appropriate to describe typical problems treated with the mixed methods which we have just described when the elastic character of the flow increases.

In most flows, it is possible to identify a characteristic shear-rate γ; in extrusion problems, γ is calculated on the wall in the upstream fully developed flow, while, for the flow through an abrupt contraction, γ is evaluated on the wall in the downstream channel. For the flow of a Maxwell fluid, the product $\lambda\gamma$ of the characteristic shear-rate times the relaxation time provides a *non-dimensional shear-rate* which is a good indicator of the elasticity of the flow. For other models, a more convenient non-dimensional parameter is the recoverable shear S_R defined by $\nu_1(\gamma)/2\eta(\gamma)\gamma$, which for the Maxwell fluid reduces to $\lambda\gamma$.#

Let us now consider the numerical simulation of the flow of a Maxwell fluid when the value of S_R is progressively increased. For low values of S_R, say less than about one, the iterative algorithm converges rapidly (about four iterations of Newton's method for a relative error of 10^{-4}) and produces a smooth solution. Beyond some value of S_R, the iterative algorithm is still converging rather quickly; however, the solution is affected by unphysical wiggles which follow the pattern of the mesh. Usually, for some critical value of S_R, the spatial oscillations prevent the convergence of the iterative algorithm towards a solution of the algebraic system; however, it may also happen that the calculation blows up without the previous development of wiggles.

Spurious wiggles are not uncommon in *Newtonian* fluid mechanics; they are known to occur when the inertia terms in the Navier-Stokes equations become important, and when the mesh is too coarse (see, for example, Gresho et al. 1981).

#For the Maxwell model, the elasticity parameter W discussed in Chapter 3 and S_R can be made equivalent by a suitable choice of the characteristic velocity and characteristic length.

A common feature between the Navier-Stokes equations and our actual problem (where we have not considered inertia terms) is the presence of convective terms which appear in the constitutive equations for the latter and in the inertia terms for the former. In solving the Navier-Stokes equations a possible cure of the spatial oscillations is a mesh refinement; in solving the flow of a Maxwell fluid, it is disturbing to find that a mesh refinement often decreases the critical value of S_R beyond which a numerical solution ceases to exist.

Most problems of practical interest, like the flow through an abrupt contraction and extrusion problems, contain an additional difficulty which is the presence of a stress singularity; it is often difficult to point out whether the degradation of a numerical solution when the elasticity of the flow increases is due to the presence of the singularity or to the numerical algorithm. In order to reach a clearer view of the problem, we will now study a specific example where there is no corner singularity (Van Schaftingen & Crochet, 1983).

We wish to study the flow of a Maxwell fluid in a wedge with perpendicular walls. The tip of the wedge is replaced by a plane channel which is connected to the walls of the wedge by means of smooth parabolic boundaries; such boundaries may be exactly represented with the use of isoparametric elements. The entry section of the wedge is also connected to a plane channel. A fully developed plane Poiseuille flow is imposed in the entry and in the exit sections. The elasticity of the flow is evaluated by the product $S_R = \lambda\gamma$ on the wall in the downstream fully developed flow. The finite element mesh and the boundary conditions are shown on Fig. 10.6; the mesh contains 116 elements and 531 nodes which produce 2805 degrees of freedom with MIX1 and MIX2, and 1662 degrees of freedom with MIX3.

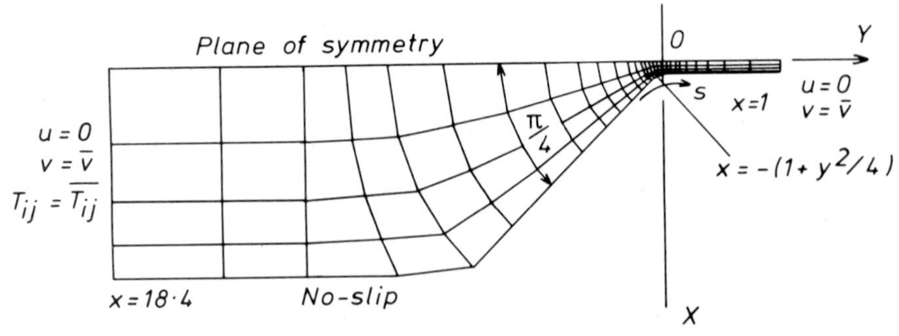

Fig.10.6 Finite element mesh and boundary conditions for the flow through a tapered contraction.

Let us now examine the behaviour of the numerical algorithms which have been described when S_R increases. This is done in practice by keeping a unit flow-rate in the wedge and by incrementing the value of λ. The maximum values of S_R for which Newton's method converges are given in Table 10.2. The highest value of S_R is reached with MIX3, where the extra-stresses are of the P^1-C^0 type, and the lowest value is reached with MIX2. The fact that MIX1 ceases to converge for a value of S_R lower than MIX3 may be related to our earlier observation that the critical value of S_R is usually lower for a fine mesh than for a coarse mesh, since MIX1 uses about four times as many nodal stresses as MIX3. In fact, it was found that in an additional calculation where all the elements of the mesh of Fig. 10.6 were divided into four small elements, the method MIX3 did not converge beyond $S_R = 4.5$.

TABLE 10.2
Critical value of S_R beyond which Newton's method did not converge

Method	MIX1	MIX2	MIX3
Maximum value of S_R	4.75	3	7

Let us study more closely the results obtained with the various algorithms. For that purpose, we will exhibit two graphs which are sensitive indicators of the behaviour of the solution, i.e. the extra-stress component T_{yy} along the wall plotted as a function of the arc length on the wall, and the y-velocity component v along the plane of symmetry plotted as a function of y. Typically, the extra-stress component T_{yy} along the wall rises sharply at some distance from the exit and, after an overshoot, settles to its constant asymptotic value on the downstream channel wall. Between the plane walls, the velocity component v along the plane of symmetry behaves like y^{-1}.

We show on Fig. 10.7 the plots of T_{yy} and v for a value of $S_R = 3$, obtained with the three mixed methods summarized on Table 10.2. The spatial oscillations of T_{yy} with MIX2 show clearly why convergence is lost for higher values of S_R. Both MIX1 and MIX3 already produce wiggles which have a growing amplitude when S_R increases; the overshoot of T_{yy} obtained with MIX1 is sharper than with MIX3, since the interval between nodes where the stresses are evaluated is twice as small. Fig. 10.7 also shows the curves of T_{yy} and v obtained with MIX3 at $S_R = 6$, just before the loss of convergence.

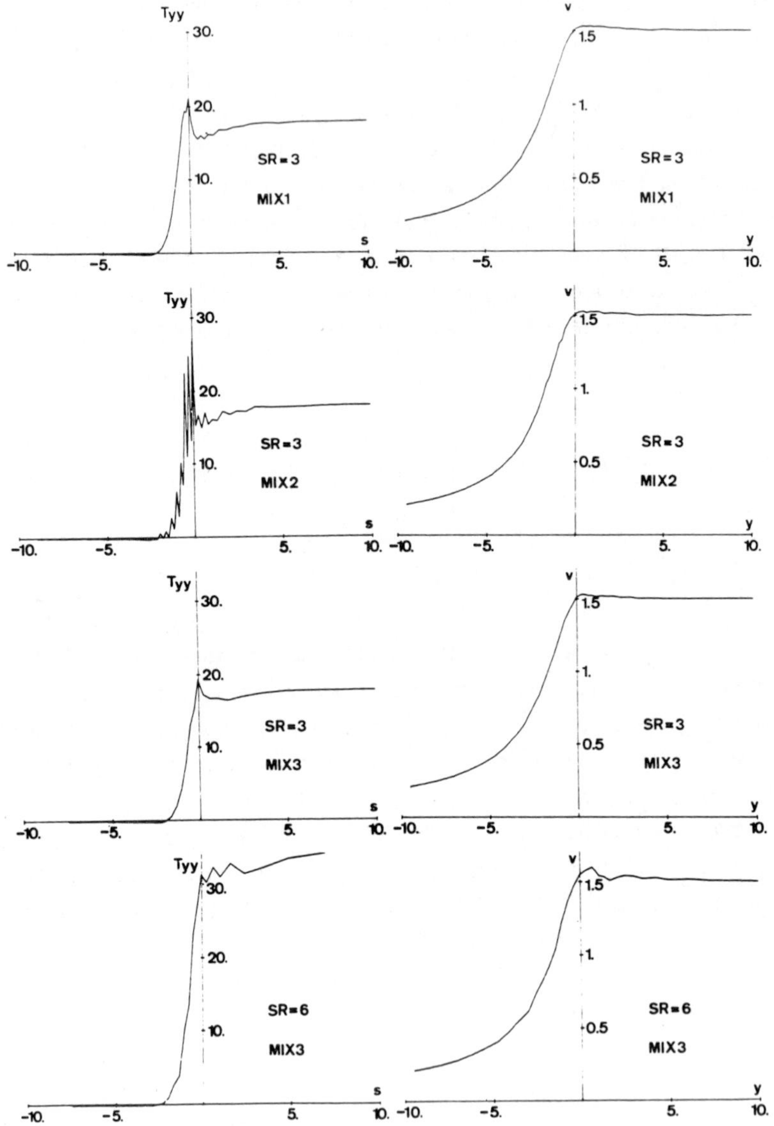

Fig.10.7 *Graphs of the extra-stress component T_{yy} along the wall of the contraction (s is a curvilinear coordinate) and of the y-velocity component v along the plane of symmetry, for a unit flow-rate and a unit shear-viscosity η_0.*

The deficiencies which have been explained above are typical of many numerical solutions for the flow of viscoelastic fluids. We will explain in the next two sections how further progress has been made by applying two different techniques.

However, before going any further, we would like to address the following question: does the spatial oscillatory behaviour of the solution have its origin in the discretized form of the constitutive equations which contain convective terms in the extra-stresses? Since the first three equations (10.19) are bi-linear forms in the extra-stress and velocity components, it is easy to think of a numerical experiment in which the smooth *Newtonian* velocity field is used for calculating the corresponding extra-stress field with at least a non-vanishing λ. On Fig. 10.8, we show the curves of T_{yy} obtained at $S_R = 7.5$ with the elements of MIX1 and MIX3, respectively. There are no wiggles, and we note that quadratic stresses produce a much sharper overshoot. We conclude that the Galerkin method used for obtaining the discretized form *of the constitutive equations* in the mixed method is not by itself responsible for the oscillatory behaviour.

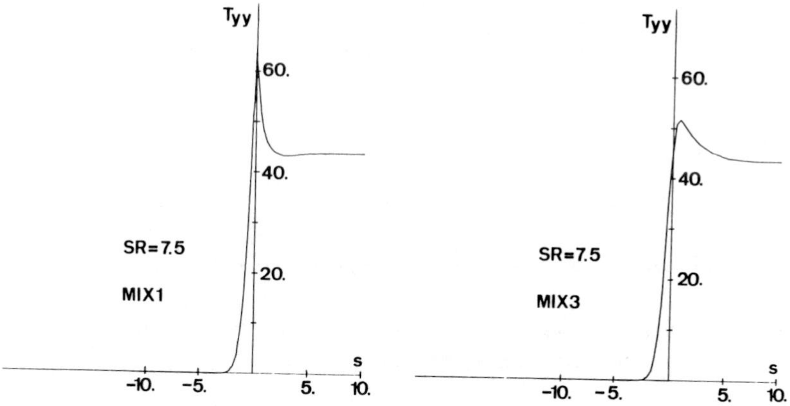

Fig. 10.8 Graphs of the extra-stress component T_{yy} along the wall of the contraction based on the Newtonian velocity field, for a unit flow-rate and a unit shear-viscosity.

10.8 THE OLDROYD-B FLUID AND RELATED MODELS

The domain of S_R over which the mixed methods do provide a converged solution can be significantly extended by means of a slight modification to the rheological model (3.5) which we have been using until now for developing numerical algorithms. In order to present the modification in a natural way, let us for a while abandon the Maxwell model and study the model (2.75) which was simultaneously introduced by Johnson and Segalman (1977) and Phan Thien and Tanner (1977) on the basis of the kinetic theory of macro-molecules. For convenience, we rewrite (2.75) with the help of (2.70) as follows :

$$T_{ik} + \lambda_1\left[\left(1 - \frac{a}{2}\right)\overset{\nabla}{T}_{ik} + \frac{a}{2}\overset{\Delta}{T}_{ik}\right] = 2\eta_0 d_{ik} \quad , \tag{10.34}$$

where a is a scalar parameter, $0 \leqslant a \leqslant 2$.

We gave in (2.75) the form of the shear viscosity η as a function of γ; it is easy to show that the shear stress T_{xy}, plotted against the shear rate γ, goes through a maximum for a finite value of γ provided $0 < a < 2$. Fig. 10.9 shows a graph of $T_{xy}\lambda/\eta_0$ as a function of $\lambda\gamma$ for various values of a.

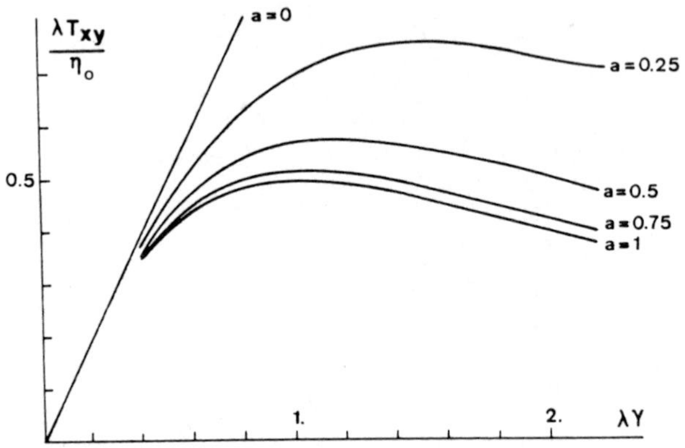

Fig.10.9 Shear stress T_{xy} plotted as a function of $\lambda\gamma$ for a Johnson-Segalman fluid.

It is clear that one cannot accept a constitutive equation which is such that, in a simple shear flow, the shear-stress T_{xy} decreases when γ increases; typically, for a given flow rate, one might obtain an infinite number of solutions for Poiseuille flow in a circular cylindrical tube.

It is however possible to remedy the situation by writing the extra-stress tensor as the sum of a purely viscous component and a term given by (10.34). To be more specific, let us consider the following constitutive equation :

$$T_{ik} = T^1_{ik} + T^2_{ik} \quad ,$$

$$T^1_{ik} + \lambda_1 \left[\left(1 - \frac{a}{2} \right) \overset{\nabla}{T}{}^1_{ik} + \frac{a}{2} \overset{\Delta}{T}{}^1_{ik} \right] = 2\eta_1 \, d_{ik} \quad , \tag{10.35}$$

$$T^2_{ik} = 2\eta_2 \, d_{ik} \quad .$$

In a simple shear experiment, it is found that

$$T_{xy} = \frac{\eta_1 \gamma}{1 + a(2-a)(\lambda_1 \gamma)^2} + \eta_2 \gamma \quad , \tag{10.36}$$

and it is easy to show that, whatever the value of a, T_{xy} is a monotonically increasing function of γ as long as (cf. Oldroyd 1958)

$$\eta_2/\eta_1 \geq 1/8 \quad . \tag{10.37}$$

While developing a numerical algorithm for solving the flow of a Phan Thien-Tanner fluid, Crochet and Keunings (1982b) found that the presence of the purely viscous component T^2_{ik} enhances the convergence properties of the algorithm MIX1, even when a vanishes in (10.35). Let us therefore examine more closely the form

$$T_{ik} = T^1_{ik} + T^2_{ik} \quad ,$$

$$T^1_{ik} + \lambda_1 \overset{\nabla}{T}{}^1_{ik} = 2\eta_1 \, d_{ik} \quad , \tag{10.38}$$

$$T^2_{ik} = 2\eta_2 \, d_{ik} \quad .$$

By eliminating T^1_{ik} and T^2_{ik} from the system (10.38), we obtain

$$T_{ik} + \lambda_1 \overset{\nabla}{T}_{ik} = 2(\eta_1 + \eta_2) \left\{ d_{ik} + \frac{\eta_2}{\eta_1 + \eta_2} \lambda_1 \overset{\nabla}{d}_{ik} \right\} \quad , \tag{10.39}$$

which we recognize as the constitutive equation of the Oldroyd-B fluid (3.13), and where $\lambda_2 = \lambda_1 \eta_2/(\eta_1 + \eta_2)$ is the retardation time.

At first sight, it might seem difficult to write an algorithm for solving the flow of a fluid given by (10.39), because of the high-order derivatives of the velocity field in $\overset{\nabla}{d}_{ik}$. However, the problem may be easily circumvented with the use of the decomposition (10.38). From (10.38) we may write

$$T_{ik} = 2\eta_2 d_{ik} + T^1_{ik} \quad , \tag{10.40}$$

while T^1_{ik} is given by the constitutive equation of the Maxwell fluid. The numerical procedure may thus consist of writing the Galerkin form of the constitutive equations for T^1_{ik}; substituting the right hand side of (10.40) for T_{ik} in the momentum equations, one obtains the following system

$$\begin{aligned}
&< -\frac{1}{\eta_1}\tilde{T}^1_{mn} - \frac{\lambda_1}{\eta_1}\overset{\nabla}{\tilde{T}}{}^1_{mn} + (\tilde{v}_{m,n} + \tilde{v}_{n,m}) \; ; \; \phi_i > = 0 \quad , \qquad 1 \leqslant i \leqslant L \quad , \\
&< \eta_2(\tilde{v}_{m,n} + \tilde{v}_{n,m}) - \tilde{p}\,\delta_{mn} + \tilde{T}^1_{mn} \; ; \; \psi_{i,m} > - < \rho F_n \; ; \; \psi_i > \\
&\qquad\qquad = \int_{\Gamma_t} \bar{t}_n \psi_i \, d\Gamma \quad , \qquad 1 \leqslant i \leqslant M \quad , \\
&< \tilde{v}_{m,m} \; ; \; \pi_i > = 0 \quad , \qquad 1 \leqslant i \leqslant N \quad .
\end{aligned} \tag{10.41}$$

The system (10.41), which was introduced by Crochet and Keunings (1982b), is very similar to (10.14); the difference is that additional purely viscous terms have been introduced into the momentum equations.

The use of the Oldroyd-B fluid rather than the Maxwell fluid, with the algorithm (10.41), extends the domain of convergence of the numerical method in a significant way,# provided one uses the shape functions of method MIX1 in Table 10.1; several examples will be shown in sections 10.11 and 10.12. It is clear that the presence of a viscous term in the momentum equation dampens the oscillatory behaviour of the solutions; however, the viscous term is not artificial and is part of the constitutive model. We recall that the Maxwell fluid and the Oldroyd-B fluid have the same behaviour in viscometric flows, i.e. a constant shear viscosity, and a first normal stress difference which is a quadratic function of γ; oscillatory experiments would allow us to distinguish between both fluids (Prilutski et al. 1983). The separation of the extra-stress tensor into a purely viscous component and a viscoelastic component is mandatory in the general case given by (10.35). The algorithm is then similar to (10.41), where $\overset{\nabla}{T}{}^1_{pq}$ is replaced by $\overset{\square}{T}{}^1_{pq}$ defined in (2.70).

The flow in a wedge considered in the previous section has been recalculated with the use of an Oldroyd-B fluid where $\eta_2/\eta_1 = 1/8$. The algorithm was still converging for $S_R = 9$, but the investigation was not pursued in view of the oscillatory behaviour of the stress field along the boundary. Fig. 10.10 shows

---

#Other advantages of the Oldroyd-B model over the Maxwell model in numerical simulation have already been discussed in connection with corner singularities in Chapter 6 (§6.2.4).

the curves of T_{yy} and v obtained at $S_R = 6.2$. As in many other problems, the size changes of the elements along the streamlines produce spatial oscillations at the element level especially on the stress field.

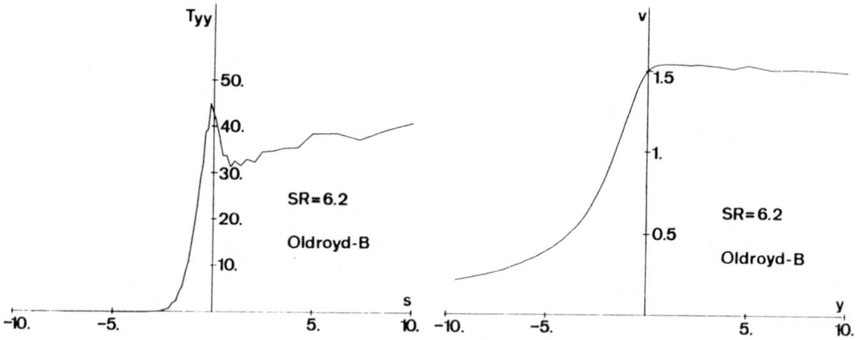

Fig.10.10 Graph of the extra stress component T_{yy} along the wall of the contraction and of the y-velocity component v along the plane of symmetry, for a unit flow rate and a unit shear viscosity, with an Oldroyd-B fluid.

We mentioned in section 3.2 that in flow problems dominated by the shear viscosity where the elastic effects do however play some part, it may be useful to use the White-Metzner fluid given by (3.4). For characterizing the White-Metzner fluid, it is necessary to know the viscosity curve as a function of γ together with the first normal stress difference. A successful method consists again of separating the extra-stress tensor as the sum of a purely viscous component and a viscoelastic component which obeys (3.4). More precisely, we write

$$T_{ik} = T^1_{ik} + T^2_{ik} \;,$$

$$T^1_{ik} + \lambda_1(I_2)\overset{\nabla 1}{T}_{ik} = 2\eta_1(I_2)d_{ik} \;, \qquad (10.42)$$

$$T^2_{ik} = 2\eta_2(I_2)d_{ik} \;,$$

where $\eta_1(I_2)$ and $\eta_2(I_2)$ may be defined as fractions of the shear viscosity $\eta(I_2)$. (10.42) is in fact a White-Metzner type extension to the Oldroyd B model.

The numerical method for solving the flow of a fluid characterized by (10.42) is again given by (10.41) where η_1, η_2 and λ_1 are now functions of I_2 which are calculated on the basis of the previous iteration.

In problems where the elongational viscosity plays a dominant role, it has been shown by Phan Thien and Tanner (1977) that the fluid properties may be well represented by the following model :

$$T_{ik} = \sum_{j=1}^{n} T_{ik}^{(j)} \quad ,$$
$$\exp[\varepsilon\lambda_j/\eta_j \; T_{mm}^{(j)}]T_{ik}^{(j)} + \lambda_j \overset{\square}{T}{}_{ik}^{(j)} = 2\eta_j \, d_{ik} \quad . \tag{10.43}$$

It is again necessary to include a purely viscous component for one of the partial stresses in (10.43). The method of calculation is similar to (10.41) (Keunings and Crochet 1983); an example of application will be shown in section 10.11.

10.9 A THIRD METHOD FOR SOLVING THE FLOW OF A MAXWELL FLUID (MIX4)

Sections 10.7 and 10.8 have taught us two important features about the behaviour of mixed methods for solving the flow of Maxwell and related fluids. First, we found in section 10.7 that the spurious wiggles which herald the loss of convergence of the iterative algorithm are not specifically due to the Galerkin form of the discretized constitutive equations; this was evidenced by the results shown on Fig. 10.8. Secondly, we found in section 10.8 that the presence of a viscous term in the momentum equations enlarges the domain of elasticity where numerical solutions may be found. In the present section, we wish to take advantage of these findings and introduce a third algorithm which appears to perform better than the other mixed methods (Van Schaftingen and Crochet 1983).

Our aim is to introduce an explicit viscous term in the discretized momentum equation, while avoiding at the same time the introduction of a convective term which appears in the second equation (10.25), and which was inevitable in MIX2 and MIX3. For the Maxwell fluid, let us introduce the following decomposition

$$T_{ik} = 2\eta_0 \, d_{ik} + S_{ik} \quad ; \tag{10.44}$$

such a decomposition has been used extensively within the framework of finite differences (see §6.1). Once we are able to use the variable S_{ik} instead of T_{ik}, the momentum equation takes an interesting form, i.e.

$$(-p_{,i} + \eta_0 d_{ki,k} + \rho F_i) + S_{ki,k} = 0 \quad . \tag{10.45}$$

Since the first parenthesis on the left hand side of (10.45) is the left hand side of Stokes' equations, one may expect that the momentum equations will be well-conditioned in view of the viscous term which was not present in MIX1; moreover, the additional term does not contain convected derivatives.

The constitutive equation for S_{ik} is obtained by inserting (10.44) into (3.5), i.e.

$$S_{ik} + \lambda_1(\overset{\nabla}{S}_{ik} + 2\eta_0 \overset{\nabla}{d}_{ik}) = 0 \quad . \tag{10.46}$$

The Galerkin method is now used with equations (10.46), (10.45) and the continuity equation in order to obtain an algebraic system in the nodal values of S_{ik}, v_i and p. After an integration by parts in the momentum equations, one obtains

$$\begin{aligned}
&< \tilde{S}_{mn} + \lambda_1(\overset{\nabla}{\tilde{S}}_{mn} + 2\eta_0 \overset{\nabla}{\tilde{d}}_{mn}) ; \phi_i > = 0 \quad , && 1 \leqslant i \leqslant L \quad , \\
&< \tilde{S}_{mn} + \eta_0(\tilde{v}_{m,n} + \tilde{v}_{n,m}) - \tilde{p}\,\delta_{mn} ; \psi_{i,m} > - < \rho F_n ; \psi_i > = \int_{\Gamma_t} \bar{t}_n \psi_i \, d\Gamma \quad , && \\
& && 1 \leqslant i \leqslant M \quad , \\
&< \tilde{v}_{m,m} ; \pi_i > = 0 \quad , && 1 \leqslant i \leqslant N \quad .
\end{aligned} \tag{10.47}$$

As it stands, the system (10.47) is not devoid of problems. Indeed, the first equation contains a term involving second-order derivatives of the velocity field which cannot be handled as long as the velocity components are C^0-continuous. It is possible, however, to obtain a weak form of the constitutive equations by performing an integration by parts on the term $< \overset{\nabla}{d}_{mn} ; \phi_i >$. In a steady flow, we may indeed write

$$\begin{aligned}
< \overset{\nabla}{d}_{mn} ; \phi_i > &= < (d_{mn} v_k)_{,k} ; \phi_i > - < v_{m,k} d_{kn} ; \phi_i > - < d_{mk} v_{n,k} ; \phi_i > \\
&= - < d_{mn} v_k ; \phi_{i,k} > - < v_{m,k} d_{kn} ; \phi_i > - < d_{mk} v_{n,k} ; \phi_i > \\
&\quad + \int_\Gamma d_{mn} v_k n_k \phi_i \, d\Gamma \quad .
\end{aligned} \tag{10.48}$$

The boundary integral in (10.48) contains the flux of the rate of deformation tensor through the boundary. The integrand vanishes identically on impervious walls, and the usual assumption of a fully developed flow in an entry section allows one to calculate the corresponding flux. In an exit section, the surface integral can be expressed in terms of the nodal velocities, and brings a contribution to the stiffness matrix.

This method, which has been programmed with the choice of shape functions given on Table 10.1, will be called MIX4. When it was applied to the wedge problem discussed in section 10.7, it was found that MIX4 was still converging when $S_R = 7.5$ where, however, the calculation was stopped because spatial oscillations developed along the element pattern. Moreover, when the density of the finite element mesh was doubled, MIX4 still converged at $S_R = 7$ against 4.5 with MIX3. Fig. 10.11 shows the behaviour of T_{yy} along the wall and of the y-velocity component on the plane of symmetry for $S_R = 3$ and $S_R = 6$; these curves should be compared with those of Fig. 10.7.

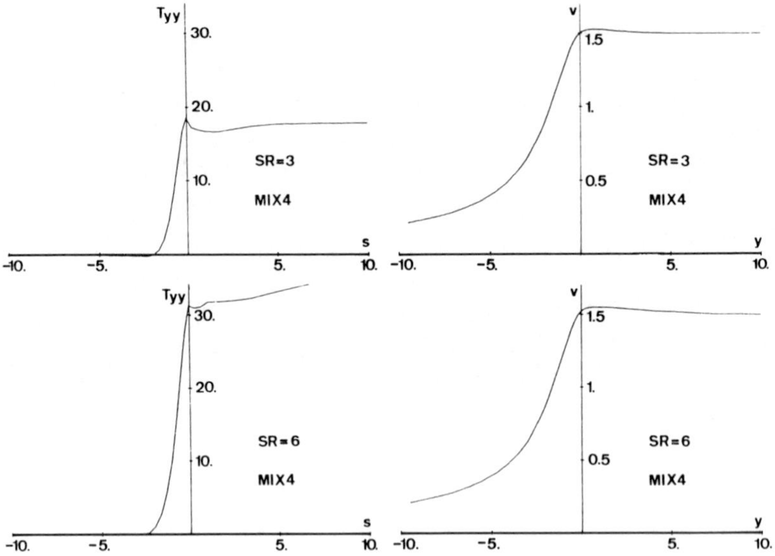

Fig.10.11 Graph of the extra-stress component T_{yy} along the wall of the contraction and of the y-velocity component v along the plane of symmetry, for a unit flow rate and a unit shear viscosity η_0, with a Maxwell fluid and method MIX4.

It is worth noting that a similar form of MIX4 was used by Mendelson et al. (1982) for solving the flow of a second order fluid, but it was disregarded in the same paper for solving the flow of a Maxwell fluid.

We have now covered most of the available mixed methods designed for calculating the flow of fluids of the differential type; their adequacy for solving some specific problems will be discussed in sections 10.11 and 10.12.

In a recent paper, Shen (1982) presented a method for integrating the flow of a Leonov fluid (see Leonov 1976). The method lies midway between the algorithms used for fluids of the differential type and for fluids of the integral type which will be considered in the next section.

10.10 THE FLOW OF VISCOELASTIC FLUIDS OF THE INTEGRAL TYPE

The last few sections have shown that the finite element calculation of the flow of fluids of the differential type is a difficult problem; the computational difficulties resemble those which are generally associated with the finite element solution of convective problems. Two important advantages of selecting fluids of the differential type are the use of the Eulerian representation of the field variables and the possibility of programming Newton's method without major difficulties.

The finite element numerical integration of the flow of fluids of the integral type is an entirely different problem which has been approached using finite elements by Viriyayuthakorn and Caswell (1980) followed by Bernstein et al. (1981); algorithms based on finite differences have already been described in Chapter 7. It may be hoped that the inherent smoothing properties of the integral operator will be helpful in solving the problems associated with fluids of the differential type. However, difficult problems are seldom endowed with miracle solutions; the obstacles encountered in discretizing the stress tensor and the convective terms are now replaced by the difficulties associated with the calculation of the strain history and the formulation of a fast iterative technique.

For the sake of simplicity, we will describe here the methods developed by Viriyayuthakorn and Caswell (1980) and Bernstein et al. (1981) for solving the flow of a Maxwell fluid in integral form and we will refer later to the possibility of extending the algorithm to more complex models of the integral type.

Let us repeat for convenience the set of constitutive and field equations for the flow of a Maxwell fluid in integral form. From (2.80) with $a = 0$, together with (2.6) and (2.3) we have, for creeping flow

$$T_{ik} = \eta_0/\lambda^2 \int_0^\infty \exp(-s/\lambda)\, H_{ik}(s)\, ds \quad ,$$

$$- p_{,i} + T_{ki,k} + \rho F_i = 0 \quad , \tag{10.49}$$

$$v_{i,i} = 0 \quad .$$

Rather than (10.49), Viriyayuthakorn and Caswell use the following form :

$$T_{ik} = \eta_0/\lambda^2 \int_0^\infty \exp(-s/\lambda)\, C^{-1}_{ik}(s)\, ds - \eta_0/\lambda\, \delta_{ik} \quad . \tag{10.50}$$

We recall that the relative strain tensor H_{ik} is given by

$$H_{ik}(s) = C^{-1}_{ik}(s) - \delta_{ik} = \frac{\partial x_i}{\partial x'_m} \frac{\partial x_k}{\partial x'_m} - \delta_{ik} \,, \tag{10.51}$$

where x'_i is the position at time $(t - s)$ of a particle which occupies the position x_j at time t,

$$x'_i = x'_i(x_j, t, t-s) \quad . \tag{10.52}$$

In steady flow, (10.52) reduces to

$$x'_i = x'_i(x_j, s) \quad . \tag{10.53}$$

The difficulty of the problem may be grasped by inspecting equations (10.49) to (10.53). Since the flow problem is formulated in its Eulerian representation, the velocity field is the primary unknown. However, the velocity field is not explicitly present in the first two equations (10.49); rather, it is implicitly used in the calculation of the strain tensor H_{ik}. Moreover, the integral on the right hand side of (10.50) must be calculated along streamlines which are not known at the outset.

A possible solution technique may therefore consist of an iterative algorithm based on the search for a velocity field and which may be summarized as follows :

i. *Tracking* : on the basis of an assumed velocity field, one calculates the streamlines and the vector function $x'_i(x_j, s)$ in (10.53).

ii. *Stress calculation* : the strain history and the extra-stress components are calculated on the basis of (10.51) and (10.50), respectively.

iii. *Velocity correction* : since a compatible pressure field is usually unavailable for satisfying the momentum equations $(10.49)_2$, it is necessary to perturb the pre-assumed velocity field; the pressure field is calculated together with the velocity correction.

We will now consider the successive steps in more detail.

The tracking procedure starts on the basis of a trial velocity field, which is the Newtonian field for a new problem, and the previous velocity field in an iterative procedure. The tracking process amounts to solving the set of differential equations,

$$\frac{Dx_i(t-s)}{D(t-s)} = v_i[x_k(t-s)] \quad , \quad \text{or} \quad -\frac{dx_i'}{ds} = v_i(x_k') \quad , \tag{10.54}$$

where we recall that s is the time variable calculated backwards.

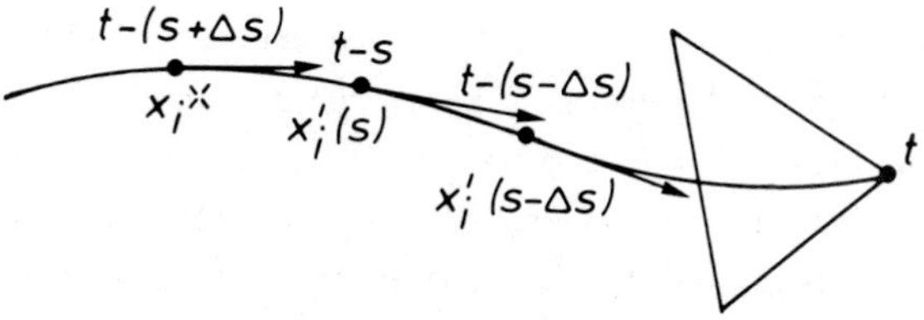

Fig.10.12 Tracking of a particle.

Viriyayuthakorn and Caswell (1980) calculate the trajectory of the nodes as a function of time (Fig. 10.12). Let us assume that the position $x_i'(s)$ of a given node at time $(t-s)$ has been found, and let Δs be a small time increment; from a second order Taylor expansion of $x_i'(s)$ one obtains

$$x_i'(s+\Delta s) = x_i'(s) - v_i[x_k'(s)]\Delta s + \dot{v}_i[x_k'(s)]\frac{\Delta s^2}{2} + O(\Delta s^3) \quad . \tag{10.55}$$

The acceleration $\mathring{v}_i[x_k'(s)]$ is calculated iteratively by following the material particle. In the first step, one uses

$$\dot{v}_i[x_k'(s)] = \{v_i[x_k'(s-\Delta s)] - v_i[x_k'(s)]\}/\Delta s \quad , \tag{10.56}$$

from which a first estimate x_i^* is found for $x_i'(s+\Delta s)$. The calculation is then refined by means of the estimate

$$\dot{v}_i[x_k'(s)] = \{v_i[x_k'(s)] - v_i[x_k^*]\}/\Delta s \quad , \tag{10.57}$$

and the calculation is repeated until no further change of position occurs within some given tolerance.

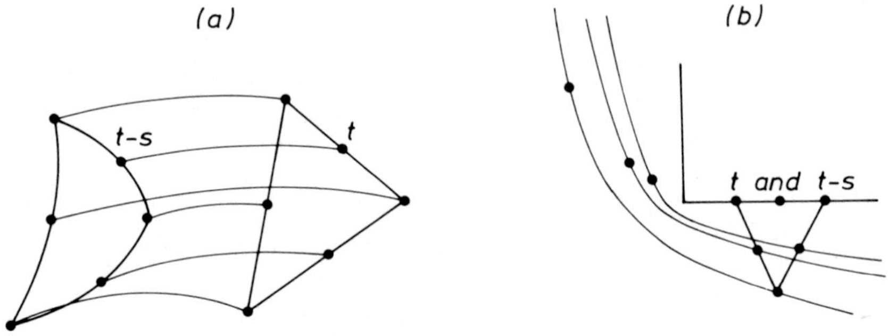

Fig. 10.13 Past configuration of an element.

Let us now assume that the tracking procedure has been completed, and that the past location of the nodes is known for a discrete set of time intervals which will be identified in a moment. Thus, at a given time s_k, the location of the nodes of an element is known (Fig. 10.13a); the deformation of the element is then described by an isoparametric transformation which, for a planar triangular element, may be written as follows :

$$x'(s_k) = \sum_{j=1}^{6} x_j'(s_k)\,\psi_j(x,y) \;, \qquad y'(s_k) = \sum_{j=1}^{6} y_j'(s_k)\,\psi_j(x,y) \;; \qquad (10.58)$$

The calculation of the deformation gradient $\partial x_i'(s_k)/\partial x_j$, of its inverse $\partial x_i/\partial x_j'(s_k)$ and of the strain tensor $H_{ij}(t-s_k)$ by means of (10.51) is then an easy task to perform.

The limitation of the procedure is obvious on Fig. 10.13b. The finite element interpolation of the displacement $x_i'(s_k)$ may, in some cases, become too coarse when s increases; important features of the flow, which would occur for example when the fluid flows around a corner, may be totally overlooked.

Special care must be taken of course when the streamline of a fluid particle being tracked crosses the boundary of the flow domain. When a problem contains an entry region, it is usually assumed that the upstream flow is known and fully developed, so that one may use the decomposition (10.21) together with the stress tensor on the boundary for avoiding calculations outside the flow region.

The method adopted by Bernstein et al. (1981, 1982) and Malkus (1981) for calculating the strain history at a given location in the mesh differs from the one which we have just described. Their basic idea is to develop a finite element which is simple enough so that the tracking procedure and the calculation of the strain history may be accomplished analytically at the element level.

The calculation of the strain history is based on the following argument: at time $(t-s)$, the relative deformation gradient $F_{ik}(t-s) = \partial x_i(t-s)/\partial x_k$ obeys the differential equation

$$\frac{D}{D(t-s)} F_{ik}(t-s) = L_{ij}(x'_m) F_{jk}(t-s) \quad ; \qquad (10.59)$$

x'_m has been defined in (10.52), and L_{ij} is the local velocity gradient. When the tensor L_{ij} is uniform within an element, it is possible to integrate (10.59) exactly.

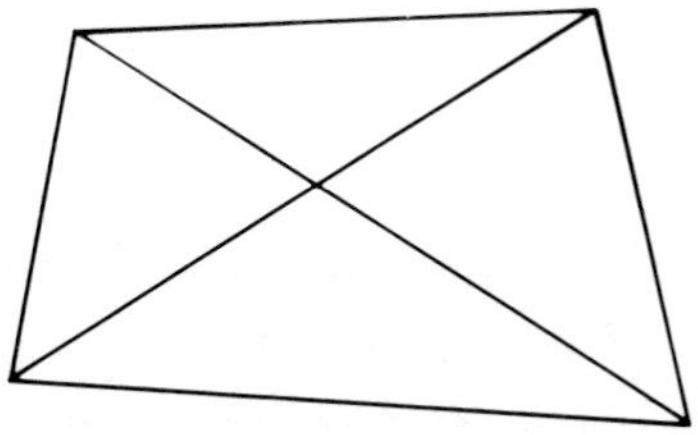

Fig.10.14 Macro-element used by Bernstein et al.(1982).

Bernstein et al. (1981) use either the bilinear quadrilateral element of Fig. 9.4, or a quadrilateral composed of four triangles shown on Fig. 9.3, which is arranged in such a way that the central node is located at the intersection of the diagonals (Fig. 10.14). The validity of this composite element for solving Newtonian flow is discussed in detail by Bernstein et al. (1982). The velocity gradient is uniform within three-node triangular elements. This is not, however, the case for the bilinear quadrilateral, since the bilinear shape functions ϕ_i^q defined in (8.78) contain a second-order term; however, the *approximation* is made that the velocity gradient is uniform and takes the value calculated at the centroid of the element.

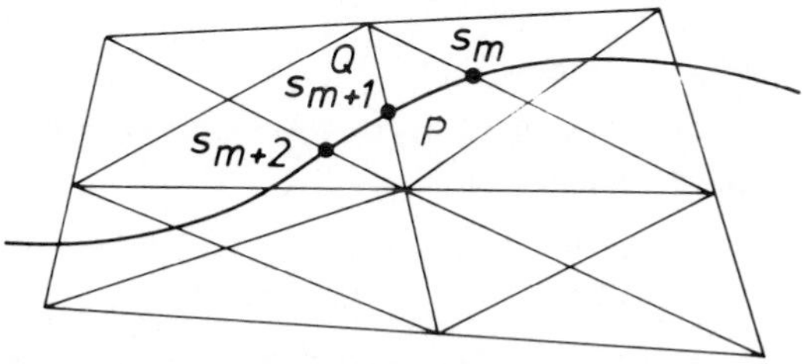

Fig.10.15 Mesh of macro-elements and path of a particle.

Let us consider on Fig. 10.15 a mesh of macro-elements and the path of a material point which occupies the position x_i at time t; we assume that the particle has left element P at time $(t-s_m)$, with a known relative deformation gradient $F_{ik}(t-s_m)$. Since the nodal velocities are known, it is possible to determine the streamline of the particle analytically within element P, together with the location where the particle crosses the boundary of the element towards the neighbouring element Q. In order to integrate (10.59), one needs to know the time at which the crossing occurs. To that effect, Bernstein and Malkus (1981) introduce the concept of a *drift function*. Let $x_i(t')$ be the location at time t' of a particle which occupies the position $x_i(t)$ at time t; the drift function $w(x_k)$ is defined by the following equation :

$$w[x_i(t)] - w[x_i(t')] = t - t' \quad . \tag{10.60}$$

The analytical calculation of the drift function in linear triangles is discussed by Bernstein et al. (1982); the axisymmetric case is considered by Chen and Bernstein (1982). Once the drift function has been calculated, the time $(t-s_{m+1})$ at which the particle left element Q is also known and, in view of (10.59), one may calculate the relative deformation gradient

$$\underline{F}(t-s_{m+1}) = e^{-\underline{L}^P(s_{m+1}-s_m)}\ \underline{F}(t-s_m) \quad , \tag{10.61}$$

where $\underline{L}^P$ is the velocity gradient in element P. Thus, the simplicity of the interpolation used for the velocity field allows for an easy calculation of the strain history.

At this stage two procedures have been described for calculating the strain history of a material particle; it is now necessary to evaluate the stress tensor given by (10.50) for the case of an upper-convected Maxwell fluid.

Consider an integral of the form

$$I = \int_0^\infty \exp(-z)\ P(z)\ dz \quad . \tag{10.62}$$

When P(z) is a polynomial of degree 2n + 1, it is possible to find n + 1 points of abscissa z_i and associated weights w_i in order to evaluate the integral $\underline{I}$ exactly by means of the formula

$$\underline{I} = \sum_{i=0}^{n} w_i\ P(z_i) \quad . \tag{10.63}$$

The numerical procedure for evaluating an integral of the type (10.62) on a semi-infinite domain is called the Gauss-Laguerre quadrature. Typical integration rules for $n=1$ and $n=2$ are given in Table 10.3; further details may be found in Carnahan et al. (1969).

TABLE 10.3
Table of Gauss-Laguerre Quadrature

z_i		w_i	
Two-point formula, n = 1			
.58578	64376	.85355	33906
3.41421	35624	.14644	66094
Three-point formula, n = 2			
.41577	45568	.71709	30099
2.29428	03603	.27851	77336
6.28994	50829	.01038	92565

With the change of variable $\sigma = s/\lambda$, equation $(10.49)_1$ may be rewritten as

$$T_{ik} = \int_0^\infty \eta_0/\lambda)\exp(-\sigma)\, H_{ik}(\sigma\lambda)d\sigma \ , \tag{10.64}$$

and, in view of (10.63), the extra-stress is evaluated as follows :

$$T_{ik} = \sum_{i=0}^{n} w_i(\eta_0/\lambda)H_{ik}(\lambda z_i) \quad . \tag{10.65}$$

The number of points of integration depends upon the complexity of the strain history. Consider for example a viscometric flow, where $H_{ik}(t-s)$ reduces to a second degree polynomial in s; in that case, the stress may be evaluated exactly with two Laguerre points given in Table 10.3. Viriyayuthakorn and Caswell use a maximum of 10 Laguerre points; when the number of points increases, so does the backward time s at which the strains must be evaluated, and the latter becomes large. The number of points is usually associated with the size of the elements. A fine grading of the mesh is necessary in regions of high velocity gradient, where a large number of integration points is also needed. Large elements are associated with regions where the flow is almost viscometric, and a small number of integration points is then associated with these elements.

Quadrature rules which are well suited for calculating the flow of fluids of the integral type which differ from the Maxwell fluid may be found in Bernstein and Malkus (1981).

The final part of the algorithm consists of calculating a correction of the velocity field in order to find the extra-stresses and a pressure field which satisfy the full set of field and constitutive equations. A very difficult problem which has not been solved as yet is the construction of Newton's procedure for calculating the velocity correction; the difficulty lies in the complex dependence of T_{ik} upon the Eulerian velocity components through (10.50). We will now describe the procedure utilized by Viriyayuthakorn and Caswell (1980), which assumes at the outset that the perturbation of the velocity field is small.

The perturbation technique is based upon the possibility of expanding the tensor $H_{ik}(s)$ in terms of Rivlin-Ericksen tensors defined by (2.53). It may indeed be shown, in view of (2.54) and (2.60) that

$$H_{ik}(t-s) = s\,A_{ik}^{(1)} + s^2[A_{ij}^{(1)}\,A_{jk}^{(1)} - A_{ik}^{(2)}/2] + O(s^3) \quad . \tag{10.66}$$

Consider two velocity fields $v_i^{(n+1)}$, $v_i^{(n)}$ and their difference $\delta v_i = v_i^{(n+1)} - v_i^{(n)}$; let $H_{ik}^{(n+1)}$, $H_{ik}^{(n)}$ be the corresponding strain histories. In view of (10.66) one has

$$H_{ik}^{(n+1)}(t-s) = H_{ik}^{(n)}(t-s) + s\,\delta A_{ik}^{(1)} + O(s^2) \quad , \tag{10.67}$$

where $\delta A_{ik}^{(1)}$ is the first Rivlin-Ericksen tensor calculated with the velocity increment δv_i. Let $T_{ik}^{(n+1)}$ and $T_{ik}^{(n)}$ be the extra-stress components calculated on the basis of $v_i^{(n+1)}$ and $v_i^{(n)}$, respectively; one has, from $(10.49)_1$,

$$\begin{aligned} T_{ik}^{(n+1)} &\simeq T_{ik}^{(n)} + \eta_0/\lambda^2 \int_0^\infty \exp(-s/\lambda)s\;\delta A_{ik}^{(1)}\,ds \\ &= T_{ik}^{(n)} + \eta_0\;\delta A_{ik}^{(1)} \quad . \end{aligned} \tag{10.68}$$

Equation (10.68) provides the relation needed for solving the system (10.49) by means of an iterative technique. Let ψ_i and π_i denote the shape functions used for the velocity field and for the pressure, respectively. After an integration by parts, the Galerkin form of the momentum and continuity equations is given by

$$\begin{aligned} &< T_{kj} - p\,\delta_{kj}\;;\;\psi_{i,j} > = F_k^i \quad , \\ &< v_{k,k}\;;\;\pi_i > = 0 \quad , \end{aligned} \tag{10.69}$$

where we recognize the second and third equations (10.8) in a slightly simpler form. Let $\tilde{v}_i^{(n)}$ be an assumed discretized velocity field, which we wish to correct by an amount $\delta\tilde{v}_i$. Using (10.68) we obtain from (10.69)

$$< \eta_0(\delta\tilde{v}_{k,j} + \delta\tilde{v}_{j,k}) - \tilde{p}\delta_{kj} \; ; \; \psi_{i,j} > = F_k^i - < T_{kj}^{(n)} \; ; \; \psi_{i,j} > \quad ,$$
$$< \delta v_{k,k} \; ; \; \pi_i > = 0 \quad . \tag{10.70}$$

The system (10.70) has been studied in detail in Chapter 9; its left hand side is precisely the set of equations used for calculating the flow of a Newtonian viscous fluid, while the right hand side contains the previously calculated stress contribution as a force term. To obtain a converging sequence of velocity fields, it may be necessary to include a relaxation parameter ω, and to replace η_0 by $\eta_0\omega$ in (10.70); the selection of the relaxation parameter, which is discussed by Viriyayuthakorn and Caswell (1980), proceeds from experience and no well established choice is available. The rate of convergence of the iterative technique is rather slow as compared to that associated with Newton's method.

We have concentrated our discussion here on solving the flow of a Maxwell fluid, but the procedure that we have just explained may be generalized easily to more complex fluids of the integral type. Bernstein et al. (1982) develop their numerical technique for solving the flow of fluids which are special cases of the model introduced by Curtiss and Bird (1981).

10.11 EXAMPLE OF THE GENERAL DEVELOPMENT - ENTRY FLOW IN A TUBULAR CONTRACTION

We discussed in section 9.9 the problem of flow through a tubular contraction; the geometry of the problem is shown on Fig. 9.12. The boundary conditions for the flow of a viscoelastic fluid are the same as those shown on Fig. 9.12, with the addition that all the extra-stress components must now be imposed in the entry section, with the assumption that the upstream flow is fully developed. The flow of a viscoelastic fluid through an abrupt contraction is a provocative problem; indeed, there is no standard flow behaviour, and the observations vary significantly from one polymeric fluid to another. Typically, Giesekus (1968) observed large vortices in the neighbourhood of the entry region when the elasticity increases, for the plane flow of a polyacrylamide solution in water. Similar large vortices have been photographed by Nguyen and Boger (1979) for the tubular entry flow of a polyacrylamide solution in glucose syrup; however, Walters and Webster (1982) observed no such vortex growth due to elasticity for the plane flow of the same fluid through a contraction. White and Kondo (1977) have reviewed several experimental results; they found that the presence of a large vortex depends upon the type of polymer melt used for the experiment.

It is hoped that numerical simulation will eventually lead to an understanding of the experimental observations. In the present section, we will limit ourselves to 4/1 contractions, for which experimental results have been made available by Nguyen and Boger (1979).

From a numerical point of view, the problem is very difficult, because the non-linearity of the viscoelastic constitutive equations is complicated by the presence of a singularity of unknown nature. The singularity is detrimental for the numerical techniques dealing with both the differential and the integral models. The mixed methods explained in this chapter for the differential models assume that the extra-stress field is continuous at the corner, which is obviously not the case. At the same time, we have seen how difficult it is to calculate the strain history around a corner when dealing with a fluid of integral type.

Numerical results for the flow of a Maxwell fluid through a 4/1 contraction were published by Crochet and Bezy (1979); their results were obtained on a coarse mesh with a type of finite element which has since been forsaken, and their results will not be exhibited here. However, they were already able to point out some disturbing factors: the absence of vortex growth, a rather low limit of S_R beyond which one cannot obtain convergence, and the lack of departure of the solution from what one might call second-order behaviour. Viriyayuthakorn and Caswell (1980) were the first to publish a comprehensive set of results for the flow through a tubular 4/1 contraction. We will now discuss some of their results, and use them as a basis of comparison with other techniques.

The finite element mesh used by Viriyayuthakorn and Caswell (1980) is shown on Fig. 10.16; we recall that these authors use quadrilaterals made of four triangles as shown on Fig. 9.6.

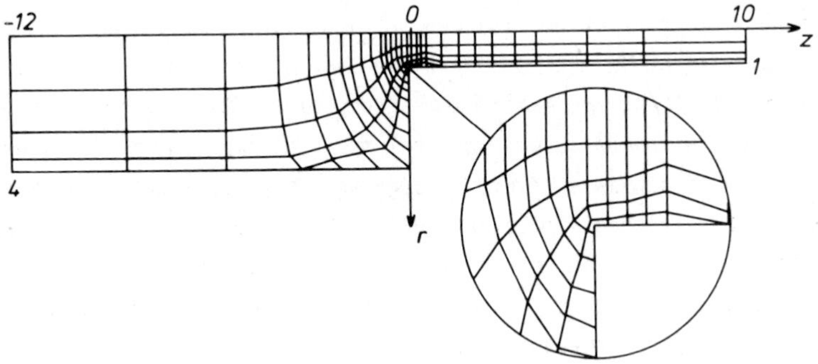

Fig. 10.16 Finite element mesh for the flow through a tubular 4/1 contraction used by Viriyayuthakorn and Caswell (1980).

In the present problem, the non-dimensional parameter which indicates the amount of elasticity of the flow is the product $\lambda\dot{\gamma}$ on the wall of the downstream tube where the flow is fully developed. Convergence is obtained by Viriyayuthakorn and Caswell (1980) up to $S_R = 2$. For $S_R = 3$, they find marginal convergence, in the sense that the velocity corrections do not decrease beyond some value which is still unacceptable for the results to qualify as a solution. The number of iterations reported by the authors is 6 for $S_R = 1$ and 10 for $S_R = 2$.

The streamlines obtained for $S_R = 0$, 1 and 2 are shown on Fig. 10.17. They show little dependence with respect to S_R; the recirculating vortex grows slightly when S_R increases, but is far from reaching the dimensions observed by Nguyen and Boger (1979).

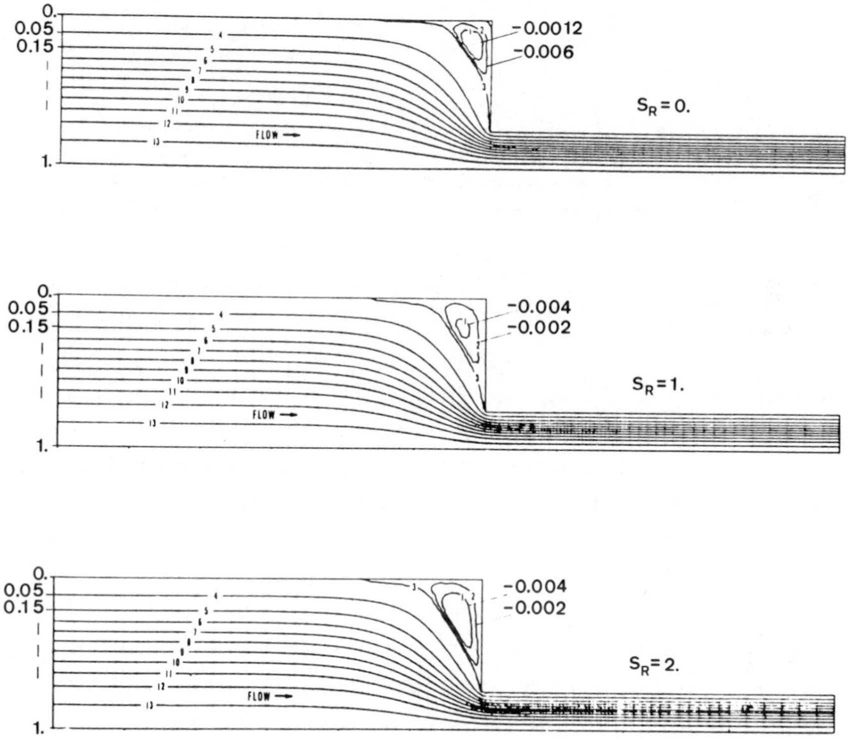

Fig.10.17 Streamlines for the flow through a tubular contraction. The stream function is normalized from 0 to 1.

A better idea of the difference between the flow of Newtonian and Maxwell fluids is obtained by comparing the centreline axial velocity distribution for $S_R = 0$ and $S_R = 2$, which is shown on Fig. 10.18. An overshoot of the axial velocity is clearly visible; its amplitude is approximately 3% and its distance from the entry section is about one radius of the downstream tube.

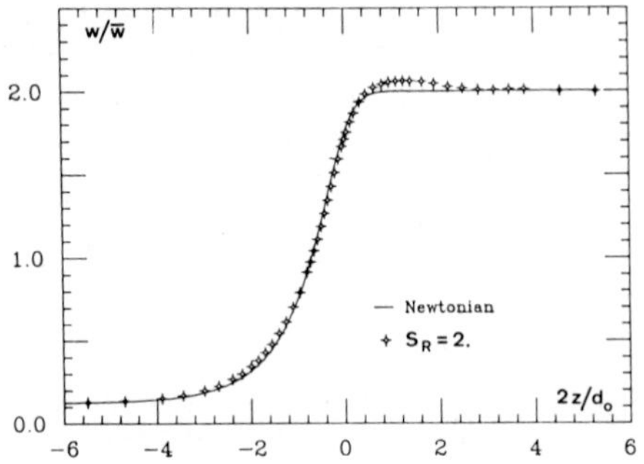

Fig.10.18 Centreline axial velocity distribution obtained by Viriyayuthakorn and Caswell (1980) for die entry of a Newtonian fluid and a Maxwell fluid ($S_R = 2$); $\bar{w}$ is the mean velocity in the downstream tube of diameter d_0.

As we might expect, another main difference between the Newtonian and the visco-elastic results is the stress-field; this will be discussed later in this section together with the excess pressure drop.

The entry flow in a tubular contraction has been studied intensively with the various mixed methods described earlier. An important observation is that the limit of convergence depends upon the mesh, as in the Hamel-flow problem. It is very disturbing however that, at the present time, even a rational description of that dependence is unavailable. Results have been obtained by Crochet and his collaborators for values of S_R as high as 3.5 with method MIX1 for a Maxwell fluid; however, the wiggles in the velocity and stress fields are sufficient to render the solution meaningless, *although the stream function, which integrates all the oscillations, is perfectly respectable.* A very good example of what happens with MIX1 beyond some value of S_R is shown on Fig. 10.19, where we show the centreline velocity obtained at $S_R = 1$ with the dense mesh of Fig. 9.18; the values of the axial velocity at the midside nodes are not aligned with those at the vertices. With the mesh of Fig. 10.16, the maximum value of S_R for convergence is 1; the axial velocity is also shown on Fig. 10.19. An overshoot is clearly visible, but the wiggles have already appeared. The method MIX2 behaves essentially like MIX1. With method MIX3, the limit of convergence with the mesh of Fig. 10.16 is 2.4, but the centreline axial velocity is unacceptable.

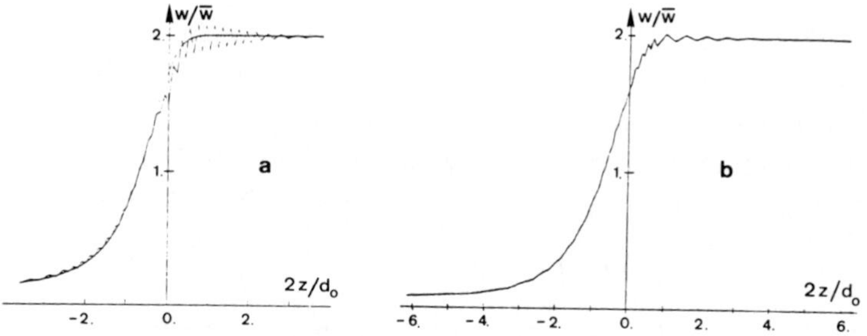

Fig.10.19 Centreline axial velocity distribution obtained with the method MIX1 at S_R=1 with the mesh of Fig.9.18(a), and the mesh of Fig.10.16(b).

The grim picture given by the mixed methods for the flow through an abrupt contraction is modified once we accept the presence of a retardation time in the constitutive equations, and consider in fact an Oldroyd-B fluid, with the method of section 10.8 (Crochet 1982b). The selected ratio of the viscosities η_1/η_2 in (10.35) is 8; it follows that the relation between S_R and $\lambda\gamma$ is now

$$S_R = 8\ \lambda\gamma/9 \quad . \tag{10.71}$$

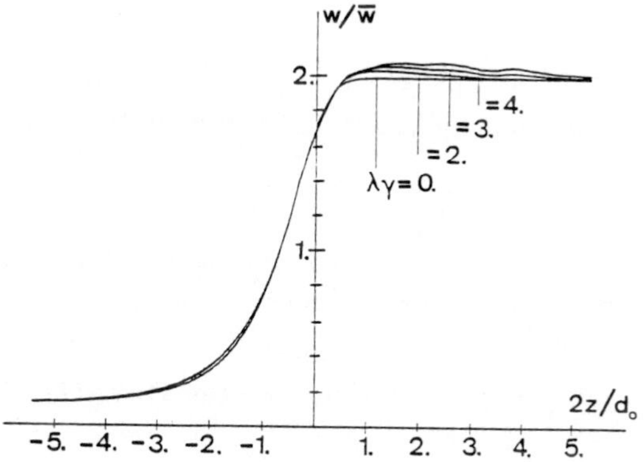

Fig.10.20 Centreline axial velocity distribution obtained for an Oldroyd-B fluid with the mesh of Fig.10.16; the value of $S_R = 8\lambda\gamma/9$.

For an easier comparison, we use again the mesh of Fig. 10.16 where the quadrilaterals represent Lagrangian elements. With an Oldroyd-B fluid, convergence was obtained with four or five iterations up to a value of $\lambda\dot{\gamma} = 4$, or $S_R = 32/9$. The calculation has not been pursued in view of the wiggles which appear in the centreline axial velocity component. Fig. 10.20 shows quite clearly that a velocity overshoot appears for all values of S_R. Table 10.4 gives the value and the location of the maximum velocity overshoot. They compare quite well with the values obtained by Viriyayuthakorn and Caswell (1980).

TABLE 10.4
Value and location of the overshoot of the centreline velocity

S_R	0	8/9	16/9	24/9	32/9
overshoot (%)	-	0.7	2.15	3.35	4.21
$2z/d_0$	-	1.	1.25	1.25	1.75

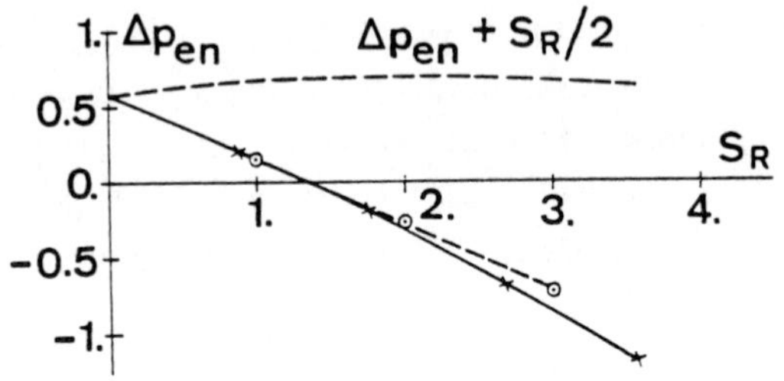

Fig.10.21 Excess pressure drop in the contraction as a function of S_R: x: Oldroyd-B fluid; ⊙ :Maxwell fluid.

The excess pressure drop in a contraction has been defined in section 9.9. Fig. 10.21 shows the values obtained for the Oldroyd-B fluid together with those obtained by Viriyayuthakorn and Caswell (1980). The fact that Δp_{en} becomes negative can be explained easily as long as the kinematics is not dramatically affected by the elasticity of the flow. Indeed, the existence of a fully developed Poiseuille flow in the exit section involves the appearance of normal stresses T_{zz} which exert a traction on the fluid. If one compares the average value of these tractions to the wall shear stress in a tube of radius r_0 one obtains

$$\left[\int_0^{r_0} T_{zz} \; 2\pi r \; dr \right] / \pi r_0^2 = S_R/2 \quad . \tag{10.72}$$

Once these tractions are taken into account in calculating the entry pressure loss, there is little difference between the Newtonian and the viscoelastic cases, as one finds on Fig. 10.21.

There is no need to show the streamlines obtained with the Oldroyd-B fluid, since for $S_R = 0$, 1 and 2, they are essentially identical to those of Fig. 10.17; for higher values of S_R, the intensity of the vortex grows slightly but its size remains essentially unchanged.

The same flow has been studied by Van Schaftingen and Crochet (1983) with the recent method MIX4 of section 10.9 for the *Maxwell fluid*, without a retardation time. Fig. 10.22 shows the centreline axial velocity for $S_R = 2.4$. The curve is smooth, and exhibits an overshoot which is compatible with the results of Table 10.4. On the same figure, we show the curve of the extra-stress component T_{zz} along the wall in the neighbourhood of the re-entrant corner. The high stress peak is due to the singularity at the corner, and it is fairly clear that the peak produces wiggles in the stress field, because the finite element representation cannot cope with such rapid variations, unless a special treatment is developed for the singularity.

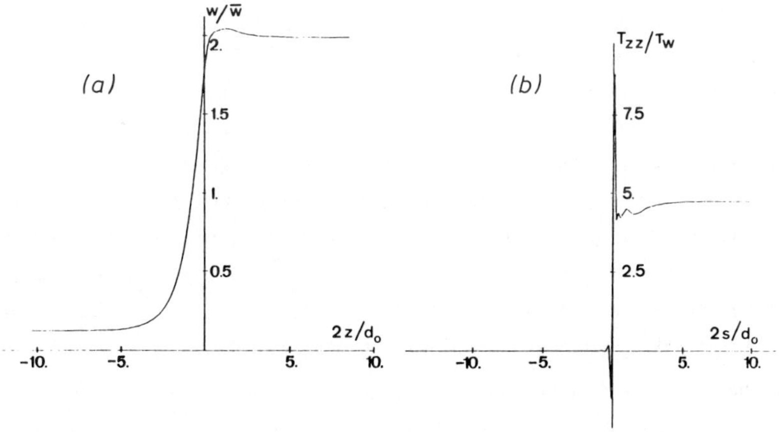

Fig.10.22 (a) Centreline axial velocity distribution obtained for a Maxwell fluid with the method MIX4 and the mesh of Fig.10.16, at $S_R = 2{\cdot}4$. (b) Extra-stress component T_{zz} along the wall in the neighbourhood of the reentrant corner; s is the arc-length along the wall.

The results presented up to now have dealt with the Maxwell fluid and the closely related Oldroyd-B fluid. Keunings and Crochet (1983) have recently studied the flow in a 4/1 tubular contraction of a Phan Thien-Tanner fluid with a constitutive equation given by (10.43). More precisely, the following form was adopted for the calculations :

$$\exp[0.015\ \lambda_1/\eta_1\ T^{(1)}_{mm}]\ T^{(1)}_{ik} + \lambda_1\ \overset{\square}{T}{}^{(1)}_{ik} = 2\eta_1\ d_{ik}\ ,$$
$$T^{(2)}_{ik} = 2\eta_2\ d_{ik}\ , \qquad (10.73)$$

with $\eta_2 = \eta_1/8$ and a = 0.2 in the convected derivative $\overset{\square}{T}{}^{(1)}_{ik}$. Such a fluid is shear-thinning and its elongational properties differ from those of a Maxwell fluid.

The non-dimensional parameter $S_R = N_1/2\tau_w$ is not a good indicator of the elasticity of the flow of the Phan Thien-Tanner fluid; when $\lambda\gamma$ increases, S_R goes through a maximum and then decreases. It is better to use the elasticity number defined by

$$W = \lambda_1\ \bar{w}/d_0\ , \qquad (10.74)$$

where $\bar{w}$ is the mean downstream velocity and d_0 is the downstream diameter.

With the finite element mesh of Fig. 10.16, it was possible to obtain a solution to the flow in the 4/1 contraction up to $W = 1.75$. It was, however, necessary to increase the length of the last few elements and to include a downstream tube which is 20 radii long; such a length is necessary to reach a fully developed flow in the exit section. This time, an important overshoot of the velocity is found on the axis of symmetry near the contraction section, which is related to a significant growth of the corner vortex. Fig. 10.23 shows the streamlines in the neighbourhood of the contraction when $W = 1$ and $W = 1.75$. It is fairly evident that a corner vortex is developing when the elasticity of the flow increases; Fig. 10.23 should be compared with Fig. 10.17, where the intensity of the vortex remains low even when $S_R = 2$.

Another important consequence of using the Phan Thien-Tanner model is the behaviour of the excess pressure drop as a function of W. While numerical experiments with the Maxwell fluid have shown repeatedly that Δp_{en} decreases when S_R increases, the situation is now reversed when $W>0.2$; Fig. 10.24 shows a graph of Δp_{en} as a function of W which exhibits a steady increase of Δp_{en} beyond $W = 0.2$. The results have been confirmed by convergence tests on several meshes.

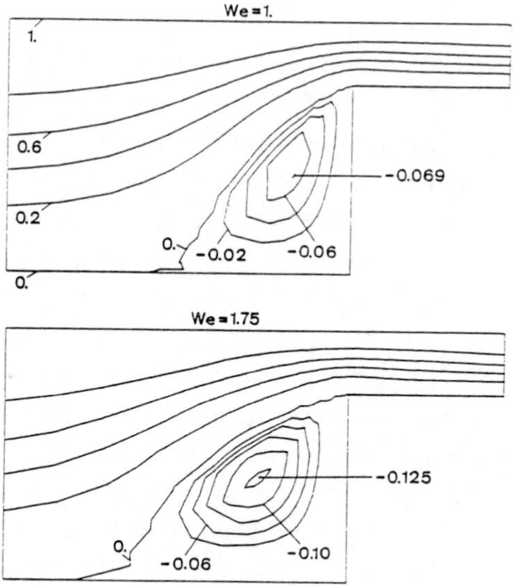

Fig.10.23 Streamlines in the neighborhood of the contraction for the flow of a Phan Thien-Tanner fluid, at W = 1 and W = 1·75; the stream function is normalized at 0 on the wall and 1 on the axis of symmetry.

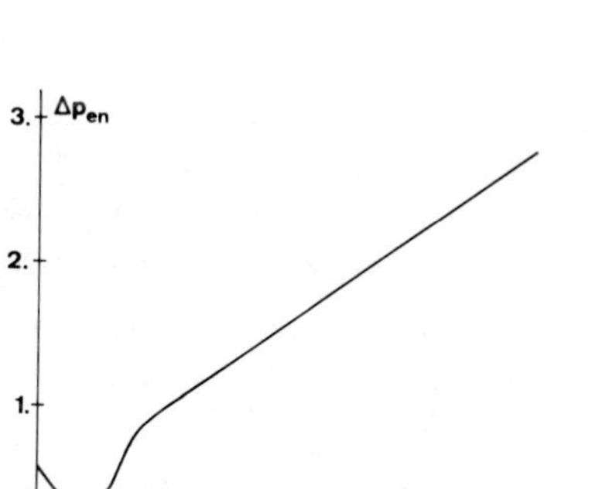

Fig.10.24 Excess pressure drop in the contraction as a function of W for the flow of a Phan Thien Tanner fluid.

10.12 EXAMPLE OF THE GENERAL DEVELOPMENT - DIE-SWELL OF A VISCOELASTIC FLUID

The problem of die-swell which was considered in section 9.10 for the case of a generalized Newtonian fluid, has attracted a great deal of attention from researchers working on the numerical simulation of viscoelastic flow. Indeed, extrudate swell is an easily observable experimental phenomenon, which is often exhibited as a characteristic feature of polymeric fluids as opposed to Newtonian fluids. Here again, there is a considerable amount of scatter among the available experimental data; the magnitude of the swelling ratio varies with the material and with the experimental conditions, and the non-dimensional groups which govern the phenomenon are not known. The problem is further complicated by the fact that in many instances the viscometric properties of the material cannot be measured in the regions of viscoelastic behaviour where the swelling ratio takes dramatic proportions. A review of the problem has been given recently by Vlachopoulos (1982) and the problem has been analyzed by Tanner (1970) and Pearson and Trottnow (1978), who considered the flow of Lodge's rubberlike elastic fluid. Tanner (1970) obtained an expression relating the swelling ratio to the recoverable shear, which is the only available non-dimensional group for the creeping flow of a Maxwell fluid out of an infinite tube; the relation is plotted on Fig. 10.30 for circular die-swell. By inspection of this curve, it is obvious that the phenomenon becomes interesting for values of S_R beyond 1; for smaller values of S_R, there is very little difference between the Newtonian case, where the swelling phenomenon already occurs, and the viscoelastic case. This remark explains why the S_R barrier for obtaining convergence of the iterative algorithm has been even more infuriating in the die-swell analysis than with other problems.

The numerical prediction of die-swell has been undertaken by Reddy and Tanner (1978b) for a second-order fluid, by Chang et al. (1979), Crochet and Keunings (1980, 1981, 1982a), Coleman (1981), Finlayson and Tuna (1982), Caswell and Viriyayuthakorn (1983) for a Maxwell fluid, and by Crochet and Keunings (1982b) for an Oldroyd-B fluid. These authors have considered either circular or slit die-swell; table 10.5 summarizes the problems treated by these authors, together with the method used and the maximum value of S_R where convergence was obtained.

Table 10.5 shows that the addition of a retardation time in the equation of a Maxwell fluid or, equivalently, the addition of a purely viscous component to the extra-stress tensor, enlarges the domain of convergence since one goes from a maximum value of $S_R = 1.5$ to $S_R = 4$. We will first discuss the results which have been obtained for the second-order fluid and the Maxwell fluid, and then consider the Oldroyd-B fluid. For simplicity, we will sometimes refer to the literature cited in this section by means of the numbers indicated in brackets in table 10.5.

TABLE 10.5 SUMMARY OF THE NUMERICAL SIMULATIONS OF THE DIE-SWELL PROBLEM

	Paper	Fluid	Method	Maximum value of S_R Slit die	Circular die
[1]	Reddy and Tanner (1978b)	2nd-order	u-v-p	0.75	
[2]	Chang et al. (1979)	Maxwell	mixed Galerkin		0.4
[3]	Crochet and Keunings (1980)	Maxwell	MIX1, MIX2	0.75	0.66
[4]	Coleman (1981)	Maxwell	MIX1	1.25	
[5]	Crochet and Keunings (1982a)	Maxwell	MIX1, MIX2	1.25	
[6]	Finlayson and Tuna (1982)	Maxwell	MIX3		1.5
[7]	Caswell and Viriyayuthakorn (1982)	Maxwell	Integral		0.75
[8]	Crochet and Keunings (1982b)	Oldroyd-B	MIX1	4.	4.

The main features that we want to examine here are the swelling ratio d/d_0 between the final diameter of the free jet and the diameter of the tube (or h/h_0 between the final thickness of the sheet and the width of the die), and the exit pressure loss Δp_{ex} defined in (9.71). These are numbers which are easy to compare between various works.

A comparison between various techniques was performed in [5] for the case of *a slit die*. Three meshes, shown on Fig. 10.25, were used for that investigation (meshes 1, 2 and 3); Fig. 10.25 also shows the finite element meshes used in [4] (mesh 4) and in [1] (mesh 5), and the type of element used in the related investigations. The number of degrees of freedom of mesh 4, for example, lies between those of mesh 2 and mesh 3, since the element used in [4] makes use of linear polynomials. Let us first compare the results obtained with method MIX1 with the meshes 1, 2 and 3; it is found that the maximum value of S_R where convergence is obtained depends upon the mesh: 0.75 with mesh 1, 0.85 with mesh 2 and 1.25 with mesh 3. This is one more case where, when the mesh is refined, the maximum value of S_R decreases; we identified the same phenomenon in sections 10.9 and 10.11. There is, however, a remarkable agreement between the values of the swelling ratio obtained with the three meshes; this is shown on Fig. 10.26, where we also give the curve of the exit pressure loss which depends slightly upon the coarseness of the mesh. The use of method MIX1 with linear polynomials leads in [4] to values of the swelling ratio which are lower than those obtained with second-order polynomials, while the exit pressure losses are in close agreement.

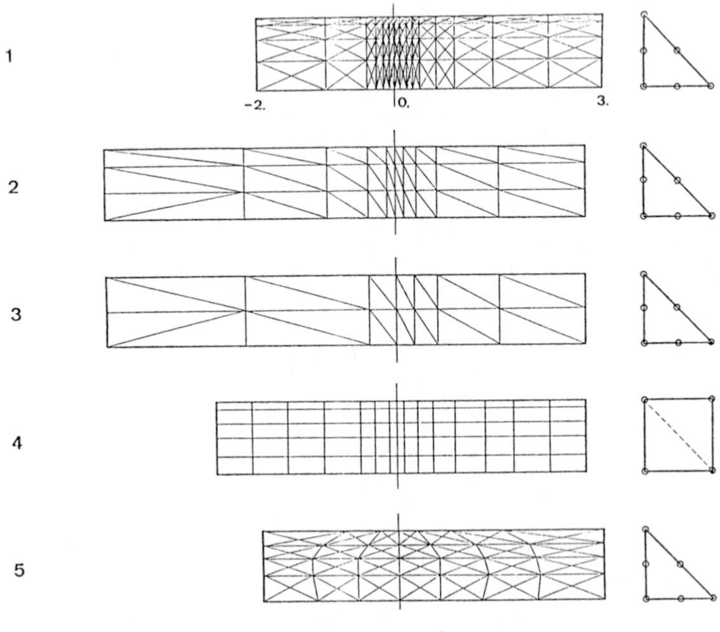

Fig.10.25 Finite element meshes used in various works mentioned in Table 10.5; 1:[2]; 2,3:[5]; 4:[4]; 5:[1].

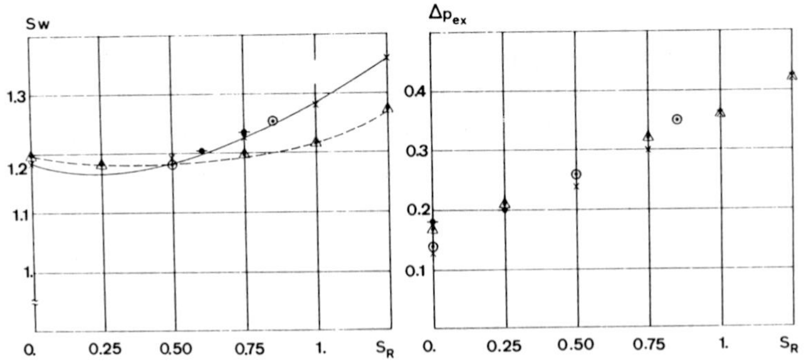

Fig.10.26 Swelling ratio and exit pressure loss for a slit die as a function of S_R, obtained with method MIX 1: ● with mesh 1, ⊙ with mesh 2, x with mesh 3; △ with mesh 4 in [4].

The method MIX2, which resembles a perturbation of the u-v-p method, is used again with meshes 1, 2 and 3; this time, the maximum values of S_R are lower: 0.6 with mesh 2, 0.70 with mesh 3, while in [1] a maximum value of 0.75 was attained for the second-order fluid with the u-v-p method. The curves of the swelling ratio and of the exit pressure loss are shown on Fig. 10.27, where we reproduce the swelling curve obtained with MIX1 and shown on Fig. 10.26. From the comparisons made in Fig. 10.26 and Fig. 10.27, we conclude that (i) the values of the swelling ratio and of the exit pressure loss are mesh dependent, in particular with MIX2; (ii) for a given value of S_R, the method MIX1 predicts lower values of the swelling ratio and of the exit pressure loss than MIX2 and the u-v-p method; (iii) the maximum value of S_R decreases when the mesh is refined; here again, the reason may be the stress singularity at the edge.

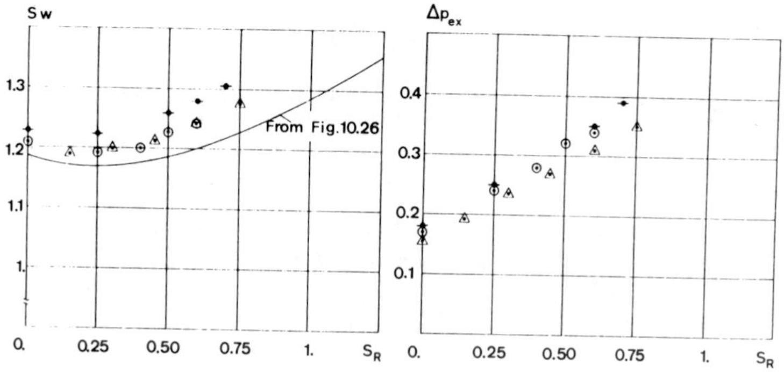

Fig.10.27 Swelling ratio and exit pressure loss for a slit die as a function of S_R, obtained with method MIX2 for a Maxwell fluid and in [1] for a second-order fluid: ⊙ with MIX2 and mesh 2, ● with MIX2 and mesh 3, △ for a second-order fluid and mesh 5 in [1].

These remarks indicate that caution is needed in the interpretation of numerical results for the swelling ratio and the exit pressure loss, since the deviations between the various results exceed several percent. On the other hand, the swelling ratios shown on Figs. 10.26 and 10.27 are so low with respect to those which we hope to attain for high values of S_R that the observed differences may be considered as insignificant at the present stage.

The same comparison is made on Fig. 10.28 for the circular die-swell, where we compare the results obtained in [3], [6] and [7] and for the sake of completeness, we have also added those of [8] for an Oldroyd-B fluid; however, the exit losses were not available in [6]. The type of agreement is similar to that found on Figs. 10.26 and 10.27. We observe again that the method MIX1 gives slightly lower results for the swelling ratio and for the exit pressure loss than MIX3 and the method for fluids of the integral type, which resembles the u-v-p method.

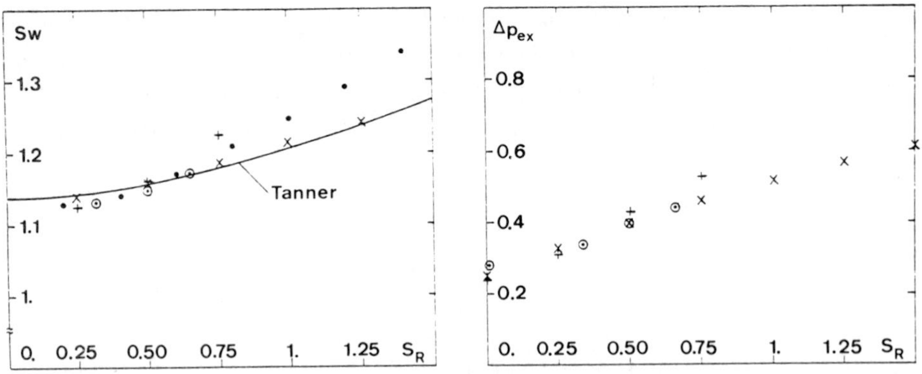

Fig. 10.28 Swelling ratio and exit pressure loss for a circular die as a function of S_R : ⊙ with MIX1 in [3], + with an integral constitutive equation in [7], • with MIX 3 in [6], and x for an Oldroyd-B fluid and MIX1 in [8].

It is comforting of course to observe that the various techniques presented in Table 10.5 offer results which are in fairly good agreement, but it is unfortunate that the maximum value of the recoverable shear S_R where convergence is obtained lies well below the range of practical applications, where large swelling ratios are commonly observed. During the course of numerical experiments performed by Crochet and Keunings (1982b) with Phan Thien-Tanner fluids, it was found that the presence of the retardation time allows one to reach much higher values of S_R than with the Maxwell fluid; results for the exit from a circular and a slit die have been obtained for values of S_R up to 4.

It is recalled that the boundary conditions for the die-swell of a viscoelastic fluid consist of the conditions given in section 9.10, together with the imposition of the extra-stress components in the entry section. In relation with the boundary conditions, Crochet and Keunings (1982b) found that the correct tube length needed for the numerical simulation with high values of S_R is an essential ingredient of a successful calculation. While going upstream from the exit inside the tube, the fully-developed velocity profile is reached within a few radii, while it takes a much longer distance before one reaches a uniform pressure profile, which is the sign of a fully-developed stress field. In order to reach a value of $S_R = 4$, it was necessary to consider a tube length of 16 radii; on the other hand, when the swelling ratio increases, the distance needed to obtain a jet of uniform radius also increases, and it was found necessary to consider 16 radii of the tube for the jet length. To reduce the cost of the calculation, three bands of Lagrangian elements were retained. The finite element mesh used for the calculations is shown on Fig. 10.29.

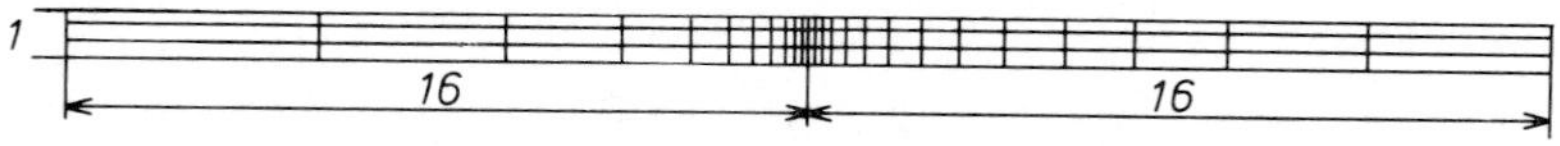

Fig.10.29 Finite element mesh used in [8] for calculating the die swell of an Oldroyd-B fluid; a $P2-C^0$ interpolation is used for the velocity and the extra-stress components.

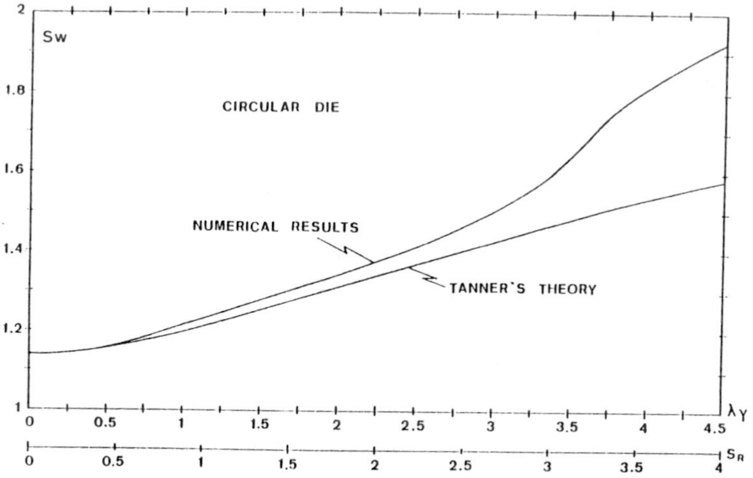

Fig.10.30 Swelling ratio of an Oldroyd-B fluid in a circular die as a function of the recoverable shear S_R.

The ratio of the viscosities in (10.38) is again $\eta_1/\eta_2 = 8$, which means that (10.71) holds. Fig. 10.30 shows the swelling ratio as a function of S_R, together with the results of Tanner's 1970 theory, and Fig. 10.31 shows the exit pressure loss for a circular die as a function of S_R. The results have been compared with those obtained with denser and shorter meshes up to $S_R = 3$ and proved to be mesh independent. The change of curvature occurring around $S_R = 3$ has not been compared with the results of other calculations.

As noted earlier in this section, there is a considerable amount of disagreement amongst the experimental data found in the literature. It is, however, worth noting in Fig. 10.32 that the order of magnitude of the results for an Oldroyd-B fluid agrees with experimental results obtained by Vlachopoulos (1981) and by Racin and Bogue (1979) with polystyrene samples.

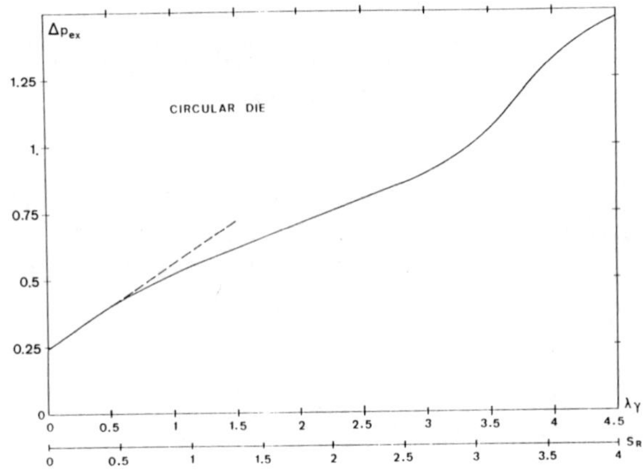

Fig. 10.31 Exit pressure loss of an Oldroyd-B fluid in a circular die as a function of the recoverable shear S_R.

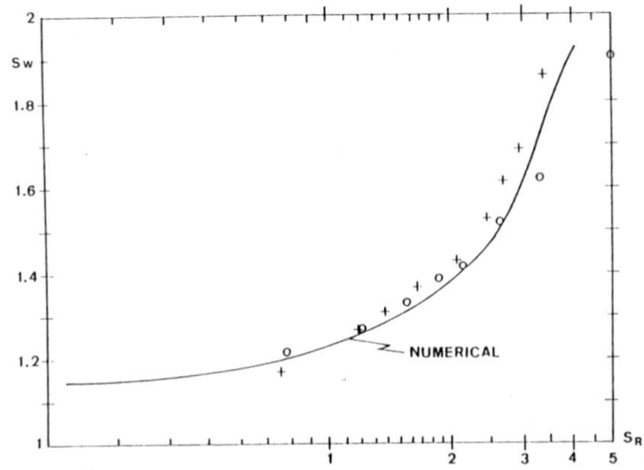

Fig. 10.32 Comparison between numerical (——) and experimental data; +: Vlachopoulos [1981]; o Racin and Bogue [1979] (TC3-30 sample).

10.13 RELATED PROBLEMS

Viscoelastic flow calculations are rather recent, and the number of flow problems solved by means of finite elements is still relatively low, the main reason being that most of the published papers deal with the development of algorithms for reaching high elasticity values rather than the solution of specific problems.

The hole pressure problem, which occupies a prominent place in rheology in view of the work by Lodge, has been considered by Richards and Townsend (1981), Jackson and Finlayson (1982) and Bernstein et al. (1982). At the present time, it seems that numerical simulation is able to give useful information for the flow in slit holes; however, three-dimensional calculations will be needed for circular holes.

Flow in contractions other than the abrupt geometry considered in §10.11 has been calculated by Bernstein et al. (1981) and Shen (1982).

The spinning of a viscoelastic fibre, which is an extension of the die-swell problem, has been studied by Keunings et al. (1983).

The slow transient flow of a Maxwell fluid contained between two squeezed parallel plates has been solved by Lee et al. (1983), who were able to identify the bouncing behaviour of the viscoelastic material.

The flow of a viscoelastic fluid around a sphere is also a problem which awaits numerical simulation. Finite elements have been used by Crochet (1982) to calculate the flow around a sphere falling along the axis of a circular cylinder with a radius fifty times the radius of the sphere. A recent paper by Hassager and Bisgaard (1983) introduces a novel technique which is also applied to the flow of a sphere falling along the axis of a narrow cylinder. While all the numerical techniques which have been explained in the present chapter use laboratory coordinates and may be qualified as Eulerian, Hassager and Bisgaard introduce Lagrangian or material coordinates. The sphere is set into motion, and the finite elements deform with the fluid; it is hoped that a steady state may be reached before the elements deform in an unacceptable way. The motion of the fluid is calculated through the application of a variational principle proposed by Hassager (1981).

Chapter 11

Outstanding Problems. Future Trends

11.1 GENERAL

We have seen in previous chapters that the basic problems associated with the numerical simulation of non-Newtonian flow are now well understood. In principle, both implicit differential and explicit integral models can be handled and there is a growing literature on the adaptation of finite difference and finite element techniques to take account of the new challenges of non-Newtonian viscosity (Chapter 9) and memory effects (Chapters 6, 7, 10). The former challenge is now resolved and existing algorithms for visco-*inelastic* models, characterized by one shear-dependent viscosity function, can be used with confidence in those situations where memory effects can be ignored. The same cannot be said about the status of algorithms presently available for memory fluids for reasons emphasized in Chapters 2, 3, 6 and 10.

Much of the previous discussion on the simulation of viscoelastic flow has concentrated on the simple upper-convected Maxwell model (2.72) and its integral equivalent (2.80) with a = 0, this preoccupation with simplicity being justified by the knowledge that most of the new problems posed by fluid-memory effects are present in the Maxwell model. Indeed, the Maxwell model may have *more* difficulties than more complicated differential models (cf. Chapters 6 and 10). Furthermore, it has already been mentioned in Chapter 10 that the basic problem of extending existing numerical algorithms to more complicated models is a relatively straightforward procedure. For example, models involving higher time derivatives (like the Oldroyd B model) can be accommodated using the decomposition techniques discussed in §10.8. Furthermore, one of the main problems associated with any complicated *integral* model is the determination of the relevant deformation tensors from a knowledge of the displacement functions. In the sense that such problems have to be faced even for the simple Maxwell model (in integral form), we may conclude that extensions to more complicated integral models will simply involve the extra detail in the model and, accordingly, we do not anticipate any new issues of principle.

This optimistic stance implies that one of the unresolved problems in the numerical simulation of non-Newtonian flow may soon be tractable. We refer to the occasional lack of agreement between numerical predictions and experimental results under conditions where converged numerical solutions can be obtained (see, for example, Crochet and Bezy 1980, Walters and Webster 1982). Before seeking more sophisticated reasons for the discrepancy between theory and experiment, it is clearly advisable to question whether the models used up to now in the simulations are too naive and simple. Already, there is some evidence to support the view that relatively minor modifications to the rheological models can alter dramatically the resulting numerical predictions (Keunings and Crochet 1983, see also §10.11).

The second unresolved problem in the numerical simulation of non-Newtonian flow concerns the degradation of the numerical solution and/or the lack of convergence of algorithms for modest values of the elasticity numbers W (or S_R) in all existing algorithms for fluids with long-range memory. Here, there is less reason for optimism and the whole question deserves special consideration.

11.2 NUMERICAL SIMULATION BREAKDOWN

There is no doubt that presently the outstanding problem in the numerical simulation of non-Newtonian flow concerns the upper limit above which the numerical algorithm fails (see, for example, Crochet and Walters 1983a,b). The limit is relatively low and often in a region where the solutions before breakdown are no more than perturbations about the Newtonian case. The following information is of relevance in this connection:

(i) A limit on W is found in all published works. It applies to finite difference and finite element techniques, to differential and integral rheological models. It even applies to the simple second-order model (2.56).

(ii) Minor changes in the constitutive equation and/or the algorithm employed can lead to higher limiting values of W (suitably modified in definition to accommodate the new constitutive equations) or S_R. For example, the use of the Oldroyd B model (2.77) in place of the upper-convected Maxwell model (2.72) can sometimes extend the range of W for which converged numerical solutions can be obtained (Chapter 10). However, it must be emphasized that such improvements, although welcome, do no more than delay the breakdown process.

(iii) As W approaches its critical value W_{crit} it is often (but not always) observed that spurious oscillations appear in the field variables; the stress components are then more severely affected than the velocity components, yielding large and erroneous stress gradients. The spurious oscillations have no physical background, and their wavelength depends upon the mesh used for the discretization.

(iv) Mesh refinement and corner strategies affect the critical conditions for breakdown, but it is difficult to discern an overall consistent trend in published works.

11.3 POSSIBLE REASONS FOR BREAKDOWN : AN EVALUATION

11.3.1 Choice of rheological model

From what has already been written, it is clear that changing the rheological model can do no more than modify the breakdown process. It is unlikely to remove it and, in any case, it would still leave unresolved the problem of why the numerical solution for one of the simplest possible models fails to converge at a very modest W value!

11.3.2 Stress evaluation

On the basis of the constitutive equations (3.5) of a Maxwell fluid, it is easy to show that the tensor $\underline{T}' = \underline{T} + \eta_0 \underline{I}/\lambda_1$ must remain positive definite throughout the flow domain. The significance of the positive definiteness of the tensor $\underline{T}'$ has been emphasized in a series of papers by Regirer and Rutkevich (1968) and Rutkevich (1969, 1970). It has been shown by Dupret and Marchal (1983) that, for a Maxwell fluid, the loss of the positive definite character of the above tensor in a subregion of the flow coincides with the breakdown of the numerical algorithm. The loss of positive definiteness of the tensor $\underline{T}'$, due to an erroneous evaluation of the stress field, produces a local change of character of the set of partial differential equations which may explain the numerical problems.

In many viscoelastic flow problems, the evaluation of the stress field is particularly difficult in view of regions of high stress gradients along solid boundaries. Along smooth boundaries with normal and tangent vectors denoted by $\underset{\sim}{n}$ and $\underset{\sim}{s}$, the shear stress T_{ns} in a Maxwell fluid is given by $\eta_0\gamma$ while the T_{ss} component is given by $2\lambda_1\eta_0\gamma^2$. The quadratic dependence of T_{ss} upon γ gives rise to high stress gradients which require a high degree of mesh refinement for a correct stress evaluation.

11.3.3 Mesh refinement

Up to the present time, the relatively small size of computers has prevented a thorough study of numerical errors and the related problem of mesh refinement. Certainly, it may be naive to simply take over established practice with the Navier-Stokes equations to the more complicated field of non-Newtonian fluid mechanics. Too much is often demanded of our numerical simulations and there is growing evidence to suggest that some workers may occasionally read too much into the finer details of their simulations. We have now reached the stage in the development of the subject when the whole question of mesh refinement and numerical errors must be given priority (cf Davies et al 1983).

Although mesh refinement is essential if inaccurate converged solutions are to be avoided, we must conclude that at the present time there is little evidence to suggest that mesh refinement *per se* would resolve the breakdown problem.

11.3.4 Corner strategies

We devoted §6.2.4 to a full discussion of re-entrant corner singularities within the finite difference context. Much work, both numerical and analytical, is still required in this important area and there is sufficient experimental evidence available (see, for example, Walters and Webster 1982) to emphasize how important it is to specify corner conditions accurately. Chapter 10 has argued that the problem also deserves serious consideration within the finite element context and the search for a suitable singular element is a pressing subject of current research.

Crochet (1982a) demonstrates that numerical breakdown occurs in flow problems *without* boundary singularities. Although a correct corner strategy is essential for accuracy, therefore, it is unlikely in itself to remove the breakdown problem.

11.3.5 Bifurcation

There is a wealth of recent literature on the general problem of bifurcation (see, for example, Iooss and Joseph 1980), and there is no doubt that sufficient evidence is available in Newtonian fluid mechanics (see, for example, Cliffe and Greenfield 1982) to warrant giving serious consideration to this possibility in the more complex field of *non*-Newtonian fluid mechanics, where we are concerned with at least two parameters R and W (see, for example, Kim-E. et al 1983).

What we say below concerning bifurcation may also be extended to include limit points in the parameter space, or points at which a solution ceases to exist.

It is essential in this context to distinguish between a bifurcation associated with the full set of *differential* equations and a bifurcation in the discretized (*algebraic*) form of these equations. We may anticipate that a bifurcation in the former will be accompanied by a similar phenomenon in the discretized algebraic equations; the converse may not be true.

Mendelson et al. (1982) examine the creeping flow ($R = 0$) of a second-order fluid in a 4/1 planar contraction using finite elements with Newton iteration and continuation with respect to the parameter W. They find that as W approaches W_{crit}, the Jacobian matrix does not become singular, thus giving no evidence of solution bifurcation in the discrete regime. This result, however, does not rule out the possibility of 3-dimensional flows arising as bifurcations from the family of 2-dimensional flows already calculated.

Another theoretical possibility is the onset of time periodic flows, which in principle should be detectable by time integration. Townsend's preliminary results for Oldroyd fluids (see Chapter 7), however, confirm the existence of steady flow solutions, with similar breakdown at W_{crit}. There is as yet no indication of time-periodicity.

Finally in this section we must mention the classical stability of the flow of any model fluid. Several authors (see, for example, Chan Man Fong and Walters 1965) have investigated the temporal and spatial response of model rheological fluids to sinusoidal perturbations in velocity for simple flow fields. Since, in computer calculations, the velocity increments will contain a range of Fourier modes, the onset of classical instability being the cause of breakdown is a possibility worthy of exploration.

11.3.6 Interaction between discretization error and nonlinear iteration

A possible cause of the breakdown can be attributed to discretization errors and their effect on the nonlinear coupling of the governing equations. For convenience of presentation, we illustrate this possibility by confining attention to the finite difference ($\psi,\omega,\underline{S}$) formulism discussed in Chapter 6.

We have to solve the following non-dimensional equations (cf. 6.1, 6.4 and 6.5):

$$T^{ik} = S^{ik} + 2d^{ik} , \tag{11.1}$$

$$\nabla^2\psi = -\omega , \tag{11.2}$$

$$\nabla^2\omega + R\left[\frac{\partial\psi}{\partial x}\frac{\partial\omega}{\partial y} - \frac{\partial\psi}{\partial y}\frac{\partial\omega}{\partial x}\right] = \frac{\partial^2}{\partial x\partial y}\left(S^{xx} - S^{yy}\right) + \left[\frac{\partial^2}{\partial y^2} - \frac{\partial^2}{\partial x^2}\right]S^{xy} , \tag{11.3}$$

together with suitable constitutive equations. In the case of planar flow of the Maxwell model (2.72), these give rise to three hyperbolic equations. The solution of these equations, themselves, may be a factor of importance in the breakdown process (cf. §11.3.2), but it is unlikely that this factor is fundamental, since for a given *fixed* velocity field it is often possible to obtain respectable converged solutions for the stress components for a high value of W. This means that the precise form of the constitutive equations *per se* does not appear to be of major significance, only the possible nonlinear interaction of these equations with (11.2) and (11.3) in the iteration process. Indeed the breakdown problem is present for a special case of the explicit second-order model given by

$$S^{ik} = -2W\overset{\nabla}{d}{}^{ik} , \tag{11.4}$$

and for illustration purposes we now concentrate on this model. Further, the breakdown is present for R = 0 (i.e. creeping flow) and there is no particular merit in retaining the inertia terms in (11.3) in the present context of attempting to locate a mechanism for breakdown.

We now consider the flow problem provided by (11.2), (11.3) with R = 0 and (11.4). For plane flows, the Tanner-Giesekus theorem (see, for example, Tanner 1966) ensures that the Newtonian velocity field for (W = 0) must satisfy (11.2)

- (11.4) with R = 0, W $\neq$ 0. It is easy to show that substitution of (11.4) into (11.3) renders the right hand side of this equation zero, consistent with the Tanner-Giesekus theorem.

Interestingly, it is found that existing finite difference algorithms adapted to the planar second-order problem also break down at some modest W value and a possible reason for the breakdown is highlighted. Analytically, the Newtonian velocity field must satisfy the second-order problem. However, any numerical noise is exaggerated in computing S^{ik} from (11.4) and even more so in taking the second derivatives of the components of S^{ik} to provide the 'forcing function' in the vorticity equation (11.3). This forcing function should be zero, as we have noted, but it is quite evidently *not zero* (numerically) when the solution fails to converge. The reason for the breakdown in this and related problems may therefore be due to nothing more than numerical noise in the forcing function of (11.3) interacting with the nonlinear iteration scheme.

If this reason for the numerical breakdown can be unequivocally associated with the numerical noise generated in the right hand side of (11.3), the next step must be the implementation of filtering techniques (see, for example, Davies 1983) to counteract the effect of noise.

With finite elements, a related result has been found by Mendelson et al. (1982) in their study of the flow of a second-order fluid in a planar contraction. Using a numerical technique similar to MIX4 (§10.9), Mendelson et al. found that the numerical noise produces a large vortex which should not be there. Further calculations with refined meshes have shown that, on the specific problem of the planar contraction, a better discretization decreases the intensity of the spurious vortex.

11.3.7 Algorithms

Ample evidence has been given in Chapter 10 that the limiting value of W depends upon the method used for discretizing the partial differential equations and upon the nonlinear iteration algorithm. Much work remains to be done in understanding how numerical convergence depends on the selection of algorithm, of finite elements, and of mesh.

11.4 CONCLUDING REMARKS

Numerical simulation of non-Newtonian flow is clearly a provocative area of current research, which is endowed with more than its usual share of interesting, often unresolved, problems.

To end this book we can do no better than advise the reader that meaningful progress in this subject must depend on careful numerical experimentation based on the most relevant guidelines that current mathematical theory can offer. We must strive to understand not only the nature of our numerical algorithms, but also the analytic properties of our governing equations. All this, of course, must be carefully balanced against our increasing knowledge of non-Newtonian fluids gleaned from laboratory experiments.

References

[1928]

COURANT, R., FRIEDRICHS, K.O., LEWY, H.: Uber die partiellen differenzengleichurgen der mathematischen physik. *Math. Ann.* 100, 32-74.

[1947]

CRANK, J., NICOLSON, P.: A practical method for numerical evaluation of solutions of partial differential equationsof the heat-conduction type. *Proc. Camb. Philos. Soc.* 43, 50-67.

[1948]

RIVLIN, R.S.: The hydrodynamics of non-Newtonian fluids Part 1. *Proc. Roy. Soc.* A193, 260-281.

[1950]

OLDROYD, J.G.: On the formulation of rheological equations of state. *Proc. Roy. Soc.* A200, 523-541.

REISSNER, E.: On a variational theorem in elasticity. *J. Math. Phys.* 29, 90-95.

[1951]

LODGE, A.S.: On the use of convected coordinate systems in the mechanics of continuous media. *Proc. Camb. Philos. Soc.* 47, 575-584.

[1952]

HESTENES, M.R., STIEFEL, E.: Methods of conjugate gradients for solving linear systems. *J. Res. natn. Bur. Stand.* 49, 409-436.

[1953]

AHLFORS, L.V.: *Complex Analysis.* New York: McGraw-Hill.

DU FORT, E.C., FRANKEL, S.P. Stability conditions in the numerical treatment of parabolic differential equations. *Math. Tables and Other Aids to Computation* 7, 135-152.

[1954]

WOODS, L.C.: A note on the numerical solution of fourth order differential equations. *Aeronaut. Q.* 5, 176.

[1955]

ALLEN, D.N. de G., SOUTHWELL, R.V.: Relaxation methods applied to determine the motion, in two dimensions, of a viscous flow past a fixed cylinder. *Quart. J. Mech. Appl. Math.* 8, 129-145.

PEACEMAN, D.W., RACHFORD, H.H.: The numerical solution of parabolic and elliptic differential equations. *J. SIAM* 3, 28-41.

RIVLIN, R.S., ERICKSEN, J.L.: Stress deformation relations for isotropic materials. *J. Rat. Mech. Anal.* 4, 323-425.

[1956]

ERICKSEN, J.L.: Overdetermination of the speed in rectilinear motion of non-Newtonian fluids. *Quart. J. Applied Math.* 14, 318-321.

LODGE, A.S.: A network theory of flow birefringence and stress in concentrated polymer solutions. *Trans. Faraday Soc.* 52, 120-130.

[1957]

GREEN, A.E., RIVLIN, R.S.: The mechanics of non-linear materials with memory. Part 1. *Arch. Rat. Mech. Anal.* 1, 1-21.
RICHTMEYER, R.D.: *Difference methods for initial-value problems*. New York: Interscience Publishers.

[1958]

CRIMINALE, W.O., ERICKSEN, J.L., FILBEY, G.L.: Steady shear flow of non-Newtonian fluids. *Arch. Rat. Mech. Anal.* 1, 410-417.
KANTOROVICH, L., KRYLOV, V.: *Approximate methods of higher analysis*. Gronigen: Noordhoff.
OLDROYD, J.G.: Non-Newtonian effects in steady motion of some idealized elastico-viscous liquids. *Proc. Roy. Soc.* A245, 278-297.
TRUESDELL, C.: Geometrical interpretation for the reciprocal deformation tensors. *Quart. J. Applied Math.* 15, 434-435.

[1959]

GREEN, A.E., RIVLIN, R.S., SPENCER, A.J.M.: The mechanics of non-linear materials with memory. Part 2. *Arch. Rat. Mech. Anal.* 3, 82-90.
JENSEN, V.G.: Viscous flow round a sphere at low Reynolds numbers (≤ 40). *Proc. Roy. Soc. London* A249, 346-366.

[1960]

CHURCHILL, R.V.: *Complex Variables and Applications*. (3rd ed.) New York: McGraw-Hill.
COLEMAN, B.D., NOLL, W.: An approximation theorem for functionals, with applications in continuum mechanics. *Arch. Rat. Mech. Anal.* 6, 355-370.
GREEN, A.E., RIVLIN, R.S.: The mechanics of non-linear materials with memory. Part 3. *Arch. Rat. Mech. Anal.* 4, 387-404.

[1961]

COLEMAN, B.D., NOLL, W.: Foundations of linear viscoelasticity. *Rev. Mod. Phys.* 33, 239-249.
OLDROYD, J.G.: Survey on Second-Order Fluid Mechanics. *Proceedings of the International Symposium on Second-Order Effects in Elasticity, Plasticity and Fluid Dynamics,* ed. D. Abir, pp. 699-711. Pergamon.
THOM, A., APELT, C.J.: *Field Computations in Engineering and Physics*. Princeton: Van Nostrand.

[1962]

BRENNER, H.: Effect of finite boundaries on the Stokes resistance of an arbitrary particle. *J. Fluid Mech.* 12, 35-48.

[1963]

BERNSTEIN, B., KEARSLEY, E.A., ZAPAS, L.: A study of stress relaxation with finite strain. *Trans. Soc. Rheol.* 7, 391-410.
VARGA, R.S.: *Matrix Iterative Analysis*. Englewood Cliffs: Prentice-Hall.
WHITE, J.L., METZNER, A.B.: Development of constitutive equations for polymeric melts and solutions. *J. Appl. Polym. Sci.* 7, 1867-1889.

[1964]

FOX, L.: *Introduction to numerical linear algebra*. Oxford: Clarendon Press.
LAX, P.D., WENDROFF, B.: Difference schemes with high order of accuracy for solving hyperbolic equations. *Comm. Pure Appl. Math.* 17, 381-398.
MOFFAT, H.K.: Viscous and resistive eddies near a sharp corner. *J. Fluid Mech.* 18, 1-18.

[1964]

PIPKIN, A.C.: Small finite deformations of viscoelastic solids. *Rev. Mod. Phys.* 36, 1034-1041.
THOMAS, R.H., WALTERS, K.: The motion of an elastico-viscous liquid due to a sphere rotating about its diameter. *Quart. J. Mech. Appl. Math.* 17, 39-53.
TRUESDELL, C.: The natural time of a viscoelastic fluid: its significance and measurement. *Phys. Fluids* 7, 1134-1142.
WHITE, J.L.: A continuum theory of nonlinear viscoelastic deformation with application to polymer processing. *J. Applied Polymer Science* 8, 1129-1146.

[1965]

BRAMBLE, J.H., HUBBARD, B.E.: Approximation of solutions of mixed boundary value problems for Poisson's equation by finite differences. *J. Assoc. Comput. Mach.* 12, 114-123.
CHAN MAN FONG, C.F., WALTERS, K.: The solution of flow problems in the case of materials with memory II. The stability of plane Poiseuille flow of slightly viscoelastic liquids. *J. de Mécanique* 4, 439-453.
HIRT, C.W.: Multidimensional fluid dynamics calculations with high speed computers. *AIAA Paper No. 65-3, AIAA 2nd Aerospace Sciences Meeting.* New York: American Institute of Aeronautics and Astronautics.
OLDROYD, J.G.: Some steady flows of the general elastico-viscous liquid. *Proc. Roy. Soc.* A283, 115-133.
TRUESDELL, C., NOLL, W.: *The Non-linear Field Theories of Mechanics.* Springer.

[1966]

GENTRY, R.A., MARTIN, R.E., DALY, B.J.: An Eulerian differencing method for unsteady compressible flow problems. *J. Comput. Phys.* 1, 87-118.
GODDARD, J.D., MILLER, C.: An inverse of the Jaumann derivative and some applications to the rheology of viscoelastic fluids. *Rheol. Acta* 5, 177-184.
WACHSPRESS, E.L.: *Iterative solution of elliptic systems.* Englewood Cliffs, New Jersey: Prentice-Hall.

[1967]

PIPKIN, A.C., OWEN, D.R.: Nearly viscometric flows. *Phys. Fluids* 10, 836-843.
TANNER, R.I., SIMMONS, J.M.: An instability in some rate-type viscoelastic constitutive equations. *Chem. Eng. Science* 22, 1079-1082.

[1968]

GIESEKUS, H.: Nicht-lineare Effekte beim Strömen viskoelastsher Flüssigkeiten durch Schlitz-und Lochdüsen. (Non-linear effects in the flow of viscoelastic fluids through slits and circular apertures.) *Rheol. Acta* 7, 127-138.
KELLER, H.B.: *Numerical methods for two-point boundary value problems.* Waltham, Massachusetts: Blaisdell.
REGIRER, S.A., RUTKEVICH, I.M.: Certain singularities of the hydrodynamic equations of non-Newtonian media. *P.M.M.* 32, 942-945.
STONE, H.L.: Iterative solution of implicit approximations of multidimensional partial differential equations. *SIAM J. Numer. Anal.* 5, 530-558.
TORRANCE, K.E.: Comparison of finite difference computations of natural convection. *J. Res. natn. Bur. Stand.* 72B, 281-301.

[1969]

CARNAHAN, B., LUTHER, H.A., WILKES, W.O.: *Applied Numerical Methods.* John Wiley.
DENNIS, S.C.R., CHANG, G.Z.: Numerical integration of the Navier-Stokes equations for steady two-dimensional flow. *Phys. Fluids* 12, Supp. II, 88-93.
FOX, L., SANKAR, R.: Boundary singularities in linear elliptic differential equations. *J. Inst. Maths Applics.* 5, 340-350.

[1969]

KAWAGUTI, M.: Numerical study of the flow of a viscous fluid in a curved channel. *Phys. Fluids* II, 101-104.
RUNCHAL, A.K., SPALDING, D.B., WOLFSHTEIN, M.: Numerical solution of the elliptic equations for transport of vorticity, heat and matter in two-dimensional flow. In *High Speed Computing in Fluid Dynamics*, eds. F.N. Frenkiel and K. Stewartson. *Physics of Fluids Supplement II*. New York: American Institute of Physics.
RUTKEVICH, I.M.: Some general properties of the equations of viscoelastic incompressible fluid dynamics. *P.M.M.* 33, 42-51.

[1970]

BUZBEE, B.L., GOLUB, G.H., NIELSON, C.W.: On direct methods for solving Poisson's equations. *SIAM J. Numer. Anal.* 7, 627-656.
DENNIS, S.C.R., CHANG, G.Z.: Numerical solutions for steady flow past a circular cylinder at Reynolds number up to 100. *J. Fluid Mech.* 42, 471-489.
GILLIGAN, S.A., JONES, R.S.: Unsteady flow of an elastico-viscous fluid past a circular cylinder. *J. Appl. Math. Phys. (ZAMP)* 21, 786-797.
IRONS, B.M.: A frontal solution program for finite element analysis. *Int. J. Num. Meth. Engng.* 2, 5-32.
ODEN, J.T.: A finite element analogue of the Navier-Stokes equations. *J. Eng. Mech. Div., Proc. Am. Soc. Civ. Eng.,* 96, 529-534.
ORTEGA, J.M., RHEINBOLDT, W.C.: *Iterative solution of nonlinear equations in several variables*. New York: Academic Press.
RUTKEVICH, I.M.: The propagation of small perturbations in a viscoelastic fluid. *P.M.M.* 34, 41-56.
TANNER, R.I.: A theory of die swell. *J. Poly. Sci.* 8, 2067-2078.

[1971]

BARNES, H.A., TOWNSEND, P., WALTERS, K.: On pulsatile flow of non-Newtonian liquids. *Rheol. Acta* 10, 517-527.
BIRKHOFF, G.: *The Numerical Solution of Elliptic Equations*. Philadelphia: SIAM.
RIVLIN, R.S., SAWYER, K.N.: Nonlinear continuum mechanics of viscoelastic fluids. *Ann. Rev. Fluid Mech.* 3, 117-146.
ULTMAN, J.S., DENN, M.M.: Slow viscoelastic flow past submerged objects. *Chem. Eng. J.* 2, 81-89.
YOUNG, D.M.: *Iterative solution of large linear systems*. New York: Academic Press.

[1972]

FINLAYSON, B.A.: *The Method of Weighted Residuals and Variational Principles*. Academic Press.
LODGE, A.S., STARK, J.H.: On the description of rheological properties of viscoelastic continua. Part 2. Proof that Oldroyd's 1950 formalism includes all 'simple fluids'. *Rheol. Acta* 11, 119-126.
REID, J.K.: The use of conjugate gradients for systems of linear equations possessing "Property A". *SIAM J. Numer. Anal.* 9, 325-332.
SPALDING, D.B.: A novel finite difference formulation for differential expressions involving both first and second derivatives. *Int. J. Num. Meth. Engng.* 4, 551-559.
WALTERS, K.: On non-Newtonian behaviour in pipe flows and the possibility of its prediction. In *Progress in Heat and Mass Transfer Vol. 5*, ed. W.R. Schowalter, pp. 217-231. Pergamon.
WILSON, J.C.: Stability of Richtmeyer type difference schemes in any finite number of space variables and their comparison with multistep Strang schemes. *J. Inst. Maths. Applics.* 10, 238-257.

[1973]

BAUDIER, F., AVENAS, P.: Calcul d'un écoulement viscoélastique dans une cavité carrée. In *Proc. Third Int. Conf. Num. Meth. Fluid Mech.*, eds. Cabannes, H., Temam, R., Vol. II, pp. 10-17. New York: Springer-Verlag.

BRANDT, A.: Multilevel adaptive technique (MLAT) for fast numerical solution of boundary value problems. In *Proc. Third Int. Conf. Num. Meth. Fluid Mech.*, eds. H. Cabannes, R. Temam, pp. 82-89. Berlin: Springer.

DUDA, J.L., VRENTAS, J.S.: Entrance flows of non-Newtonian fluids. *Trans. Soc. Rheol.* 17, 89-108.

FOX, L., SANKAR, R.: The regula-falsi method for free-boundary problems. *J. Inst. Maths. Applics.* 12, 49-54.

KESTIN, J., SOKOLOV, M., WAKEHAM, W.: Theory of capillary viscometers. *Appl. Sci. Res.* 27, 241-264.

STRANG, G., FIX, G.J.: *An analysis of the finite element method.* Prentice-Hall.

TANNER, R.I.: Die-swell reconsidered: some numerical solutions using a finite element program. *Applied Polymer Symposium* 20, 201-208.

TAYLOR, C., HOOD, P.: A numerical solution of the Navier-Stokes equations using the finite element technique. *Comput. Fluids* 1, 73-100.

THOMPSON, E.G., HAQUE, M.I.: A high order finite element for completely incompressible creeping flow. *Int. J. Num. Meth. Engng.* 6, 315-321.

TOWNSEND, P.: Numerical solutions of some unsteady flows of elastico-viscous liquids. *Rheol. Acta* 12, 13-18.

VAN ES, H.E., CHRISTENSEN, R.M.: A critical test for a class of nonlinear constitutive equations. *Trans. Soc. Rheol.* 17, 325-330.

VELDMAN, A.E.P.: The numerical solution of the Navier-Stokes equations for laminar incompressible flow past a paraboloid of revolution. *Comput. Fluids* 1, 251-271.

VRENTAS, J.S., DUDA, J.L.: Flow of a Newtonian fluid through a sudden contraction. *Appl. Sci. Res.* 28, 241-260.

[1974]

ASTARITA, G., MARRUCCI, G.: *Principles of non-Newtonian Fluid Mechanics.* McGraw-Hill.

BAIOCCHI, C., COMINCIOLI, V., MAGENES, V., POZZI, G.A.: Fluid flow through porous media: a new theoretical and numerical approach. *Publication No. 69, Laboratorio di Analisi Numerica, Pavia.*

BRILEY, W.R.: Numerical method for predicting three-dimensional steady viscous flow in ducts. *J. Comput. Phys.* 14, 8-28.

DODSON, A.G., TOWNSEND, P., WALTERS, K.: Non-Newtonian flow in pipes of non-circular cross section. *Comput. Fluids* 2, 317-338.

HENRICI, P.: *Applied and Computational Complex Analysis. Vol. I.* New York: Wiley.

HOOD, P., TAYLOR, C.: Navier-Stokes equations using mixed-interpolation. In *Finite Elements in Flow Problems*, pp. 55-56, Huntsville: UAH Press.

LADEVÈZE, J., PEYRET, R.: Calcul numérique d'une solution avec singularité des équations de Navier-Stokes: écoulement dans un canal avec variation brusque de section. *J. de Mécanique* 13, 367-396.

LODGE, A.S.: *Body-Tensor Fields in Continuum Mechanics.* Academic Press.

NICKELL, R.E., TANNER, R.I., CASWELL, B.: The solution of viscous incompressible jet and free surface flows using finite element methods. *J. Fluid Mech.* 65, 189-206.

RHEINBOLDT, W.C.: *Methods for solving systems of nonlinear equations.* Philadelphia: SIAM.

ZIENKIEWICZ, O.C., GODBOLE, P.N.: Flow of plastic and visco-plastic solids with special reference to extrusion and forming processes. *Int. J. Num. Meth. Engng.* 8, 3-16.

[1975]

CROCHET, M.J., PILATE, G.: Numerical study of the flow of a fluid of second grade in a square cavity. *Comput. Fluids* 3, 283-291.
HUILGOL, R.R.: *Continuum Mechanics of Viscoelastic Liquids*. John Wiley and Sons.
NICOLAIDES, R.A.: On multiple grid and related techniques for solving discrete elliptic systems. *J. Comput. Phys.* 19, 418-431.
OLSON, M.D.: Variational finite element methods for two-dimensional and axi-symmetric Navier-Stokes equations. In *Finite Elements in Fluids, Vol. 1*, eds. R.H. Gallagher *et. al.*, pp. 57-72. London: Wiley.
SCHUMANN, U.: Linear stability of finite difference equations for three-dimensional flow problems. *J. Comput. Phys.* 18, 465-470.
TANNER, R.I., NICKELL, R.E., BILGER, R.W.: Finite element methods for the solution of some incompressible non-Newtonian fluid mechanics problems with free surfaces. *Comput. Methods Appl. Mech. Engng.* 6, 155-179.
WALTERS, K.: *Rheometry*. Chapman and Hall.
ZANA, E., TIEFENBRUCK, F., LEAL, L.G. A note on the creeping motion of a visco-elastic fluid past a sphere. *Rheol. Acta* 14, 891-898.

[1976]

ASTARITA, G.: Is non-Newtonian fluid mechanics a culturally autonomous subject? *J. non-Newtonian Fluid Mech.* 1, 203-206.
CROCHET, M.J., PILATE, G.: Plane flow of a fluid of second grade through a contraction. *J. non-Newtonian Fluid Mech.* 1, 247-258.
KELLER, H.B.: *Numerical solution of two-point boundary value problems*. Philadelphia: SIAM.
LEONOV, A.I.: Non equilibrium thermodynamics and rheology of viscoelastic polymer media. *Rheol. Acta* 15, 85-98.
ODEN, J.T., REDDY, J.N.: *An introduction to the mathematical theory of finite elements*. New York: Wiley.
ROACHE, P.J.: *Computational fluid dynamics*. Albuquerque: Hermosa Publishers.
TANNER, R.I.: Some experiences using finite element methods in polymer processing and rheology. In *Proc. of the VIIth Int. Cong. on Rheology*, eds. C. Klason and J. Kubàt, pp. 140-145. Gothenburg.

[1977]

AMES, W.F.: *Numerical Methods for Partial Differential Equations*. New York: Academic Press.
BIRD, R.B., ARMSTRONG, R.C., HASSAGER, O.: *Dynamics of Polymeric Liquids: Vol. I Fluid Mechanics*. John Wiley and Sons.
BOGER, D.V.: A highly elastic constant-viscosity fluid. *J. non-Newtonian Fluid Mech.* 3, 87-91.
BRANDT, A.: Multi-level adaptive solutions to boundary value problems. *Math. Comp.* 31, 333-390.
CHIEN, J.C.: A general finite difference formulation with application to Navier-Stokes equations. *Comput. Fluids* 5, 15-31.
CRYER, C.W.: A bibliography of free boundary problems. *MRC Tech. Summ. Rept. 1793*. Mathematics Research Center, University of Wisconsin, Madison.
FURZELAND, R.M.: A survey of the formulation and solution of free and moving boundary (Stefan) problems. *Tech. Rept. TR/76*. Dept. of Mathematics, Brunel University, Uxbridge.
GHIA, U., GHIA, K.N., STUDERUS, C.J.: Three-dimensional laminar incompressible flow in straight polar ducts. *Comput. Fluids* 5, 205-218.
IRONS, B.M., ELSAWAF, A-F.: The conjugate Newton algorithm for solving finite element equations. In *Formulations and Computational Algorithms in Finite Element Analysis*, eds. K-J. Bathe, J.T. Oden, W. Wunderlich, pp. 665-672. Cambridge, Massachusetts: MIT Press.
JOHNSON, M.W., Jr., SEGALMAN, D.: A model for viscoelastic fluid behaviour which allows non-affine deformation. *J. non-Newtonian Fluid Mech.* 2, 255-270.
KAWAHARA, M., TAKEUCHI, N.: Mixed finite element method for analysis of visco-elastic fluid flow. *Comput. Fluids* 5, 33-45.

[1977]

MEIJERINK, J.A., VAN DER VORST, H.A.: An iterative solution method for linear systems of which the coefficient matrix is a symmetric M-matrix. *Maths. Comp.* 31, 148-162.

PERERA, M.G.N., WALTERS, K.: (a) Long-range memory effects in flows involving abrupt changes in geometry. Part I. Flows associated with L-shaped and T-shaped geometries. *J. non-Newtonian Fluid Mech.* 2, 49-81.

PERERA, M.G.N., WALTERS, K.: (b) Long-range memory effects in flows involving abrupt changes in geometry. Part II. The expansion/contraction/expansion problem. *J. non-Newtonian Fluid Mech.* 2, 191-204.

PHAN THIEN, N., TANNER, R.I.: A new constitutive equation derived from network theory. *J. non-Newtonian Fluid Mech.* 2, 353-365.

PILATE, G., CROCHET, M.J.: Plane flow of a second-order fluid past submerged boundaries. *J. non-Newtonian Fluid Mech.* 1, 247-258.

REDDY, K.R., TANNER, R.I.: Finite element approach to die-swell problems of non-Newtonian fluids. *6th Australian Hydraulics and Fluid Mechanics Conference,* 431-434.

SIGLI, D., COUTANCEAU, M.: Effect of finite boundaries on the slow laminar isothermal flow of a viscoelastic fluid around a spherical obstacle. *J. non-Newtonian Fluid Mech.* 2, 1-21.

THAMES, F.C., THOMPSON, J.F., MASTIN, C.W., WALKER, R.L.: Numerical solutions for viscous and potential flow about arbitrary two-dimensional bodies using body fitted coordinate systems. *J. Comput. Phys.* 24, 245-273.

THOMPSON, J.F., THAMES, F.C., MASTIN, C.W.: TOMCAT - A code for numerical generation of boundary-fitted curvilinear coordinate systems on fields containing any number of arbitrary two-dimensional bodies. *J. Comput. Phys.* 24, 274-302.

WESSELING, P.: Numerical simulation of the stationary Navier-Stokes equations by means of a multiple grid method and Newton iteration. *Tech. Rept. NA-18,* Dept. of Mathematics, Delft University of Technology, The Netherlands.

WHITE, J.L., KONDO, A.: Flow patterns in polyethylene and polystyrene melts during extrusion through a die entry region: measurement and interpretation. *J. non-Newtonian Fluid Mech.* 3, 41-64.

ZIENKIEWICZ, O.C.: *The finite element method.* 3rd Edition. McGraw-Hill.

[1978]

BOGER, D.V., GUPTA, R., TANNER, R.I.: The end-correction for power-law fluids in the capillary rheometer. *J. non-Newtonian Fluid Mech.* 4, 239-248.

CAMPION-RENSON, A., CROCHET, M.J.: On the stream function-vorticity finite element solutions of Navier-Stokes equations. *Int. J. Num. Meth. Engng.* 12, 1809-1818.

CASWELL, B., TANNER, R.I.: Wirecoating die design using finite element methods. *Polym. Eng. Sci.* 18, 416-421.

CIARLET, P.G.: *The finite element method for elliptic problems.* Amsterdam: North Holland.

CRANK, J., FURZELAND, R.M.: The numerical solution of elliptic and parabolic partial differential equations with boundary singularities. *J. Comput. Phys.* 26, 285-296.

DOI, M., EDWARDS, S.F.: Dynamics of concentrated polymer systems Parts 1-4. *J. Chem. Soc. Faraday Trans.* II, 74, 1789-1801, 1802-1817, 1818-1832; 75, 38-54.

GATSKI, T.B., LUMLEY, J.L.: (a) Steady flow of a non-Newtonian fluid through a contraction. *J. Comput. Phys.* 27, 42-70.

GATSKI, T.B., LUMLEY, J.L.: (b) Non-Newtonian flow characteristics in a steady two-dimensional flow. *J. Fluid Mech.* 86, 623-639.

HUYAKORN, P.S., TAYLOR, C., LEE, R.L., GRESHO, P.M.: A comparison of various mixed-interpolative finite elements in the velocity-pressure formulations of the Navier-Stokes equations. *Comput. Fluids* 6, 25-35.

KERSHAW, D.S.: The incomplete Cholesky conjugate gradient method for the iterative solution of systems of linear equations. *J. Comput. Phys.* 26, 43-65.

[1978]

MALKUS, D.S., HUGHES, T.J.R.: Mixed finite element methods - reduced and selective integration techniques: a unification of concepts. *Comp. Meth. in Appl. Mech. and Engng.* 15, 63-81.
NOYE, J.: (ed.) *Numerical simulation of fluid motion*. Amsterdam: North-Holland.
PEARSON, J.R.A., TROTTNOW, R.: On die swell: some theoretical results. *J. non-Newtonian Fluid Mech.* 4, 195-216.
PHAN THIEN, N. On pulsatile flow of polymeric fluids. *J. non-Newtonian Fluid Mech.* 4, 167-176.
REDDY, K.R., TANNER, R.I.: (a) Finite element solution of viscous jet flows with surface tension. *Comput. Fluids* 6, 83-91.
REDDY, K.R., TANNER, R.I.: (b) On the swelling of extruded plane sheets. *Trans. Soc. Rheol.* 22, 661-665.
RICHARDS, C.W., CRANE, C.M.: An economical central difference algorithm for Navier-Stokes equations convergent for high mesh Reynolds numbers. *Appl. Math. Modelling* 2, 59-61.
SCHOWALTER, W.R.: *Mechanics of non-Newtonian Fluids*. Pergamon Press.
SMITH, G.D.: *Numerical solution of partial differential equations: finite difference methods*. Oxford University Press.
WACKER, H.: (ed.) *Continuation methods*. London: Academic Press.
WAGNER, M.H.: A constitutive analysis of uniaxial elongational flow data of a low-density polyethylene melt. *J. non-Newtonian Fluid Mech.* 4, 39-55.
ZIENKIEWICZ, O.C., JAIN, P.C., ONATE, E.: Flow of solids during forming and extrusion: some aspects of numerical solutions. *Int. J. Solids Structures* 14, 15-38.

[1979]

AKAY, G.: Non-steady two-phase stratified laminar flow of polymeric liquids in pipes. *Rheol. Acta* 18, 256-267.
ASTARITA, G.: Scale-up problems arising with non-Newtonian fluids. *J. non-Newtonian Fluid Mech.* 4, 285-298.
BEN-SABAR, E., CASWELL, B.: A stable finite element simulation of convective transport. *Int. J. Num. Meth. Engng.* 14, 545-565.
CHANG, P.W., PATTEN, T.W., FINLAYSON, B.A.: (a) Collocation and Galerkin finite element methods for viscoelastic fluid flow. I. Description of method and problems with fixed geometries. *Comput. Fluids* 7, 267-283.
CHANG, P.W., PATTEN, T.W., FINLAYSON, B.A.: (b) Collocation and Galerkin finite element methods for viscoelastic fluid flow. II. Die swell problems with a free surface. *Comput. Fluids* 7, 285-293.
CROCHET, M.J., BEZY, M.: Numerical solutions for the flow of viscoelastic fluids. *J. non-Newtonian Fluid Mech.* 5, 201-218.
DAVIES, A.R., WALTERS, K., WEBSTER, M.F.: Long-range memory effects in flows involving abrupt changes in geometry. Part III. Moving boundaries. *J. non-Newtonian Fluid Mech.* 4, 325-344.
DUFF, I.S.: Conjugate gradient methods and similar techniques. *Harwell Report AERE-R 9636*, Computer Science & Systems Division, AERE Harwell, Didcot, U.K.
GIRAULT, V., RAVIART, P-A.: Finite Element Approximation of the Navier Stokes Equations. *Lecture Notes in Mathematics*, Vol. 749. Heidelberg: Springer-Verlag.
GLADWELL, I., WAIT, R. *A survey of numerical methods for partial differential equations*. Oxford University Press.
GODDARD, J.D.: Polymer fluid mechanics. *Advances in Applied Mechanics* 19, 143-219.
GRESHO, P.M., LEE, R.L.: Don't suppress the wiggles - they're telling you something! In *Finite Element Methods for Convection Dominated Flows*, ed. T.J.R. Hughes, AMD Vol. 34, pp 37-62, The American Scty of Mech. Engineers.

[1979]

GUPTA, M.M., MANOHAR, R.P.: Boundary approximations and accuracy in viscous flow computations. *J. Comput. Phys.* 31, 265-288.
HEINRICH, J., ZIENKIEWICZ, O.C.: The finite element method and upwinding techniques in the numerical solution of convection dominated flow problems. In *Finite Element Methods for Convection Dominated Flows*, ed. T.J.R. Hughes, AMD Vol. 34, pp 105-136. The American Scty of Mech. Engineers.
HUGHES, T.J.R.: (ed.) *Finite Element Methods for Convection Dominated Flows.* The American Scty of Mech. Engineers, AMD, Vol. 34.
HUGHES, T.J.R., LIU, W.K., BROOKS, A.: Finite element analysis of incompressible viscous flow by the penalty function formulation. *J. Comput. Phys.* 30, 1-60.
LEAL, L.G.: The motion of small particles in non-Newtonian fluids. *J. non-Newtonian Fluid Mech.* 5, 33-78.
LEONARD, B.P.: A survey of finite differences of opinion on numerical muddling of the incomprehensible defective confusion equation. In *Finite Element Methods for Convection Dominated Flows*, ed. T.J.R. Hughes, AMD, Vol. 34, pp 1-17. The American Scty of Mech. Engineers.
NGUYEN, H., BOGER, D.V.: The kinematics and stability of die entry flows. *J. non-Newtonian Fluid Mech.* 5, 353-368.
PADDON, D.J.: The numerical solution of some rheological flow problems. Ph.D. Thesis, University of Wales.
PETRAVIC, M., KUO-PETRAVIC, G.: An ILUCG algorithm which minimizes in the Euclidean norm. *J. Comput. Phys.* 32, 263-269.
PETRIE, C.J.S.: *Elongational Flows.* Pitman.
RACIN, R., BOGUE, D.C.: Molecular weight effects in die swell and in shear rheology. *J. Rheol.* 23, 263-280.
RICHARDS, C.W., CRANE, C.M.: The accuracy of finite difference schemes for the numerical solution of Navier-Stokes equations. *Appl. Math. Modelling* 3, 205-211.
STURGES, L.D.: Die swell: the separation of the free surface. *J. non-Newtonian Fluid Mech.* 6, 155-159.
TEMAM, R.: *Navier-Stokes Equations* (2nd ed.). Amsterdam: North-Holland.
TEMPERTON, C.: Direct methods for the solution of the discrete Poisson equation: some comparisons. *J. Comput. Phys.* 31, 1-20.
WAIT, R.: *The Numerical Solution of Algebraic Equations.* Chichester: John Wiley.
WALTERS, K.: Developments in non-Newtonian fluid mechanics - A personal view. *J. non-Newtonian Fluid Mech.* 5, 113-124.
WEBSTER, M.F.: The numerical solution of rheological flow problems. Ph.D. Thesis, University of Wales.

[1980]

BOGER, D.V., DENN, M.M.: Capillary and slit methods of normal stress measurements. *J. non-Newtonian Fluid Mech.* 6, 163-185.
CHHABRA, R.P., TIU, C., UHLHERR, P.H.T.: Shear-thinning effects in creeping flow about a sphere. In *Rheology*, eds. G. Astarita, G. Marucci, L. Nicolais, Vol. 2, pp. 9-16. Plenum Press, New York.
COLEMAN, C.J.: A finite element routine for analysing non-Newtonian flows. Part I: Basic method and preliminary results. *J. non-Newtonian Fluid Mech.* 7, 289-301.
COURT, H.: Computational rheological fluid dynamics. Ph.D. Thesis, University of Wales.
CROCHET, M.J., BEZY, M.: Elastic effects in die entry flow. In *Rheology, Vol. 2: Fluids*, ed. G. Astarita, G. Marrucci, L. Nicolais, pp 53-58. New York: Plenum.
CROCHET, M.J., KEUNINGS, R.: Die swell of a Maxwell fluid: numerical prediction. *J. non-Newtonian Fluid Mech.* 7, 199-212.
HESTENES, M.R.: *Conjugate Direction Methods in Optimization.* New York: Springer-Verlag.
HIEBER, C.A., SHEN, S.F.: A finite-element/finite difference simulation of the injection-molding filling process. *J. non-Newtonian Fluid Mech.* 7, 1-32.
IOOSS, G., JOSEPH, D.D.: *Elementary stability and Bifurcation Theory.* New York: Springer-Verlag.

[1980]

JONES, I.P., THOMPSON, C.P.: On the use of nonuniform grids in finite difference calculations. *Harwell Report AERE-R 9765*. London: H.M.S.O.
KEENTOK, M., GEORGESCU, A.G., SHERWOOD, A.A., TANNER, R.I.: The measurement of the second normal stress difference for some polymer solutions. *J. non-Newtonian Fluid Mech*. 6, 303-324.
MANERO, O.: Problems in non-Newtonian fluid mechanics. Ph.D. Thesis, University of Wales.
MANERO, O., WALTERS, K.: On elastic effects in unsteady pipe flows. *Rheol. Acta* 19, 277-284.
MITCHELL, A.R., GRIFFITHS, D.F.: *The Finite Difference Method in Partial Differential Equations*. Chichester: John Wiley.
PHUOC, H.B., TANNER, R.I.: Thermally-induced extrudate swell. *J. Fluid Mech*. 98(2), 253-271.
RUSCHAK, K.J.: A method for incorporating free boundaries with surface tension in finite element fluid-flow simulators. *Int. J. Num. Meth. Engng*. 15, 639-648.
SILLIMAN, W.J., SCRIVEN, L.E.: Separating flow near a static contact line: slip at a wall and shape of a free surface. *J. Comput. Phys*. 34, 287-313.
TOWNSEND, P.: (a) A computer model of hole-pressure measurement in Poiseuille flow of visco-elastic liquids. *Rheol. Acta* 19, 1-11.
TOWNSEND, P.: (b) A numerical simulation of Newtonian and visco-elastic flow past stationary and rotating cylinders. *J. non-Newtonian Fluid Mech*. 6, 219-243.
TROGDON, S.A., JOSEPH, D.D.: The stick-slip problem for a round jet. 1. Large surface tension. *Rheol. Acta* 19, 404-420.
VIRYAYUTHAKORN, M., CASWELL, B.: Finite element simulation of viscoelastic flow. *J. non-Newtonian Fluid Mech*. 6, 245-267.
WALTERS, K.: (ed.) *Rheometry: Industrial Applications*. J. Wiley and Sons.
WALTERS, K., BARNES, H.A.: Anomalous extensional flow effects in the use of commercial viscometers. In *Rheology, Vol. 1: Principles,* eds. G. Astarita, G. Marrucci and L. Nicolais, pp 45-62. Plenum.
ZIENKIEWICKZ, O.C., TAYLOR, R.L.: Some developments of the finite element methods for fluid mechanics. *3rd Int. Conf. on Finite Elements in Flow Problems,* Banff, Canada, Vol. 1, 1-10.

[1981]

BERNSTEIN, B., KADIVAR, M.K., MALKUS, D.S.: Steady flow of memory fluids with finite elements: two test problems. *Comput. Methods Appl. Mech. Engng*. 27, 279-302.
BERNSTEIN, B., MALKUS, D.S.: Steady flow of memory fluids with finite elements: a progress report. *I.I.T. Report*.
COCHRANE, T., WALTERS, K., WEBSTER, M.F.: On Newtonian and non-Newtonian flow in complex geometries. *Philos. Trans. Roy. Soc. London Ser. A* 301, 163-181.
COLEMAN, C.J.: A finite element routine for analysing non-Newtonian flows. Part 2: The extrusion of a Maxwell fluid. *J. non-Newtonian Fluid Mech*. 8, 261-270.
COURT, H., DAVIES, A.R., WALTERS, K.: Long-range memory effects in flows involving abrupt changes in geometry. Part IV. Numerical simulation using integral rheological models. *J. non-Newtonian Fluid Mech*. 8, 95-117.
CROCHET, M.J., KEUNINGS, R.: Numerical simulation of die swell: geometrical effects. *Proc. 2nd World Congr. Chem. Engng., Montreal* 6, 285-288.
CURTISS, C.F., BIRD, R.B.: A kinetic theory for polymer melts. Parts 1, 2. *J. Chem. Phys*. 74, 2016-2025, 2026-2033.
GUPTA, M.M., MANOHAR, R.P., NOBLE, B.: Nature of viscous flows near sharp corners. *Comput. Fluids* 9, 379-388.
HAGEMAN, L.A., YOUNG, D.M.: *Applied Iterative Analysis*. New York: Academic Press.
HASSAGER, O.: Variational principle for the KBKZ rheological equation of state with potential function. *J. non-Newtonian Fluid Mech*. 9, 321-328.

[1981]

HEINRICH, J.C., MARSHALL, R.S.: Viscous incompressible flow by a penalty function finite element method. *Comput. Fluids* 9, 73-83.
HOLSTEIN, H.: The Numerical Solution of Some Rheological Flow Problems. Ph.D. Thesis, University of Wales.
HOLSTEIN, H., PADDON, D.J.: A singular finite difference treatment of re-entrant corner flow. Part I. Newtonian fluids. *J. non-Newtonian Fluid Mech.* 8, 81-93.
MALKUS, D.S.: Functional derivatives and finite elements for the steady spinning of a viscoelastic filament. *J. non-Newtonian Fluid Mech.* 8, 223-237.
MANERO, O., MENA, B.: On the slow flow of viscoelastic liquids past a circular cylinder. *J. non-Newtonian Fluid Mech.* 9, 379-387.
MEIS, T., MARCOWITZ, U.: *Numerical Solution of Partial Differential Equations.* New York: Springer-Verlag.
RICHARDS, G.D., TOWNSEND, P.: A finite element computer model of the hole pressure problem. *Rheol. Acta* 20, 261-269.
RYAN,M.E.,DUTTA, A.: A finite difference simulation of extrudate swell. *Proc. 2nd World Congr. Chem. Engng., Montreal,* 6, 277-280.
SANI, R.L., GRESHO, P.M., LEE, R.L., GRIFFITHS, D.F.: (a) The cause and cure (?) of the spurious pressures generated by certain FEM solutions of the incompressible Navier-Stokes equations. Part I. *Int. J. Num. Meth. Fluids* 1 (1981) 17-43.
SANI, R.L., GRESHO, P.M., LEE, R.L., GRIFFITHS, D.F., ENGELMAN, M.: (b) The cause and cure (?) of the spurious pressures generated by certain FEM solutions of the incompressible Navier-Stokes equations. Part II. *Int. J. Num. Meth. Fluids* 1, 171-204.
SHIMAZAKI, Y., THOMPSON, E.G.: Elasto visco-plastic flow with special attention to boundary conditions. *Int. J. Num. Meth. Engng.* 17, 97-112.
STURGES, L.D.: A theoretical study of extrudate swell. *J. non-Newtonian Fluid Mech.* 9, 357-378.
TELIONIS, D.P.: *Unsteady Viscous Flows.* New York: Springer-Verlag.
THOMASSET, F.: *Implementation of Finite Element Methods for Navier-Stokes Equations.* New York: Springer-Verlag.
VLACHOPOULOS, J.: Extrudate swell in polymers. In *Reviews on the Deformation Behavior of Materials Vol. III,* 219-248.

[1982]

BAKER, C.T.H., MILLER, G.F.: (eds.) *Treatment of integral equations by numerical methods.* London: Academic Press.
BERNSTEIN, B., MALKUS, D.S. Finite elements for steady flows of memory fluids. In *Numerical Methods in Industrial Forming Processes,* eds. J.F.T. Pittman, R.D. Wood, J.M. Alexander, O.C. Zienkiewicz, pp 611-620. Pineridge Press.
BERNSTEIN, B., MALKUS, D.S., OLSEN, E.T.: A finite element for incompressible plane flows of fluids with memory. *I.I.T. Report.*
BEZY, M.: Simulation numérique de l'écoulement de fluides polymériques dans les convergents. Ph.D. Thesis, Louvain-la-Neuve.
BOGER, D.V.: Circular entry flows of inelastic and viscoelastic fluids. In *Advances in Transport Processes Vol. 2,* eds. A.S. Mujundar and R.A. Mashelkar, Wiley Eastern Ltd., pp 43-104.
BRANDT, A. Introductory remarks on multigrid methods. In *Numerical Methods for Fluid Dynamics,* eds. K.W. Morton, M.J. Baines, pp 127-134. London: Academic Press.
CHEN, F., BERNSTEIN, B.: The artificial time-drift function method for finite element techniques for axially symmetric flows of memory fluids. *I.I.T. Report.*
CLIFFE, K.A., GREENFIELD, A.C.: Some comments on laminar flow in symmetric 2-dimensional channels. *Technical Report TP 939,* Theoretical Physics Division, AERE Harwell.
COCHRANE, T., WALTERS, K., WEBSTER, M.F.: Newtonian and non-Newtonian flow near a re-entrant corner. *J. non-Newtonian Fluid Mech.* 10, 95-114.

[1982]

CROCHET, M.J.: (a) The flow of a Maxwell fluid around a sphere. In *Finite Elements in Fluids, Vol. 4,* ed. R.H. Gallagher, pp 573-597, New York: Wiley.
CROCHET, M.J.: (b) Numerical simulation of die-entry and die-exit flow of a viscoelastic fluid. In *Numerical Methods in Industrial Forming Processes,* eds. J.F.T. Pittman, R.D. Wood, J.M. Alexander, O.C. Zienkiewicz, pp 85-94, Pineridge Press.
CROCHET, M.J., KEUNINGS, R.: (a) On numerical die-swell calculation. *J. non-Newtonian Fluid Mech.* 10, 85-94.
CROCHET, M.J., KEUNINGS, R.: (b) Finite element analysis of die swell of a highly elastic fluid. *J. non-Newtonian Fluid Mech.* 10, 339-356.
FINLAYSON, B.A., TUNA, N.Y.: Mathematical modeling of polymer flows. In *Finite Element Flow Analysis,* ed. T. Kawai. Proc. of 4th Int. Symp. on Finite Element Methods in Flow Problems, Tokyo, pp 363-370, University of Tokyo Press.
HACKBUSCH, W., TROTTENBERG, U. *Multigrid Methods.* New York: Springer-Verlag.
HOLSTEIN, H., PADDON, D.J.: A finite difference strategy for re-entrant corner flow. In *Numerical Methods for Fluid Dynamics,* eds. K.W. Morton, M.J. Baines, pp 341-358. London: Academic Press.
JACKSON, N.A., FINLAYSON, B.A.: Calculation of hole pressure. II. Viscoelastic fluids. *J. non-Newtonian Fluid Mech.* 10, 71-84.
LEE, S.J., DENN, M.M., CROCHET, M.J., METZNER, A.B.: Compressive flow between parallel disks: Part 1. Newtonian fluid with a transverse viscosity gradient. *J. non-Newtonian Fluid Mech.* 10, 3-30.
MENDELSON, M.A., YEH, P-W., BROWN, R.A., ARMSTRONG, R.C.: Approximation error in finite element calculation of viscoelastic flow. *J. non-Newtonian Fluid Mech.* 10, 31-54.
NAKAZAWA, S., PITTMAN, J.F.T., ZIENKIEWICZ, O.C.: Numerical solution of flow and heat transfer in polymer melts. In *Finite Elements in Fluids, Vol. 4,* ed. R.H. Gallagher, pp 251-284, New York: Wiley.
RICHARDS, G.D., TOWNSEND, P.: Computer modelling of flows of elastic liquids through complex vessels and with forced convection. *J. non-Newtonian Fluid Mech.* 10, 175-183.
SHEN, S.F.: Simulation of non-isothermal polymeric flows in the injection molding process. In *Finite Element Flow Analysis,* ed. T. Kawai. Proc. of 4th Int. Symp. on Finite Element Methods in Flow Problems, Tokyo, pp 337-356, University of Tokyo Press.
THOMPSON, C.P., WILKES, N.S.: Experiments with higher-order finite difference formulae. *Harwell Report AERE-R 10493.* London: H.M.S.O.
THOMPSON, J.F.: (ed.) *Numerical Grid Generation.* Amsterdam: Elsevier.
THOMPSON, J.F., WARSI, Z.U.A., MASTIN, C.W.: Boundary-fitted coordinate systems for numerical solution of partial differential equations - A review. *J. Comput. Phys.* 47, 1-108.
TIEFENBRUCK, G., LEAL, L.G.: A numerical study of the motion of a viscoelastic fluid past rigid spheres and spherical bubbles. *J. non-Newtonian Fluid Mech.* 10, 115-155.
WALTERS, K., WEBSTER, M.F.: On dominating elastico-viscous response in some complex flows. *Philos. Trans. Roy. Soc. London Ser. A* 308, 199-218.

[1983]

BELYTSCHKO, T., HUGHES, T.J.R.: (eds.) *Computational Methods for Transient Analysis.* Amsterdam: North-Holland.
CASWELL, B., VIRYAYUTHAKORN, M.: Finite element simulation of die swell for a Maxwell fluid. *J. non-Newtonian Fluid Mech.* 12, 13-30.
CROCHET, M.J., WALTERS, K.: (a) Numerical methods in non-Newtonian fluid mechanics. Annual Reviews of Fluid Mechanics 15, 241-260.
CROCHET, M.J., WALTERS, K.: (b) Computational techniques for viscoelastic fluid flow. In *Computational analysis of polymer processing,* eds. J.R.A. Pearson, S.M. Richardson, pp 21-62. Applied Science Publishers.
DAVIES, A.R.: Numerical filtering and the high Weissenberg number problem. *J. non-Newtonian Fluid Mech.* To appear.

[1983]

DAVIES, A.R., LEE, S.J., WEBSTER, M.F.: Numerical simulation of viscoelastic flow: the effect of mesh size. *J. non-Newtonian Fluid Mech.* To appear.
DAVIES, A.R., MANERO, O.: Finite difference solution of viscoelastic flows by preconditioned conjugate gradients. To be published.
DUPRET, F., MARCHAL, J.M.: Characteristic surfaces for a Maxwell fluid: theory and numerical consequences. To be published.
HASSAGER, O., BISGAARD, C.: A Lagrangian finite element method for the simulation of flow of non-Newtonian liquids. *J. non-Newtonian Fluid Mech.* 12, 153-164.
KEUNINGS, R., CROCHET, M.J.: Numerical simulation of the flow of a viscoelastic fluid through an abrupt contraction. *J. non-Newtonian Fluid Mech.*, to appear.
KEUNINGS, R., CROCHET, M.J., DENN, M.M.: Profile development in continuous drawing of viscoelastic liquids. *I.& E.C. Fundamentals 22, 347-355.*
KIM-E, M., YEH, P.W., ARMSTRONG, R.C., BROWN, R.A.: Multiple solutions in the calculation of axisymmetric contraction flow of an upper convected Maxwell fluid. *J. non-Newtonian Fluid Mech.* To appear.
LEE, S.J., DENN, M.M., CROCHET, M.J., METZNER, A.B., RIGGINS, G.J.: Compressive flow between parallel disks II: oscillatory behaviour of viscoelastic materials under a constant load. *J. non-Newtonian Fluid Mech.* To appear.
LODGE, A.S.: A classification of constitutive equations based on stress relaxation predictions for the single-jump shear strain experiment. *J. non-Newtonian Fluid Mech.* To appear.
PRILUTSKI, G., GUPTA, R.K., SRIDHAR, T., RYAN, M.E.: Model viscoelastic liquids. *J. non-Newtonian Fluid Mech.* 12, 233-242.
RIVLIN, R.S.: Integral representations of constitutive equations. *Rheol. Acta* 22, 260-267.
SAUT, J.C., JOSEPH, D.D.: Fading memory. *Arch. Rat. Mech. Anal.* 81, 53-95.
TANNER, R.I.: Extrudate swell. In *Computational analysis of polymer processing.* eds. J.R.A. Pearson, S.M. Richardson, pp 63-91. Applied Science Publishers.
TOWNSEND, P.: On the numerical simulation of two-dimensional time-dependent flows of Oldroyd fluids. Part 1: Basic method and preliminary results. *J. non-Newtonian Fluid Mech.* To appear.
VAN SCHAFTINGEN, J.J., CROCHET, M.J.: A comparison of mixed methods for solving the flow of a Maxwell fluid. *Int. J. of Numerical Methods in Fluids*, to appear.
WILKES, N.S., THOMPSON, C.P.: An evaluation of higher-order upwind differencing for elliptic flow problems. *Harwell Report* CSS 137, Engineering Sciences and Computer Science and Systems Divisions, AERE Harwell, Oxfordshire.

Author Index

Subject Index